MODELLING PERCEPTION WITH ARTIFICIAL NEURAL NETWORKS

Studies of the evolution of animal signals and sensory behaviour have more recently shifted from considering 'extrinsic' (environmental) determinants to 'intrinsic' (physiological) ones. The drive behind this change has been the increasing availability of neural network models. With contributions from experts in the field, this book provides a complete survey of artificial neural networks. The book opens with two broad, introductory level reviews on the themes of the book: neural networks as tools to explore the nature of perceptual mechanisms, and neural networks as models of perception in ecology and evolutionary biology. Later chapters expand on these themes and address important methodological issues when applying artificial neural networks to study perception. The final chapter provides perspective by introducing a neural processing system in a real animal. The book provides the foundations for implementing artificial neural networks, for those new to the field, along with identifying potential research areas for specialists.

COLIN R. TOSH is a postdoctoral researcher currently based at the Institute of Integrative and Comparative Biology, University of Leeds. He began his career as an experimental behavioural biologist, specialising in the host utilisation behaviour of insects. More recently he has extended his interests to theoretical biology and is currently interested in applying neural network models to study the impact of information degradation and bias between trophic levels (predator–prey, herbivore–plant, etc.). He is the author of numerous papers in international journals of ecology and evolution and recently published a major review on insect behaviour.

GRAEME D. RUXTON is Professor of Theoretical Ecology at the University of Glasgow. He began academic life as a physicist, but ended up in behavioural ecology after a detour into statistics. His research focuses on the use of mathematical models as tools for understanding animal behaviour, with particular interest in cognitive aspects of predator–prey interactions. He has co-authored over 200 peer-reviewed papers, one textbook and two monographs. Ruxton and Tosh have several years' experience of fruitful collaboration centred on the use of neural networks as representations of the sensory and decision-making processes of predators.

MODELLING PERCEPTION WITH ARTIFICIAL NEURAL NETWORKS

COLIN R. TOSH
Faculty of Biological Sciences, University of Leeds

GRAEME D. RUXTON
Faculty of Biomedical Life Sciences, University of Glasgow

CAMBRIDGE
UNIVERSITY PRESS

CAMBRIDGE UNIVERSITY PRESS
Cambridge, New York, Melbourne, Madrid, Cape Town, Singapore,
São Paulo, Delhi, Dubai, Tokyo

Cambridge University Press
The Edinburgh Building, Cambridge CB2 8RU, UK

Published in the United States of America by Cambridge University Press, New York

www.cambridge.org
Information on this title: www.cambridge.org/9780521763950

First published 2010

Printed in the United Kingdom at the University Press, Cambridge

A catalogue record for this publication is available from the British Library

Library of Congress Cataloguing in Publication data
Modelling perception with artificial neural networks / [edited by] Colin R. Tosh,
Graeme D. Ruxton.
p. cm.
Includes index.
ISBN 978-0-521-76395-0 (hardback)
1. Perception–Computer simulation. 2. Neural networks (Computer science) I. Tosh, Colin.
II. Ruxton, Graeme D. III. Title.
QP441.M63 2010
612.8′2–dc22 2010010418

ISBN 978-0-521-76395-0 Hardback

Contents

Contributors

Nihat Ay
Max Planck Institute for
Mathematics in the Sciences,
Inselstrasse 22–26,
D-04103 Leipzig,
Germany 13

Gianluca Baldassarre
Laboratory of Computational
Embodied Neuroscience,
Istituto di Scienze e Tecnologie
della Cognizione,
Consiglio Nazionale delle Ricerche
(LOCEN-ISTC-CNR),
Via San Martino della Battaglia 44,
I-00185 Roma,
Italy

Alexander Borst
Max Planck Institute for Neurobiology,
Systems and Computational
Neurobiology,
Am Klopferspitz 18,
82152 Martinsried-Planegg,
Germany

Raffaele Calabretta
Institute of Cognitive Sciences and
Technologies,
Italian National Research Council,
Rome,
Italy

Abdul Chrachri
Faculty of Science,
University of Plymouth,
Drake Circus,
Plymouth PL4 8AA,
UK

Dara Curran
Cork Constraint
Computation Centre,
Western Gateway Building,
University College Cork,
Cork,
Ireland

L. Douw
Department of Neurology,
VU University Medical Centre,
PO Box 7057,
1007 MB Amsterdam,
the Netherlands

Magnus Enquist
Group for Interdisciplinary
Cultural Research/Zoology Institution,
Stockholm University,
SE-106 91 Stockholm,
Sweden

Jessica Flack
Living Links,
Yerkes National Primate Research Center,

Emory University,
201 Dowman Drive,
Atlanta,
Georgia 30322,
USA

Daniel W. Franks
York Centre for Complex Systems Analysis (YCCSA),
Department of Biology,
& Department of Computer Science,
University of York,
YO10 5YW,
UK

Wayne M. Getz
Department of Environmental Sciences, Policy and Management,
University of California at Berkeley,
201 Wellman Hall,
CA 94720–3112,
California,
USA

Stefano Ghirlanda
Group for Interdisciplinary Cultural Research,
Stockholm University,
SE-106 91 Stockholm,
Sweden

Kevin Gurney
Adaptive Behaviour Research Group,
Department of Psychology,
University of Sheffield,
Sheffield S10 2TP,
UK

J. J. Heimans
Department of Neurology,
VU University Medical Centre,
PO Box 7057,
1007 MB Amsterdam,
the Netherlands

Noél M. A. Holmgren
Department of Life Sciences,
University of Skövde,
P.O. Box 408,
SE-541 46 Skövde,
Sweden

M. Klein
Department of Medical Psychology,
VU University Medical Centre,
PO Box 7057,
1007 MB Amsterdam,
the Netherlands

David C. Krakauer
Santa Fe Institute,
1399 Hyde Park Road,
Santa Fe,
NM 87501,
USA

Francesco Mannella
Laboratory of Computational Embodied Neuroscience,
Istituto di Scienze e Tecnologie della Cognizione,
Consiglio Nazionale delle Ricerche (LOCEN-ISTC-CNR),
Via San Martino della Battaglia 44,
I-00185 Roma,
Italy

Sami Merilaita
Environmental and Marine Biology,
Åbo Akademi University,
Biocity,
Tykistökatu 6 A,
FIN-20520,
Turku,
Finland

Marco Mirolli
Laboratory of Computational Embodied Neuroscience,

Istituto di Scienze e Tecnologie
della Cognizione,
Consiglio Nazionale delle Ricerche
(LOCEN-ISTC-CNR),
Via San Martino della Battaglia 44,
I-00185 Rome,
Italy

Niclas Norrström
Department of Life Sciences,
University of Skövde,
P.O. Box 408,
SE-541 46 Skövde,
Sweden

Colm O'Riordan
Department of Information
Technology,
National University of Ireland,
Galway,
Ireland

Takashi Odagaki
Department of Physics,
Kyushu University,
Fukuoka 812–8581,
Japan

Julian D. Olden
School of Aquatic and Fishery
Sciences,
Box 355020,
University of Washington,
Seattle,
Washington 98195,
USA

Hiraku Oshima
Department of Physics,
Kyushu University,
Fukuoka 812–8581,
Japan

Richard A. Peters
Centre for Visual Sciences,
Research School of Biological
Sciences,
Australian National University,
Canberra ACT 0200,
Australia

Karin S. Pfennig
Department of Biology,
CB #3280,
University of North Carolina,
Chapel Hill,
NC 27599,
USA

Steven M. Phelps
P.O. Box 118525,
Department of Zoology,
University of Florida,
Gainesville,
FL 32611,
USA

J. C. Reijneveld
Department of Neurology,
VU University Medical Centre,
PO Box 7057,
1007 MB Amsterdam,
the Netherlands

Michael J. Ryan
Section of Integrative
Biology C0930,
University of Texas,
Austin,
TX 78712,
USA

Graeme D. Ruxton
Division of Environmental and
Evolutionary Biology,
Faculty of Biomedical and Life
Sciences,
University of Glasgow,
Glasgow G12 8QQ,
UK

Lawrence H. Snyder
Department of Anatomy and Neurobiology,
Box 8108,
Washington University School of Medicine,
660 S. Euclid Ave.,
St. Louis,
MO 63110,
USA

C. J. Stam
Department of Clinical Neurophysiology,
VU University Medical Centre,
PO Box 7057,
1007 MB Amsterdam,
the Netherlands

Colin R. Tosh
Institute of Integrative and Comparative Biology,
Faculty of Biological Sciences,
University of Leeds,
Leeds LS2 9JT,
UK

Robert L. White III
Department of Anatomy and Neurobiology,
Washington University School of Medicine,
660 S. Euclid Ave.,
St. Louis,
MO 63110,
USA

Roddy Williamson
Faculty of Science,
University of Plymouth,
Drake Circus,
Plymouth PL4 8AA,
UK

Introduction: Modelling perception with artificial neural networks

Colin R. Tosh and Graeme D. Ruxton

This book represents a substantial update of a theme issue of the *Philosophical Transactions of the Royal Society B Journal*, 'The use of artificial neural networks to study perception in animals' (*Phil Trans R Soc B* 2007 March 29; 362(1479)). Most of the 14 papers in that theme issue have been significantly updated and we include a further five entirely new chapters, reflecting emerging themes in neural network research. Our reasons for undertaking the theme issue and this book were not entirely altruistic. Having a young but growing interest in the use of artificial neural networks, we hoped that the publications would be an excuse for us to learn about areas in neural network research that seemed interesting to us and of potential application to our research. The people who will get most from the book are, therefore, ecologists and evolutionary biologists, perhaps with a notion of using neural network models of perception, but with little experience of their use. That said, the content of this book is extremely broad and we are confident that there is something in it for any scientist with an interest in animal (including human) perception and behaviour.

We organise the book into four fairly loose categories. The chapters by Kevin Gurney and Steve Phelps are broad reviews and introduce the two main themes of the book: neural networks as tools to explore the nature of perceptual processes, and neural networks as models of perception in ecology and evolutionary biology. Kevin Gurney's chapter is an excellent general introduction to the theory and use of neural networks and tackles the question: what can simple neural network models tell us about real neural circuits and the brain? Steve Phelps's chapter is a 'where it's at and where it's going' of artificial neural network models used to explore perceptual allocation and bias, and the models and ideas in it can be applied to many other areas of ecology and evolutionary biology. Like most of the articles in the book, both of these chapters can be appreciated by those with little or no mathematical expertise.

The next six chapters are research or focused review articles on neural network models and their use in elucidating the nature of perceptual processes in animals. Axel Borst's chapter describes and compares the properties of different neural models of motion

Modelling Perception with Artificial Neural Networks, ed. C. R. Tosh and G. D. Ruxton. Published by Cambridge University Press.

detection: specifically Reichardt and gradient detectors. We (the editors) are excited about the potential of applying such models to issues in predator–prey interactions, to address how predator-targeting accuracy is affected by the speed and number of moving prey items. Robert White and Larry Snyder use a recurrent neural network model to investigate how accurate internal representations of visual space are formed in primates. Francesco Manella *et al.* use a novel computational model which is strongly rooted in the anatomy and physiology of the mammalian brain to investigate the role of the amygdala in the phenomenon of devaluation in an instrumental conditioning task. Raffaele Calabretta explores the concept of 'genetic interference': a phenomenon that can reduce the evolvability of both modular and nonmodular visual neural networks but can be alleviated by 'sexual reproduction' in neural networks. The last two chapters of this section represent a distinct sub-theme: the relationship between connective architecture of neural networks and their functioning. Hiraku Oshima and Takashi Odagaki investigate the influence of regular, small world, random network structures on the storage capacity and retrieval time of Hopfield networks. Linda Douw *et al.* consider whether the neural and behavioural consequences of brain tumours are due to disruption of the small world properties of whole brain networks. The issue of the relationship between network structure and functioning in a burgeoning theme in wider network theory (e.g. social and communication networks) should be of interest to anyone interested in how animal behaviour evolves in response to the environment.

The next five chapters are by ecologists and evolutionary biologists and apply neural networks to classic questions in these disciplines. Karin Pfennig and Michael Ryan apply Elman networks to study the evolution of character displacement and mate choice using the calls of tundra frogs as network input. Sami Merilaita reviews recent work on the anti-predator benefits of prey colouration that uses simple neural network models. Noél Holmgren *et al.* review recent work on the use of neural networks to study ecological specialisation and sympatric speciation: an interesting approach that offers a potentially powerful alternative to traditional mathematical simulation models in these areas. All of these papers additionally use or discuss genetic algorithms, an optimisation framework also applicable to models other than neural networks that tune model parameters through a selective process analogous to natural selection. This powerful 'organic' selection method can be applied to a variety of systems. David Krakauer *et al.* use analytical mathematics with simple feedforward neural networks to show that multimodal signals (animal signals that exploit multiple sensory organs) can increase the robustness of signals through multiple channels (e.g. frequencies in vocalisation). Finally, Richard Peters investigates the difficulties involved in signal recognition by a species of lizard using a saliency map and a winner-take-all neural network of leaky integrate-and-fire neurons. This model is based on some of the known properties of visual processing in primates and will appeal to ecologists who want to explore what the most salient object is in a particular visual scene, but are discouraged by the abstraction of simple connectionist approaches.

The next five chapters are generally on methodological issues in the use simple feedforward networks. Chapter 14 by Stephano Ghirlanda and Magnus Enquist and

Chapter 15 by Colin Tosh and Graeme Ruxton (the editors) are on the phenomenon coined 'path dependence' by the former authors. This is the tendency of certain neural networks with commonly used architectures and training methods to vary in predictive properties, depending on the order of presentation of training inputs, or with stochastic variation in the starting properties of networks. This effect could have important biological as well as methodological implications. In their chapter, Dan Franks and Graeme Ruxton argue that training methods such as back propagation that researchers have used with feedforward nets to study learning in animals are inappropriate as normally applied because learning is too slow. They offer a modified protocol for the application of training procedures that better replicates the tempo of learning in real animals. Dara Curran and Colm O'Riordan review methods used to effect adaptive evolution in both the weights and architecture of artificial neural networks. We also place the chapter by Julian Olden in this section. This chapter, as well as being an interesting research paper on the relationship between landscape properties and animal movements, applies methods that allow one to dissect the functioning of neural networks. These methods should help to dispel the common myth that neural networks are 'black boxes' that produce interesting results but whose functioning and action cannot be analysed. Finally, the chapter by Roddy Williamson and Abdul Chrachri does not fit into any of the aforementioned categories and describes a real neural network: the cephalopod vestibular system. This chapter emphasises the fact that real neural networks are considerably more complex than most of the simple artificial ones described in the book, and in some (perhaps many) neural systems this complexity must be embraced in order to fully understand the system.

One of our loftier objectives in putting together this book was to attract readers from a broad and disparate range of disciplines and so foster cross-fertilisation of ideas. Papers in the book should interest readers from psychology, neurobiology, mathematics, ethology, ecology and evolutionary biology. It is hoped that readers from each of these disciplines might find something from another discipline that interests them and gives them new ideas for their own research. For example, many psychologists and neurobiologists could undoubtedly benefit from an increased appreciation of the evolutionary context of their study system, while many ecologists and evolutionary biologists could benefit from a greater appreciation of the neural mechanisms underlying phenomena at the level of the whole organism. We also hope that greater use of artificial neural networks might reduce the need for invasive animal experimentation. The study of nervous systems, using artificial models or otherwise, will always be founded on experiments with real nervous systems, but models can reduce the need for experimentation at particular stages of a research programme. A reliable model can simulate multiple scenarios and inform researchers on which areas of endeavour are likely to be most rewarding, thereby reducing the need for experimentation in areas that could lead up 'blind alleys'.

We keep this introduction short and leave the job of covering broad scientific themes in the use of neural network models to study animal perception to the first two chapters of the book.

Part I

General themes

1

Neural networks for perceptual processing: from simulation tools to theories

Kevin Gurney

1.1 Introduction

This paper has two main aims. First, to give an introduction to some of the construction techniques – the 'nuts-and-bolts' as it were – of neural networks deployed by the authors in this book. Our intention is to emphasise conceptual principles and their associated terminology, and to do this wherever possible without recourse to detailed mathematical descriptions. However, the term 'neural network' has taken on a multitude of meanings over the last couple of decades, depending on its methodological and scientific context. A second aim, therefore, given that the application of the techniques described in this book may appear rather diverse, is to supply some meta-theoretical landmarks to help understand the significance of the ensuing results.

In general terms, neural networks are tools for building models of systems that are characterised by data sets which are often (but not always) derived by sampling a system input-output behaviour. While a neural network model is of some utility if it mimics the behaviour of the target system, it is far more useful if key mechanisms underlying the model functionality can be unearthed, and identified with those of the underlying system. That is, the modeller can 'break into' the model, viewed initially as an input-output 'black box', and find internal representations, variable relationships, and structures which may correspond with the underlying target system. This target system may be entirely non-biological (e.g. stock market prices), or be of biological origin, but have nothing to do with brains (e.g. ecologically driven patterns of population dynamics). In these instances, we can ask whether the internal network machinations are informative of specific relationships between system inputs and outputs, and any internal variables. However, the mechanistic elements of a network have names which are evocative of processing in the animal brain; there is talk of 'artificial neurons', their interconnection strengths and 'learning'. If, therefore, a neural network is a model of part of the brain, the problem of interpretation of internal mechanisms is particularly acute. For, if these mechanisms are based on those in the brain, is it the case that they reflect genuine, biological neural mechanisms? These and related questions are explored in the second half of the chapter.

Modelling Perception with Artificial Neural Networks, ed. C. R. Tosh and G. D. Ruxton. Published by Cambridge University Press.

1.2 Neural network principles

This section gives a high-level view of some of the principles and techniques used in this book. A more comprehensive treatment at this level can be found in Gurney (1997) while the books by Haykin (1999) and Bishop (1996) take a more mathematical approach.

We start with a pragmatic, working definition of a neural network: A neural network is an interconnected assembly of simple processing elements, *units* or *nodes* whose functionality is loosely based on the animal neuron. The processing ability of the network is stored in the inter-unit connection strengths, or *weights*, obtained by a process of adaptation to, or *learning* from, a set of *training patterns*. The rest of this section is devoted to unwrapping these terms with special emphasis on those networks that appear in subsequent chapters in this book.

1.2.1 Artificial neurons

Figure 1.1 is a graphical description of a typical neural network node.

Input signals $x_1, x_2, \ldots x_n$ are combined to form an output y via an *activation* variable a. The latter is formed by taking a weighted sum of inputs. That is,

$$a = \sum_i w_i x_i \tag{1}$$

The weights w_i may be positive or negative. The activation is then usually transformed by some kind of squashing function which limits the output y to a specified range (usually the interval [0,1]) and introduces a nonlinearity; this latter feature proves to be crucial in endowing neural nets with their powerful functionality (see next section). In the figure, the squashing function has been chosen to be the logistic sigmoid

$$y = \frac{1}{1 + \exp(-(a - \theta))} \tag{2}$$

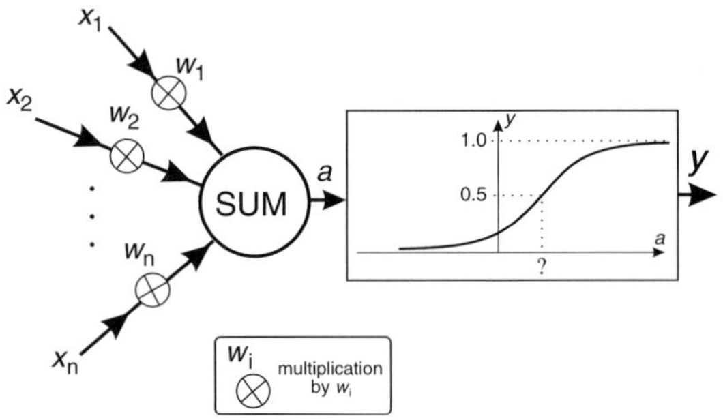

Figure 1.1. Simple model neuron.

although other, similar functions are occasionally used. The constant θ defines the point at which y takes its mid-point value. Moreover, it is the point where the function is changing most rapidly and is therefore the value of the activation at which the node is most sensitive to small changes in the inputs. The negative of θ is therefore sometimes referred to as the *bias*. Notice that y approaches 0 and 1 asymptotically as the activation decreases and increases respectively (so, y is never equal to 0 or 1, but may be made as close to these as we please).

The basic node described above has a long lineage. The first artificial neural node was the Threshold Logic Unit (TLU) introduced by McCulloch & Pitts (1943). This was also a two-stage device with the first stage given by (1) but with the output nonlinearity defined by a discontinuous step function, rather than the smooth ramp described by (2). Thus, the output of the TLU had only two values, 0 or 1, depending on whether the activation was less than or greater than the threshold θ, respectively. A more complex node – the *Perceptron* – was introduced by Rosenblatt (1958) which retained the Boolean (0,1) output of the TLU, but allowed pre-processing of Boolean input variables with arbitrary functions (so-called 'association units') whose outputs then formed the variables x_i in (1). The TLU is therefore a special case of the Perceptron when 'association units' just pass a single input through to each weight.

The neurobiological inspiration for the structure of Figure 1.1 is as follows. The input x_i corresponds to the presynaptic input on afferent i, while the weight w_i encapsulates the corresponding synaptic strength. The product $w_i x_i$ is akin to the post-synaptic potential (PSP) which is inhibitory/excitatory according to whether w_i is negative/positive. The integration of PSPs over the dendritic arbour and soma is represented by simple arithmetic addition, and the quantity a corresponds to the somatic membrane potential. This is then transformed by the squashing function to give a firing rate y. Clearly some of these correspondences are, at best, merely qualitative analogues. The issue of realism is revisited in the second half of the chapter.

1.2.2 Feedforward networks and classification

A ubiquitous problem in perception is that of classification or pattern recognition. As an example, consider the problem of identifying letters of the alphabet. Humans are able to recognise letters in many sizes, orientations and fonts (including handwritten variations) with ease. Any individual person can never see all possible letter variants, but, instead, will learn idealised letter shapes from a very small set of possibilities (usually a plain font in children's reading books). This latter point demonstrates that *generalisation* is a key component in the classification process. That is, the ability to generalise knowledge of specific pattern exemplars to a wide variety of related cases.

Based on this example we now formalise the general problem of classification as follows. Given an arbitrary sensory input pattern drawn from some universal set of patterns, is it possible to place that pattern in its appropriate class or category, where there are generally many fewer classes than the patterns themselves? Further, we suppose that we do not have an exhaustive list of the entire universe of patterns; rather, we only have immediate access to some subset of patterns P, and knowledge of the category that each member of P belongs to. By way of terminology P is the *training set* referred to in the motivational definition of a neural network at the start of this section. The problem is to construct an input-output model, based on this limited knowledge, which will generalise so that, if it is presented with a pattern not in P, it will elicit the correct classification for that pattern. Notice that a 'model' which simply classifies P but does not generalise is easy to construct but of no real interest – it is just a lookup table of pattern-class pairings. We will return to the relationship between neural processing and generalisation later. In the meantime we will look at how the classification problem may be solved in principle by a neural network.

Figure 1.2 shows a *feedforward network* which consists of a layered structure with information flowing from the inputs, at the bottom of the diagram, to the outputs at the top.

The inputs have no functionality as such, but are simply points which receive pattern information and distribute this information to the first layer of neural nodes per se (of the type described above). In the example, there are four inputs, and so all patterns for classification would have to be defined by a list of four numbers. In more formal analyses, these lists of numbers are properly referred to as *vectors* with numeric *components*, and we sometimes speak of *pattern vectors*. This first layer of functional nodes is sometimes referred to as a *hidden layer* since we are not supposed to inspect or control the output values on these nodes (y in Eq. 2) during the process of setting the network weights; that is during *training* or *learning*. The outputs of the hidden layer are subsequently processed by an *output layer* which is used to read out the category in which the input pattern is placed. There are several ways of doing this depending on the way information is represented in the network. We will refer to a network of the kind shown in Figure 1.2 as a two-layer network since it contains two layers of processing nodes. Some authors include the input layer in the layer count so that the network in Figure 1.2 would constitute a 3-layer net. The final point to make here is that networks of the kind depicted

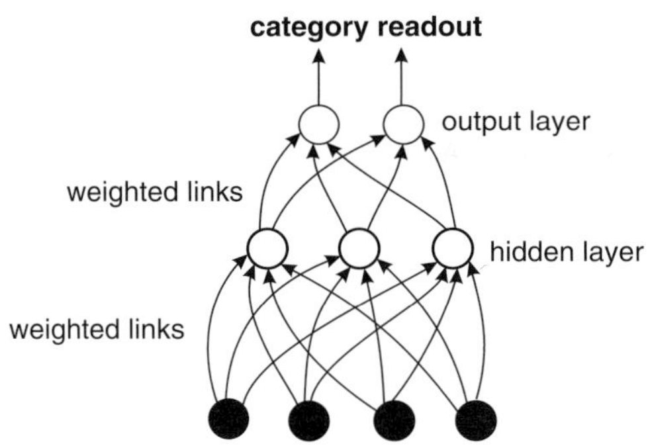

Figure 1.2. Simple two-layer feedforward neural network.

in Figure 1.2 are sometimes called *multi-layer perceptrons* or *MLPs*, in deference to the important role played by Rosenblatt's original perceptron in shaping the theory of neural network learning (Minsky & Papert, 1969).

Notice that processing by any particular node can be performed independently of that in any other. Thus, processing could, in principle, be performed in parallel if we had the necessary hardware resources to assign to each node. In spite of this, most networks find their implementation in software simulation in a conventional computer in which each node has to be visited serially to compute its output.

There is a mathematical framework which is particularly useful for describing quantitatively the process of classification in networks. It is based on the notion that patterns reside in some *pattern space* and is evocative of geometric analogies that enable the problem to be visualised. Suppose, for example, we have patterns belonging to two classes, A and B. If each pattern was defined by only two numerical components, then it could be represented quantitatively as a point in Cartesian axes as shown in Figure 1.3a.

If, in fact, each pattern is a vector with $n > 2$ components, Figure 1.3 is just a cartoon schematic which is simply illustrative of the case in n-dimensions. In Figure 1.3a, the patterns are shown as being separated by a straight line. In 3-D this situation implies a

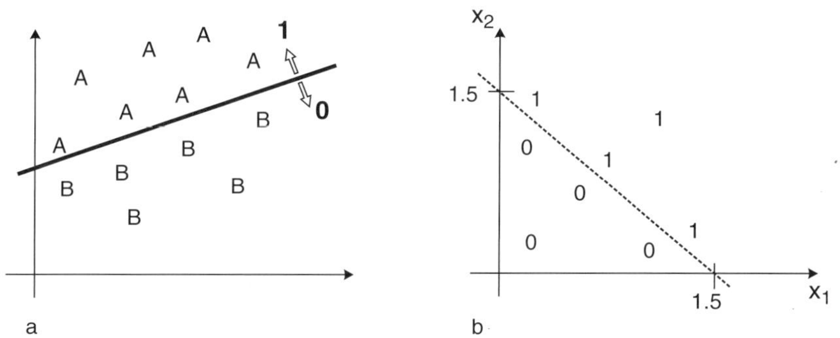

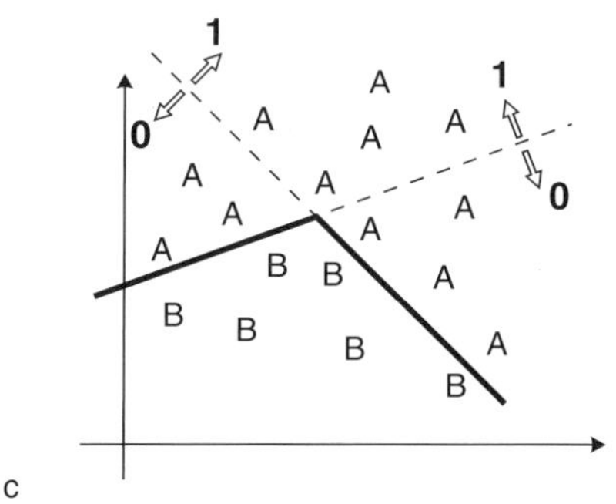

h_1	h_2	y	class
0	0		B
0	1	1	A
1	0	1	A.
1	1	1	A

Figure 1.3. Geometric view of pattern classification.

plane, and in n-D ($n > 3$) a hyperplane. In all these cases we say that the patterns are *linearly separable*, and the straight line is schematically indicative of this.

Suppose we have a single artificial neuron with n-inputs, then it could attempt to solve the classification problem in Figure 1.3a by indicating output values of 1,0 for classes A,B respectively. This could occur exactly if the node was a TLU, and approximately using a node of the form shown in Figure 1.1 (since, in this case, the output approaches 0 and 1 asymptotically). It may be shown that linearly separable problems can indeed be solved by a single artificial neuron; a result which follows from the linearity of signal combination in Eq. 1. To see this, consider, the two-input case and assume, for simplicity, a TLU. The critical condition that defines classification occurs when the activation a equals the threshold θ, since small changes in a around this value cause the node to switch its output between 0 and 1. Putting $a = \theta$, gives $w_1x_1 + w_2x_2 = \theta$. This may be solved for x_2 in terms of x_1 to give

$$x_2 = -\left(\frac{w_1}{w_2}\right)x_1 + \left(\frac{\theta}{w_2}\right)$$

which is a straight line with slope $-w_1 / w_2$ and intercept θ / w_2. Now put, for example, $w_1 = w_2 = 1$, and $\theta = 1.5$. This defines a line $x_2 = -x_1 + 1.5$ as shown in Figure 1.3b. Here, pairs of values (x_1,x_2) defining points on the same side of the line as the origin give TLU outputs of 0, while values defining points on the other side of this line give TLU outputs of 1. In particular, the Boolean inputs (1,1) give an output of 1, while the other three Boolean input pairs give an output of 0 (in this case the TLU is acting as a classical logic AND gate).

Figure 1.3c shows a harder problem in pattern space which may only be solved by a *decision line* (in n-D, a *decision surface*) consisting of two straight, but non-colinear segments (shown by the solid line in the figure). The dotted lines show the extension of the line segments which make each of them a continuous straight line throughout pattern space (similar to the line in Figure 1.3a). Each extended straight line then defines a linearly separable problem which may be solved by nodes with outputs *h1* and *h2*. While each of these separate classifications mixes patterns A and B together, the table in the figure shows how the original classification problem may now be solved by taking suitable combinations of *h1* and *h2*; that is, class B is signalled if and only if both *h1* and *h2* are zero. This 2-component classification problem is linearly separable and may be solved with a single 2-input neural node. The original *A/B* classification problem has therefore been decomposed into two stages which may be solved by a two-layer net with two hidden nodes (yielding *h1* and *h2*) and a single output node.

As the classification becomes more complex, we may now ask the following question: is it possible to solve an arbitrary classification problem with a two-layer net – or do we need to resort to more complex structures? That is, in an analogous way to the example above, can we describe the decision surface of the problem in a piecewise linear way, solve the resulting decomposition using hidden units, and then combine their outputs in a linearly separable fashion which solves the original problem? A series of general results (Lippmann, 1987; Wieland & Leighton, 1987; Makhoul *et al.*, 1989) has answered this in the affirmative so that, in principle, we never need a network with more than two layers. Further, the two-layer network capability may be extended to more than two classes. There is at least one solution to such problems, since each class X may be used to define a binary classification between X and non-X classes at a single output node. It is worth noting that the nonlinearity of the squashing function is essential to these results, for it may be shown that a two-layer net with linear nodes reduces to a single-layer net (which may only solve linearly separable problems).

While this result appears to confer enormous power on neural networks as models of classification, it is a double-edged sword as far as interpreting the network in biological terms is concerned. Thus, we are not forced to explore more biologically realistic architectures simply to achieve a desired functionality (these issues are discussed further below). Further, the result is only an existence proof; no method is supplied for identifying the detailed structure of the net (specifically how many hidden nodes one needs), how to find the weights (training procedure), or how to specify the size or nature of the training set P needed to find the solution. These problems are the subject of the next two sections.

1.2.3 Training feedforward networks

In this section we describe methods for determining the weights in a network, given that it should attempt to classify a training set P, with each member p belonging to a known category t_p. The general idea is to iteratively supply members of the training set as input to the network, compare the network's output with the desired target performance t_p, and make adjustments to the weights to gradually bring the network's output ever closer to the target. Since we have access to the desired output for each pattern, this general paradigm is referred to as *supervised training* (or *supervised learning*).

We consider first the simple case of a 'network' consisting of a single node of the form described by Eqs 1 and 2. In this case the output in response to pattern p is just a single scalar number y_p, as is also the target t_p. The latter will, in general, be 0 or 1 to flag one of two possible classes. A quantitative way to compare the network's output and the target is to take the square of the difference $y_p - t_p$ which guarantees a positive measure. Thus, we define a pattern error $e_p = (y_p - t_p)^2$, and an overall *error* E to be the sum of the e_p over all members of the training set. That is

$$E = \sum_{p \in P} (y_p - t_p)^2 \tag{3}$$

The problem can now be restated as one of attempting to minimise E by making changes in the network weights, a process which is illustrated schematically in Figure 1.4a.

The dependence of E on the weights $E(w)$ is shown in cartoon form in Figure 1.4a by the 'U'-shaped line. In general there will be n inputs ($n > 1$) each with its own weight w_i,

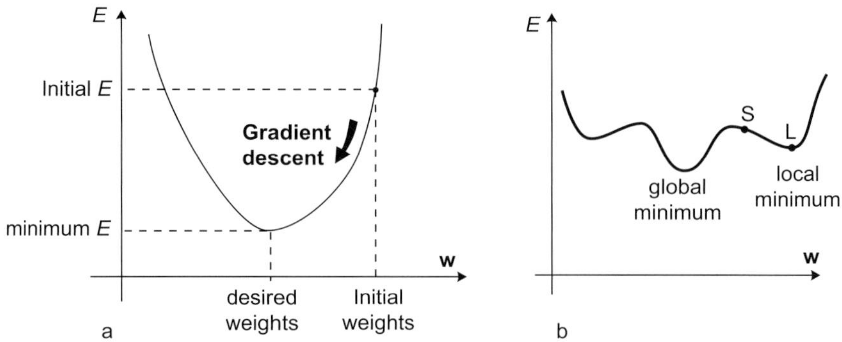

Figure 1.4. Gradient descent.

that helps determine the error; that is, the error-weight function is described by a surface in $n+1$ dimensions. In spite of this, the schematic form in Figure 1.4 gives a useful insight into the training process. Thus, starting with an initial weight set and error, training proceeds by altering the weights in a series of small steps until the ideal or desired weight set is found which corresponds to the minimum error the net can achieve. Figure 1.4 illustrates the point that, if we could always guarantee each step giving a 'downhill' move along the error-weight function $E(w)$, then we would eventually reach the target weight set. It is worth noting one proviso to this, however, which arises because we have assumed that the path from initial to final weights is continually decreasing. In general, the error surface may contain many 'bumps' and not just the global minimum as shown in Figure 1.4a. In this case, moving downhill may result in the network becoming trapped in a local (rather than the desired global) minimum. This phenomenon is illustrated in Figure 1.4b in which, starting at point *S*, training which induces movement down the error surface results in a suboptimal network at the local minimum *L*.

Notwithstanding these potential difficulties, the required process is one of *gradient descent* (moving 'downhill') and is enabled if we can compute the gradient or slope of the function $E(w)$ for each weight w_i. This is indeed possible using processes in the differential calculus. Using these techniques, it is possible to verify the intuitively plausible result that, since E has contributions from all patterns (see Eq. 3), so too do the error gradients. In principle therefore, we should evaluate all these contributions and sum them to find the true gradients for each weight. However, it transpires that it is possible to sidestep this rather compute-intensive process and train the network 'online' using estimates for the gradients found by calculating them for the current input pattern only. Under this regime, the required change in the ith weight Δw_i is given by

$$\Delta w_i = \alpha\sigma'(t_p - y_p)x_p^i \tag{4}$$

where x_p^i is the component of pattern p on input i, σ' is the slope of the squashing function (with the current input) and α is a constant referred to as the *learning rate* which controls how large the weight updates are at each step. Training then consists of repeatedly supplying input patterns from P (together with associated targets) and updating the weights according to the *learning rule* in Eq. 4. The difference term $(t_p - y_p)$ is sometimes referred to as the 'delta', so that this particular method is referred to as the *delta rule*. Nominally, the iterative process continues until there is no appreciable change in the error or until all the outputs are a reasonable match with the target. However, we will see in the next section why it might be sensible to curtail training before this point.

The method described above for a single node may be immediately extended to several nodes forming a single-layer network by simply applying Eq. 4 to each node. Extending to the case of a two-layer network is, however, non-trivial. The problem is that an error on a particular output node could be due to the weights on that node being poorly trained, or to the hidden nodes which supply its inputs being poorly trained; where does the blame lie? This *credit assignment problem* is in fact solvable using regular techniques in the calculus and, because the solution involves computations at the hidden nodes which make use of error information originally at the output nodes, it is referred to as *error back-propagation* or simply *back-propagation*. Several of the chapters in this book make use of this learning technique. Back-propagation has been criticised on the grounds that it is not biologically plausible but, unless we are specifically interested in developmental processes (rather than simply the final, fully functional network), this is not a relevant issue. Further, recent evidence (Fitzsimonds *et al.*, 1997) has shown that synaptic plasticity (the biological analogue of weight changes in neural nets) is indeed able to propagate in local neuronal circuits in ways not dissimilar to that envisaged in back-propagation.

The delta rule was first reported in a meeting abstract by Widrow & Hoff (1960); it was therefore occasionally referred to as the Widrow–Hoff rule in earlier work. For a derivation and overview of the delta rule see Widrow & Stearns (1985). Back-propagation was discovered by Werbos (1974) who reported it in his PhD thesis. It was later rediscovered by Parker (1982) but this version languished in a technical report that was not widely circulated. It was discovered again and made popular by Rumelhart *et al.* (1986) in their seminal book *Parallel Distributed Processing*. Many variants and improvements have since been made to the back-propagation algorithm since these first formulations (for a review see Tveter, 1996).

What happens if we don't have access to individual target information for each output node, but simply have, instead, information as to whether the net has performed 'well' or 'badly'; that is, we have a single scalar reinforcement signal which flags 'reward' if the network performs better than expected in approximating the target, and 'penalty' if the net performs worse than expected. The key to using this approach is to allow the network to 'explore' the space of possible solutions to the problem by adding noise to the node activation. Training then proceeds by increasing the magnitude of weights during learning trials that result in 'reward', and decreasing their magnitude (and eventually changing their sign) during 'penalty' trials. In this way the network learns by 'trial and error'. Andrew Barto and co-workers have developed a range of algorithms for training both single nodes (Barto, 1985; Barto & Anandan, 1985) and networks, (Williams, 1987). More complex variants make use of *temporal difference* learning (Sutton, 1988) and *actor-critic* models (Barto *et al.*, 1983). All these methods fall within the remit of the general theory of *reinforcement learning* (Sutton & Barto, 1998). These methods are also not intrinsically supervised – that is, the reward signal is not necessarily derived from knowledge of what the 'right answer' should be, although reward signals *may* be derived from error functions of the form in Eq. 3.

Neural network learning is an example of the class of problems known as 'function optimisation' or 'parameter search'. As such, neural net learning may take advantage of any of the general purpose techniques developed therein. In particular those methods which do not suffer from becoming trapped in local minima are especially attractive. These include simulated annealing and evolutionary computing methods such as genetic algorithms; for a review of some of these techniques see Shang & Wah (1996).

1.2.4 Networks and generalisation

As noted previously, models of perceptual classification are only useful if they can generalise and discover the underlying model that gave rise to the training set. It is therefore crucial to determine whether neural networks are able to exhibit this key property. To explore this question, consider again the model neuron defined in Eq. 1 and 2. The first thing to note is that the functionality is defined by the processing of continuously graded signals using continuous, smoothly varying relationships. This means that small changes in the inputs will, in general, result in small changes in the output. Moreover, suppose a model neuron (with continuous output function) is responding decisively to a particular pattern so that its output lies close to the asymptotes (0 and 1) of the squashing function. In this regime, the slope of this function is much less than one, and any change in activation will make a negligible change in the output. Since variations in the inputs cause activation changes, we conclude that a model neuron which is responding decisively is relatively insensitive to its input. This will, in turn, foster generalisation, since patterns similar to that currently being applied will not significantly alter the neuron's output.

In summary, it is the 'smooth' or analogue-style signal processing paradigm of artificial neurons, together with their intrinsic nonlinearity, which promotes generalisation. In a full, well-trained network (one which has clear responses to each pattern) these properties work synergistically across individual model neurons, so that similar input patterns will evoke similar patterns of activity over the hidden layer which will, in turn, elicit similar network outputs. Further, while the particular artificial neuron we have studied so far is a rather impoverished model of the real animal neuron, the nonlinear combination of analogue signals is almost certainly found in real neurons. We therefore conjecture that generalisation is an intrinsic property of real brain circuits arising through some quite general properties of biological neural processing.

Generalisation in a network can, however, be severely reduced if the network is not well suited to modelling the underlying pattern data. In particular, if there are too many

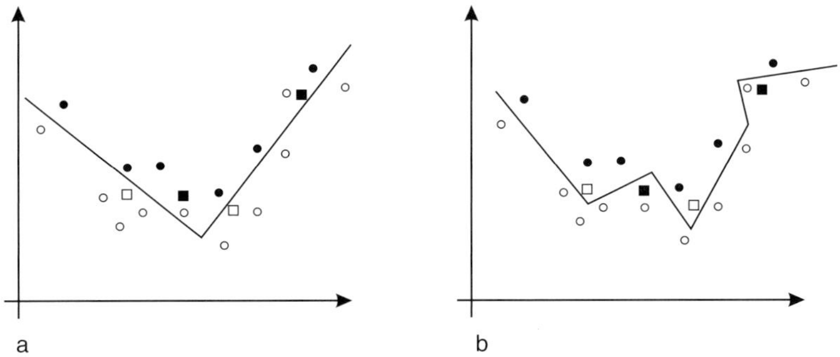

Figure 1.5. Overtraining in pattern space. Two pattern classes are denoted by open and filled symbols; training and test data are shown by circular and square symbols, respectively.

hidden nodes, then generalisation can be curtailed – a point which is best made in the context of pattern space. Consider the classification problem with two classes, shown schematically in Figure 1.5. In Figure 1.5a, there are two line segments indicating two hyperplane fragments in the decision surface, implemented using two hidden units. Two members of the training set have been misclassified, apparently indicating poor performance. However, suppose that four previously unseen *test patterns* (i.e. patterns not in P), are presented, as shown by the square symbols. These have been classified correctly and the net has generalised from the training data. Thus, if we interpret the two misclassified training patterns as outliers or noisy data, the net has implemented a good model of the data which captures its essential characteristics in pattern space.

Consider now Figure 1.5b, in which there are six line segments associated with the use of six hidden units. The training set is identical to that used in the previous example and each one has been successfully classified. However, all four test patterns have been incorrectly classified so that, even though the training data are all dealt with correctly, there may be many patterns which are misclassified. The problem here is that the net has too much freedom (via its abundance of hidden nodes) to choose its decision surface and has *overfitted* it to accommodate any noise and intricacies in the data without regard to the underlying model. Thus, while a two-layer net with an essentially unlimited supply of hidden nodes can tackle arbitrary tasks, inferring the underlying model with a limited training set P is a problem of some delicacy.

Some of the techniques for dealing with the problem of overfitting are outlined in Gurney (1997, section 6.10), however one prominent technique deserves mention here, and is based on the method of *cross-validation* found in conventional statistical modelling. Consider again the network with too many hidden units whose decision surface is shown schematically in Figure 1.5b. The diagram shows the decision surface after exhaustive training, but what form does this take in the early stages of learning? It is reasonable to suppose that a smoother form (something more like that in Figure 1.5a) would be developed at this time, before the net has had a chance to learn the details in the

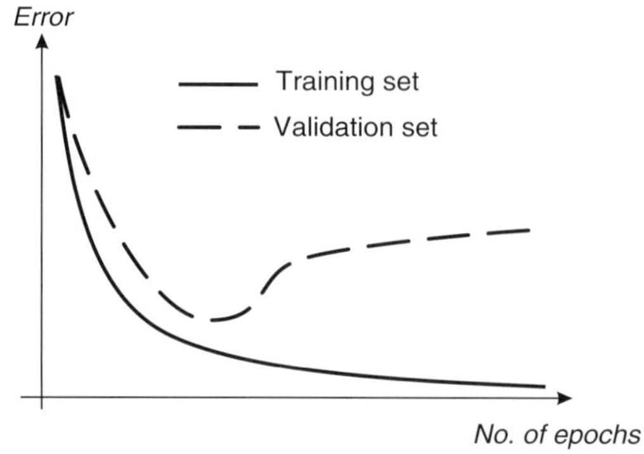

Figure 1.6. Cross-validation.

training set. If we then curtail the training at a suitable stage it may be possible to 'freeze' the net in a form that generalises more appropriately. Rosin & Fierens (1995) showed that this is exactly what does happen in a simple example with two inputs.

Assuming that this process of gradual approximation is a general one when training feedforward nets, how are we to know when to stop training the net? One approach is to divide the available training data into two sets: one training set proper P, and one so-called *validation set* V. The idea is to train the net in the normal way with P but, every so often, to determine the error with respect to the validation set V. The typical behaviour of a network under this process of *cross-validation* is shown in Figure 1.6.

One criterion for stopping training therefore, is to do so when the error over V reaches a minimum, since this is indicative that generalisation with respect to patterns not in the nominal training set P is optimal. Cross-validation is a technique borrowed from regression analysis in statistics and has a long history (Stone, 1974). That such a technique should find its way into the 'toolkit' of supervised training in feedforward neural networks should not be surprising because networks are also fitting models to statistical data. While we have emphasised the pattern space viewpoint of network function in the paper, it is also possible to conceive of networks as performing function approximation (Gurney, 1997, section 6.7.2) and so they can be thought of as performing nonlinear regression. These similarities are explored further in the review article by Cheng & Titterington (1994).

1.2.5 Knowledge representations in neural networks

At the end of training a network, any knowledge or long-term memory is stored in the weights or connection strengths. This has given rise to the term *connectionism* to describe the neural network modelling approach in biological areas (especially in psychology). However, while knowledge is stored in the weights, there is an additional way in which knowledge representation occurs in networks. On applying an external input and allowing it to be processed, a characteristic pattern of activity will be developed across the net that may be thought of as representing knowledge about the current input. For feedforward nets it is the intermediate hidden layers that provide a particular focus of interest here, for it is across these nodes that an *internal representation* of the training set occurs. In this view, these representations are then operated on, or decoded by, the output layer into a form which is to be interpreted as the 'answer' or response to the input. As noted above, the role of the hidden layer(s) may be thought of as ensuring that perceptually similar inputs are re-represented so as to be close together in the pattern vector sense, a perspective of hidden unit functionality emphasised by Rumelhart & Todd (1993). A related viewpoint conceives of hidden units as 'feature detectors' that extract the underlying aspects of the training set while ignoring irrelevant clutter or noise.

Whatever interpretation is adopted in respect of the hidden layer, there are essentially three different types of activity profile that can occur over any neural layer (hidden or otherwise). In a *localist* representation, each semantically discrete item, concept or idea is associated with the activity of a single node. For example, in a classification task, the occurrence of a particular class would be signalled in a neural layer by a single node of that layer being active (output close to 1) while all others are inactive (output close to 0). Figure 1.7a shows an example of this scheme using four nodes to categorise objects into one of four classes – horizontal rectangle/ellipse, vertical rectangle/ellipse. Activities which are close to one/zero are shown by filled/open circles respectively.

Figure 1.7b shows a *semilocalist* or *feature based* representation of the same classes. Each node now stands for a feature of the object class – 'horizontal', 'vertical', 'rectangle', 'ellipse'. Classes are now designated by suitable combinations of active feature nodes. Notice that in both localist and semilocalist representations, a minimum of four nodes is required in each case. Finally, Figure 1.7c shows a *distributed* representation, in which every node within the layer plays some role in the activity profile representing each class. Here, nodes may be partially active (indicated by shades of grey)

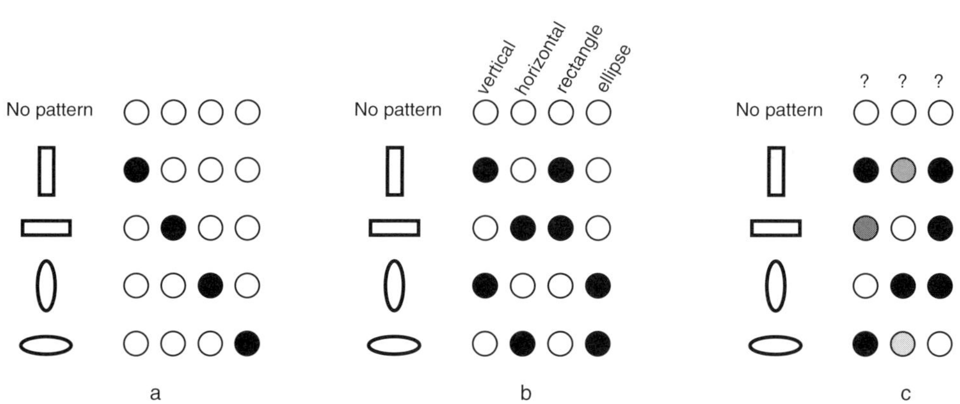

Figure 1.7. Representations.

and we don't necessarily need four nodes to obtain a unique pattern of activity for each class (the example uses three). There is now no clear interpretation of the significance of activity on any particular node (as indicated by the question marks) and we talk of the nodes designating *micro-features* or *sub-symbolic* entities.

The emphasis by some on distributed representations, combined with the fact that, in principle, artificial neurons can process their information in parallel, has led to the term *Parallel Distributed Processing* (PDP) to be applied to the computational paradigm established using neural networks (Rumelhart *et al.* 1986).

1.2.6 Learning temporal sequences

The networks we have dealt with so far are unable to deal with temporal dependencies; they respond immediately to their input with an output response. However, many perceptual processes rely on discovering temporal sequences in the input. For example, language, and animal noises and calls in general, may be thought of as a sequence of aural stimulus primitives, concatenated to produce an overall temporal pattern. To learn this patterning requires that a network stores some kind of 'memory' or previous state information that can be combined with its present input, thereby conditioning that input on its temporal context within a sequence.

Figure 1.8 shows a simple example of a network that can perform this kind of task. Structurally, it consists of a two-layer network of hidden and outputs nodes as in Figure 1.2, but it is augmented by another set of nodes – the *state* or *context* nodes – which supply input to the hidden layer. The state nodes are also hidden in the sense that they do not interact with the environment, and receive a copy of the activity on the hidden layer via fixed weights with value 1. The net is also endowed with a clock which sequences operations in the network over a series of discrete time steps as follows. At the start of all operations, the context nodes are initialised to output value 0.5. At the

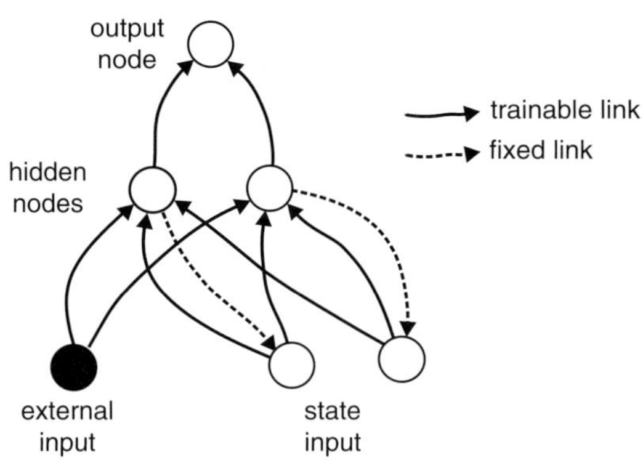

Figure 1.8. Elman network.

first time step, the first input pattern in the sequence is supplied to the external input, the network produces an output, and the hidden layer copies its pattern of activity to the state nodes. If the network is being trained, then back-propagation could be applied at this point. At the second time step, the next input pattern is presented but, since the context nodes have a record of the previous hidden layer activity, the hidden nodes also receive their own previous state as input. In this way, the network can learn temporal context and sequential dependencies.

The first example of this kind of network appeared in a technical report by Jordan (1986) and it was Elman (1990) who popularised the technique by showing how this general architecture could learn sequences in a variety of abstract temporal patterns as well as simple examples in the English language. Chapter 9 by Pfennig & Ryan describes a network of the kind described by Elman.

Quite generally, any network with feedback or *recurrent* connections (like those between the state and hidden nodes in the Elman net) will support 'memory' of some kind. Indeed, networks with massively recurrent interconnections have been extensively studied as models of associative memory in both abstract (e.g. Hopfield, 1982) and psychological (e.g. McClelland & Rumelhart, 1985) settings.

This completes the technical exposition part of the paper. In the next section we move on to discuss the theoretical status of network models and their structural and functional components.

1.3 Meta-theoretical issues

It has already been noted that a two-layer net can, in principle, perform any input–output mapping. The question to ask when building a neural network model is therefore not 'will it work?' but, rather, 'does the model shed any light on the target system?'. More formally we ask whether the model gives any theoretical insight into the mental and neural processes underlying perception and cognition. The relationship between neural network models and theories of mental processing has been hotly debated – see for example Smolensky (1988), McCloskey (1991) and Green (2001). However, as a prelude to our discussion, it useful to distinguish, quite generally, between two rather different types of computational model, without specific reference to neural networks.

1.3.1 Description versus governance

A phenomenological or *descriptive* model of a system is one which replicates the behaviour of the system but, in doing so, does not make recourse to the mechanisms that are believed to *govern* the system's behaviour. This distinction is scientifically very general, but has been explored in relation to brain modelling by Dayan (2002). For example, the time course of the voltage of a neural membrane when an action potential is produced may be described reasonably accurately by some high-order polynomial function of distance along the axon, and time. However, this makes no reference to the sodium and potassium currents which are supposed to underlie action potential generation. In contrast, the account of Hodgkin & Huxley (Hodgkin & Huxley, 1952; Koch, 1999) invokes just these mechanisms to provide an account of action potential generation. As another example, this time from economics, consider financial indicators such as the Dow-Jones or FTSE indices. It may be possible to fit their time series approximately by arbitrarily complex functional forms (indeed, a linear trend with positive slope is a good first approximation) but these quantities are ultimately governed by the action of thousands of independent 'agents' – stockbrokers who buy and sell shares on the financial markets. A mechanistic model might attempt to invoke the dynamic interaction of these agents and the markets they generate.

These two examples differ in that one (the neural membrane) is deterministic, while the other (stock prices) is subject to noise and is a problem in statistical pattern analysis. Neural networks are a class of tools that are well suited to tackling problems of the latter kind. For example, using training data generated by recording market prices over the recent past, a network could take as inputs a set of raw share prices and other financial indicators, and attempt to generate the next day's prices or FTSE index. The resulting model will make no reference to agent-based mechanisms but will describe, phenomenologically, the trends in the data.

Neural networks, considered as general statistical modelling tools, share similarities with other techniques in this general area. Thus, the outputs of the hidden layer in a feedforward network – the internal representation of the data – are analogous to the factors in factor analysis, and the hidden unit weights akin to the loadings on the variables. Moreover, in the same way that it is possible to try and interpret the factors of factor analysis (under 'rotation' of the loading vectors) it makes sense to try and understand what the hidden nodes in a network are representing. That is, rather than conceive of the network as a 'black box' that simply replicates behaviour (which is nevertheless still useful) one can probe the model to see what combinations of input are primarily responsible for determining the output. In this way, neural networks as statistical models are endowed with explanatory power. Olden provides an example of a neural network as a statistical model in Chapter 18, and goes on to examine methods for discovering causal relationships amongst the model variables.

1.3.2 The problem of biological realism

Suppose we have a neural network model which takes inputs that could directly represent sensory stimuli of some kind, and whose outputs represent a perceptual classification. The function of the network is, therefore, potentially mirroring a processing task performed in the brain. We might then reasonably ask to what extent any explanatory power of the neural network relates to the mechanisms in the brain that underlie the perceptual task. The problem here however is that, while biological perception is mediated by the brain, and neural networks are somewhat brain-like, they are in many respects biologically implausible. The functionality of real neurons is enormously more complex than that implied in our model neuron (Eqs 1 and 2) and brain circuits are usually more complex than the simple homogeneous feedforward nets described here. Neural networks might therefore appear *prima facie* to have little or no explanatory power in regards of the computations performed by brain circuits underlying perception (Crick, 1989).

There are two possible ways to tackle this problem. First, we could choose to meet the demands of biological realism head on, as it were, and constrain our models to be biologically plausible. This route takes us into the newly established discipline of computational neuroscience in which it is acknowledged that, unlike the stereotypical, and homogeneous networks described so far, brain systems typically make use of complex circuits with many layers, together with both inter- and intra-layer feedback connections (Shepherd & Koch, 1998). Further, the functionality of real neurons is more powerful than that of the artificial nodes in neural networks, and shows a much richer diversity (Koch *et al.*, 2002). Thus, the output of real neurons is best characterised as a series of discrete voltage 'spikes' (action potentials) rather than a continuous-valued variable. Information may therefore be encoded, not only in the mean firing rate (one interpretation of the node output y), but also in the specifics of inter-spike timing (Rieke *et al.*, 1996). Neural behaviour is mediated by a profusion of different ionic currents traversing the neural membrane, and real neurons have an extended morphology allowing complex computation to be performed over the dendritic arbour. The possible variations of membrane function and morphology give rise to a huge diversity of cell types in the brain with a corresponding diversity of computations. This may include the simple 'weight and add' of artificial neural net nodes, but also extend to include nonlinear combinations of inputs (Shepherd & Koch, 1998). Models which are explicitly constrained by detailed biological data form the subject of Chapter 3 by Borst, and Chapter 19 by Williamson & Chrachri.

An alternative approach to tackling the problem of biological realism and neural networks is to step back and try to define the problem of 'brains and computation' more precisely. This was the remit of Dror & Gallogly (1999) when they explored the problem of biological plausibility in the general arena of cognitive modelling. Here, we focus specifically on neural network models of brain function but, nevertheless, start with a general contextual framework for our analysis.

1.3.3 Marr's hierarchical analysis

In a technical report (Marr & Poggio, 1976), and later in his seminal book on vision (Marr 1982), Marr described a hierarchical framework for understanding computational processes in perception and cognition. At the top of the hierarchy is the *computational* level. This attempts to answer the questions – what is being computed and why? It describes the essential characteristics of the input–output transforms and any constraints that apply therein. The next level is the *algorithmic* which describes precisely how the computation is being carried out, and finally, there is the *implementation* level which gives a detailed description of what hardware the algorithm makes use of. Marr's original example in his book *Vision* (Marr 1982) provides a very clear illustration of this framework. Consider the computation of the bill in a supermarket with a cash register. In answer to the top-level question of 'what' is being computed, it is the arithmetical operation of addition. As to 'why' this is being done, it is simply that the laws of addition reflect or model the way we should accumulate prices together from piles of goods in a trolley; it is incorrect, for example, to multiply the prices together. Next, we wish to know exactly how this arithmetic operation is performed. The answer is that it is done by the normal procedure taught at school where we add individual digits in columns and carry to the next column if required. Further, in cash registers, this will be done in the decimal representation rather than binary (normally encountered in machine arithmetic) because rounding errors are incurred when converting between the normal (decimal) representation of currency and binary. As for the implementation, this occurs using logic gates made out of silicon, silicon-oxide and metal. Notice that choices at different levels are, in principle, independent of each other. For example, we could have chosen to use a binary representation, and alternative implementations might make use of mechanical machines or pencil and paper. The importance of discovering good representations for solving the problem is crucial: the ancient Romans failed to develop a positional number system and so struggled to fully develop arithmetic.

We now consider an example in neural networks models of perception. It concerns the computation of the apparent velocity of moving objects in the visual field. This is ethologically useful since we might want to know how fast a car, or falling piece of fruit, is travelling in order to avoid it, or catch it, respectively. In Marr's scheme then, the computation being performed is: find the velocity of the object in retinotopic coordinates (find how fast the object is moving with respect to the eye).

One algorithm for doing this computation is based on Fourier analysis of the image. Just as 1D temporal audio waveforms may be decomposed into Fourier (sinusoidal) components, so too can 2D spatial images be decomposed into components which consist of gratings with sinusoidally varying luminance. Figure 1.9a shows an example of such a grating which is supposed to be moving in a direction indicated by the velocity vector (arrow) $\mathbf{v}_g$. This direction is defined by the perpendicular to the grating bars, and the length of this vector is proportional to the speed of the grating.

However, if the grating is the only component in the image, it nominally extends infinitely in both directions in the plane. Therefore, any motion component of the grating along its bars is undetectable. Even if the grating were attached to some large but finite surface, any realisable motion detection system will only be able to sample a small part of this image (indicated in Figure 1.9 by the circular aperture through which the grating is being viewed). Thus, both theoretically and practically, the only observable motion component is that which is perpendicular to the grating bars (Wallach, 1976; Vallortigara & Bressan, 1991). The grating motion is therefore indistinguishable from that of any other grating which has the same perpendicular velocity component (for example, the one with velocity $\mathbf{v}^*$ in Figure 1.9a) and there is a family of gratings, compatible with the moving image, having velocities whose vectors lie on a *constraint line* (shown as a dotted line in the

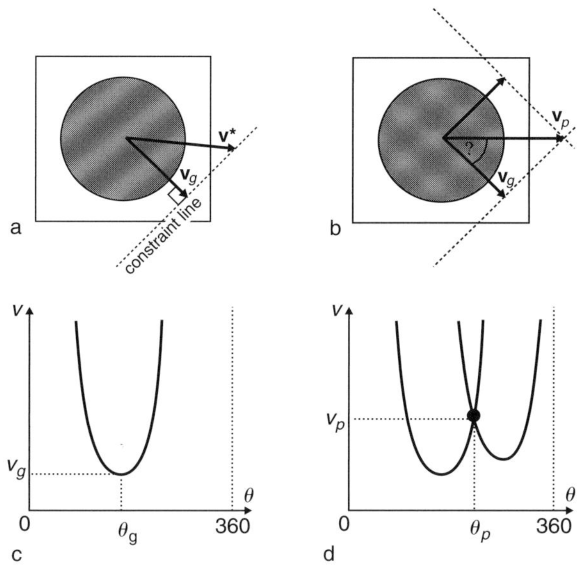

Figure 1.9. Intersection of constraints for motion detection.

figure). Of course, real moving surfaces consist of the superposition of many sinusoidal components, the simplest of which consists of a combination of two gratings. This results in a moving plaid, an example of which is shown in Figure 1.9b. The velocity of the plaid must be the same as that of its components and so its vector must lie at the intersection of the two constraint lines associated with the two component gratings. The algorithm for computing the image velocity is therefore a two-stage process: evaluate the perpendicular motion vectors of each spatial component and then compute the velocity consistent with the intersection of constraints (IOC) of these components (Adelson & Movshon, 1982).

In Marr's scheme, there are several possibilities at the implementational level for executing the IOC algorithm. It could be done, for example, with paper and pencil using a drawing of the geometry (as in Figure 1.9b), or algebraically using the associated trigonometric relations. We now show that there is a neural network implementation, which can solve IOC (Gurney & Wright, 1992). The network has a layer of inputs that encode preferential responses to specific gratings and motions. Thus, each input responds best to a grating of given direction, speed and spatial frequency. The output layer is trained to preferentially encode true image velocity in which each node responds most strongly when encoding a particular image speed v, and direction θ. For example, the plaid of Figure 1.9b would give rise to two inputs responding to the component gratings, and a single output encoding the plaid velocity.[1] Suppose we arrange the output nodes on a pair

[1] In fact the net uses so-called 'course coding' (Touretzky, 1995). So small clusters of nodes are active rather than individuals, but this is not crucial to the argument.

of Cartesian axes for speed, and direction, so the location of each node is determined by its preferred values for these quantities (see Figure 1.9c). By dint of the arguments above, the response of the network to an input representing a single grating with speed v_g and direction θ_g is ambiguous; the grating does not uniquely define an image velocity and it will stimulate many output nodes. The resulting pattern of activity in the output layer is shown schematically in Figure 1.9c by the 'U'-shaped curve (the geometry in Figure 1.9b shows that this function takes the form $v_g/\cos(\theta_p - \theta_g)$ where θ_p is the associated pattern direction). When two gratings are presented, as input, they each give rise to a 'U'-shaped pattern of activity but, at the point where these function overlap, the response of the network is enhanced since it sees twice the input (Figure 1.9d). There is therefore a unique 'hump' of activity in the network corresponding to a velocity defined by the Intersection of Constraints, and the network implements the IOC algorithm.

1.3.4 Mapping from networks to biological neural circuits

The image velocity network, as well as illustrating an application of Marr's hierarchy to perceptual processing, also shows that it is possible to think of networks as implementing algorithms rather than simply combining signals in apparently unstructured ways. This is important because it points to the possibility that network models could be thought of as quantitative tests of computational hypotheses about perceptual and cognitive processing in the brain. This idea is predicated on the following assumption: that neural network models have sufficient points of contact with real brain circuits that the hypotheses they are founded on can allude directly to the brain, albeit at a fairly high level of abstraction. Thus, in the example of the neural network model of IOC, we demonstrated that, since an abstract neural network can implement the proposed algorithm, a real brain system could also *in principle*, do the same thing. That is, we hypothesise that the brain solves the computational problem of determining image motion velocity using a two-stage process with IOC applied to analysis of local Fourier components.

To test this hypothesis, we are now charged with trying to *map* the abstract neural network onto real brain circuits. Gurney & Wright (1992) advanced the argument that there is evidence for correspondence between the input layer and visual cortical area V1, and the output layer and visual are MT in primates. This evidence is based on V1 and MT expressing sensitivity to moving gratings (Hubel & Wiesel, 1968) and image motion (Movshon *et al.*, 1985) respectively, which is consistent with the representations used in the network. The project of mapping the neural network for IOC onto biological neuronal circuits is, however, far from complete. Neocortex (including MT) is a complex six-layered structure with a variety of neural cell types and microcircuits that our impoverished single layer of simplified neurons cannot hope to emulate. In addition, while we know V1 innervates MT, we do not know if the precise connectivity required by the model is to be found anatomically. However, generalising the foregoing analysis from the IOC example, implies that, in so far as neural networks are used to model brain processes, they occupy a position between the algorithmic level in the Marr hieracrchy, and the implementational level (i.e. biological neural tissue). The abstract neural network is an additional *mechanistic* level, and we refer to the process of attempting to realise the neural network in brain circuits as one of *mechanism mapping* (Gurney *et al.*, 2004).

1.3.5 Networks and neural representations

We now explore further the role played by representations in linking neural networks with brain circuits. The first step in attempting to map the IOC net onto the visual system in the brain depended on an identification of the representations used in the net with counterparts in the visual system. In this model the input representation was explicitly based on knowledge of the way in which gratings are known to be represented in area V1. However, the network was trained in a way rather different to those outlined in the first half of this article. Rather than supplying the net with a supervisory target output, the net had to 'discover' structure in the training set under a process of *self-organisation* using learning rules that had biological plausibility.

The function of this kind of network has similarities with statistical techniques like principal component analysis (PCA) since the network performs a *dimension reduction* or compression of the patterns in the input space. Thus, the network consists of a single layer of artificial neurons and, like PCA, discovers features of the training set that are sufficient to describe the essentials of this set of patterns. In the IOC net for example, these features are speed and direction. Each node becomes maximally responsive to patterns with a particular speed and direction, and tends to ignore other aspects of the input. To this extent it has 'discovered' that the input set is essentially two-dimensional.

The learning takes place using rules based on the principles outlined by Hebb (1949) that are supposed to govern synaptic change in real neurons. Hebb proposed that if a pre-synaptic neuron was simultaneously active with a post-synaptic partner, then the synapse between these neurons would increase its strength. In the network, the amount of change on a weight is governed by the size of the corresponding input, and the output of the neuron to which the weight belongs; the weight grows if there is a strong correlation between input and output, and decay if there is little or none. Self-organising networks of this kind were developed and popularised by Kohonen (1984); an introduction to these ideas may also be found in Gurney (1997).

Based on the observations above, self-organising networks can be used in situations where we suspect that the input patterns have too many components and that they may essentially be described more naturally using only a few parameters or dimensions. The utility of the IOC network model was, however, motivated by a more biologically grounded developmental perspective. Thus, it aimed to demonstrate that the representation of image velocity, observed in area MT, could emerge naturally from a biologically plausible developmental process using a prior stage of processing whose encoding was also biologically plausible. The use of self-organising neural networks to demonstrate the emergence of known, neural representations has a long history associated especially with topographic maps of retinotopy (Willshaw & von der Malsburg, 1976), orientation selectivity (von der Malsburg, 1973), and ocular dominance (Swindale, 1980). For a more recent review of the field see Price & Willshaw (2000).

In contrast to this, Zipser (1992) has shown how feedforward networks, trained with supervised methods to perform a perceptual task, often develop representations in their hidden layer which appear similar to those observed experimentally in the neural circuits believed to implement the same computational task in the brain. For example, neurons in posterior parietal area 7a in monkeys are believed to compute a head-centred representation of object location in space, by combining retinotopic information about object spatial location, together with gaze angle (Andersen *et al.*, 1985). Zipser & Andersen (1988) constructed a network model with two layers to perform the same computation. They discovered that the hidden layer yielded nodes whose response properties mimicked well that of certain neurons in area 7a.

Zipser (1992) supplies further examples of this kind and conjectures an explanation for why the representations found in networks might mirror the biology. First, he notes that the setting of parameters (the weights) in a network, in order to achieve some given input–output behaviour, is an example of the general process of system identification. In the general case, forcing a model to show the same behaviour as a target system will not necessarily force the behaviour of the model's *constituents* to mimic those of the system. Indeed, it may not even be sensible to make that comparison because the constituents in each case will be too dissimilar. However, in the case of neural networks and biological neural circuits there is enough similarity to at least attempt a comparison, and Zipser (1992) speculates that this similarity is, in fact, sufficient to force a close correspondence between the model representations and those in the target neural system.

The ability of neural networks to exhibit biologically plausible representations leads to hypotheses that can guide experimental programmes, and can help explain complex patterning in physiological data from cells that have non-trivial receptive field properties as a result of taking part in complex representations.

1.3.6 Networks and neural architectures

We now turn to another way in which neural networks can make test computational hypotheses about perceptual processing. In this instance, it concerns the gross structural properties or architecture of the network and the target brain system; we illustrate it with an example based on the connectionist model of Stroop processing by Cohen *et al.* (1990). In the Stroop task, the subject has to name the colour used to render stimulus words which are colour names (e.g. 'red', 'blue', etc.). In some instances the word and the colour ink are congruent (e.g. 'red' in red ink) in others they conflict (e.g. 'red' in blue ink). The subject has to speak aloud the name of the ink colour as quickly as possible and reaction times and error rates are used as performance measures. There is also a control condition in which no word is presented but, rather, a meaningless pattern like 'XXXX'. In addition, it is possible to contrast the colour-naming task with another one in which the subject has to read the word. The main result is that the mean reaction time in the colour-

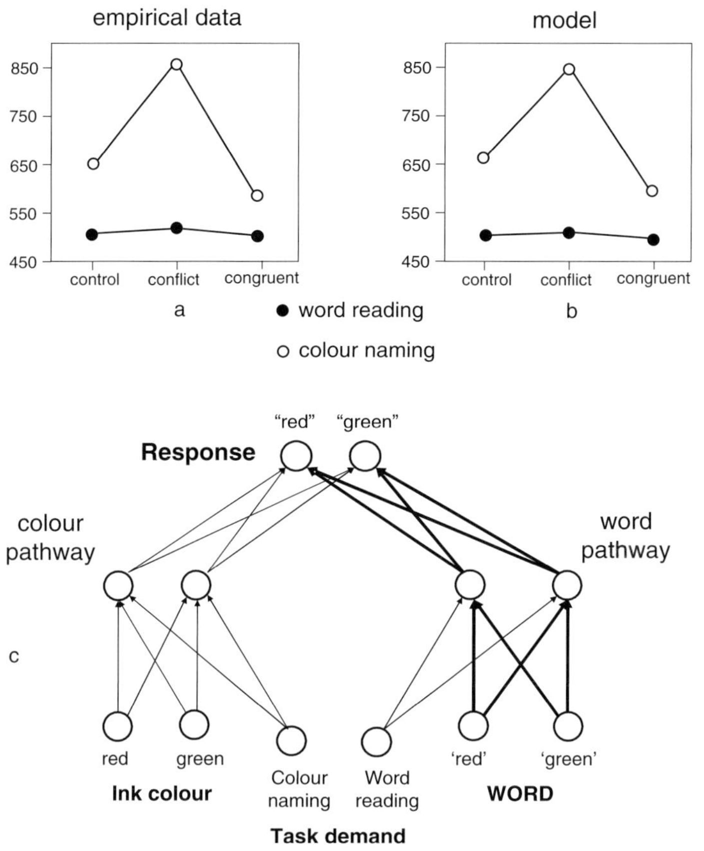

Figure 1.10. The Cohen model of the Stroop task.

naming task for the conflict condition is usually longer than that for the congruent condition – a phenomenon known as the Stroop effect (Stroop, 1935; MacLeod, 1991). Typical reaction-time data are shown in Figure 1.10a.

In their model, Cohen *et al.* (1990) advanced the hypothesis that the Stroop effect may be explained by the difference in processing strength between the pathways in the brain that process colour information and word information. Note that this hypothesis is couched in terms of neural processing because it refers to signal 'pathways' and the relative strength of transmission within these pathways. It therefore makes sense to test this hypothesis with a neural network. The model consisted of two subnetworks – one each for colour and word information (see Figure 1.10c). The pathways had an identical minimal architecture with two inputs (for two colours), two hidden nodes and a common output layer. Crucially, the pathways for colour and word information differed in their connection strengths with the word pathway having the stronger links (as indicated by thicker lines in Figure 1.10c). The task (word reading or colour naming) was specified by sensitising or biasing the hidden nodes using task-demand nodes. When the latter became active, the hidden nodes required less input from elsewhere to generate significant output.

The model was successful in several respects, including its ability to replicate the basic pattern of Stroop data (see Figure 1.10b). At the very least it is therefore a successful phenomenological model of the Stroop task. However, the model also provides evidence for the differential strength of processing hypothesis for, while there is clearly more to the biological neural processing of colour and word than the simple subnetworks used in this model, the hypothesis is not contingent on the details within those pathways. Thus, any arbitrarily complex pair of pathways would give the same result, so long as that used for word reading transmitted information more efficiently than that for colour processing.

In a similar way to the use of networks in discovering neural representations, the Stroop example highlights the utility of the core similarity between abstract nets and real neuronal circuits. In the Stroop network, this similarity (connection strengths, pathways, signal combinations, etc.) is sufficient to frame a hypothesis about biological neural architectures (different 'strengths' in two pathways) in a simplified setting. Indeed, the simplification implicit in the abstract network endows the model with explanatory power that it would not have, if encumbered with too much detail. Of course, in going further and attempting to map the colour and word reading pathways onto their biological counterparts, we would discover whether the pathways really were different in the way suggested, and a more stringent test of the hypothesis would ensue.

The Stroop network of Cohen *et al.* (1990) is but one example of many where neural networks have been used to explore architectural issues. Jacobs *et al* (1991) examined networks that could simultaneously determine the location and identity ('where' and 'what') of objects in visual space. Undifferentiated networks did not perform as well as split or modular networks that dealt with each sub-problem independently, thereby giving insight into why these problems also appear to be solved independently in the animal visual system. Hinton & Shallice (1991) described a network model which was able to exhibit many of the phenomena associated with deep dyslexia in humans. More significantly, in the current context, they experimented with a variety of architectures and discovered that many of the key results were contingent only on there being a layer of nodes with internal recurrent (feedback) connections able to support 'memory' or state information. Finally, we draw attention to the work of McClelland *et al.* (1995) which used a neural network-based analysis of the need for two memory systems in animals: a short-term memory system which can learn 'isolated' or episodic items rapidly, and a long-term system that integrates each episode into a wider knowledge base in a prolonged process of consolidation. Once again, the arguments used by the authors of the model were quite general, making reference only to fundamental properties of neural network development and learning.

1.4 Concluding remarks

In the first half of the chapter we outlined some of the technical issues in neural networks. We focused largely on feedforward nets (and variants) because these form the basis of many of the models described in this book. However, there are a plethora of network architectures we have not covered and the reader is referred to the references at the beginning of this chapter for more information. The concept of generalisation leads to a better understanding of what neural networks might buy us in terms of modelling power, and enables us to develop principled ways of training them to take advantage of this power (using cross-validation and the like). The theory of knowledge representation is an important one if we are to understand network mechanisms and their relation to any corresponding features in the target system (be it the animal brain or an ecosystem).

Our aim in the second part of the chapter was to demonstrate that neural networks are not limited to 'mere simulation' of input–output behaviour, but that they have a role to play in developing theories of cognition and perception. Some neural network models are purely phenomenological descriptions of a target system and make no claim to establish links with internal mechanisms in that system. Rather the networks provide a statistical model description of the system which, nevertheless, provide explanatory power by highlighting relationships between variables, and suggesting new internal (hidden layer) variable combinations, features or 'factors'. If a network models some computational task performed by the animal brain, it is tempting to make correspondences between the network and the corresponding brain mechanisms underlying the computation. At first glance, this approach appears flawed because neural networks lack the biological realism to directly model real neural tissue. However, careful examination of a principled approach to computational modelling due to Marr, suggests otherwise. Thus, neural networks appear to occupy a place in Marr's hierarchy somewhere between the abstract algorithmic level and implementation in biological neural circuits. Construction of neural networks at this 'abstract mechanistic' level is therefore one part of a much larger modelling strategy that seeks to understand high-level computational questions, and algorithms, as well as details of implementation in real neural tissue. As such, neural networks seem to have sufficient core similarities with biological neural circuits to offer insights in two general areas: first, in discovering and understand the role of neural representations; and second, in testing hypotheses about large-scale neural connectivity or architectures.

Acknowledgements

I would like to thank Tom Stafford for reading a draft of the chapter. This work was supported in part by EPSRC grant EP/C516303/1.

References

Adelson, E. H. & Movshon, J. A. 1982. Phenomenal coherence of moving visual patterns. *Nature* **300**, 523–525.

Andersen, R. A., Essick, G. K. & Siegel, R. M. 1985. Encoding of spatial location by posterior parietal neurons. *Science* **230**, 456–458.

Barto, A. G. 1985. Learning by statistical cooperation of sele-interested neuron-like computing elements. *Hum Neurobiol* **4**, 229–256.

Barto, A. G. & Anandan, P. 1985. Pattern-recognizing stochastic learning automata. *IEEE Syst Man Cybernetics* **SMC-15**, 360–375.

Barto, A. G., Sutton, R. S. & Anderson, C. W. 1983. Neuronlike elements that can solve difficult learning control problems. *IEEE Trans Systems Man Cybernetics* **13**, 835–846.

Bishop, C. M. 1996. *Neural Networks for Pattern Recognition*. Oxford University Press.

Cheng, B. & Titterington, D. M. 1994. Neural networks: a review from a statistical perspective. *Statist Sci* **9**, 2–54.

Cohen, J. D., Dunbar, K. & McClelland, J. L. 1990. On the control of automatic processes – a parallel distributed-processing account of the Stroop effect. *Psychol Rev* **97**, 332–361.

Crick, F. 1989. The recent excitement about neural networks. *Nature* **337**, 129–132.

Dayan, P. 2002. Levels of analysis in neural modeling. In *Encyclopedia of Cognitive Science* (ed. L. Nadel). Nature Publishing Group; John Wiley & Sons Ltd.

Dror, I. E. & Gallogly, D. P. 1999. Computational analyses in cognitive neuroscience: in defense of biological implausibility. *Psychon Bull Rev* **6**, 173–182.

Elman, J. L. 1990. Finding structure in time. *Cogn Sci* **14**, 179–211.

Fitzsimonds, R. M., Song, H. J. & Poo, M. M. 1997. Propagation of activity-dependent synaptic depression in simple neural networks. *Nature* **388**, 439–448.

Green, C. D. 2001. Scientific models, connectionist networks, and cognitive science. *Theory Psychol* **11**, 97–117.

Gurney, K. 1997. *An Introduction to Neural Networks*. UCL Press (Taylor and Francis group).

Gurney, K., Prescott, T. J., Wickens, J. R. & Redgrave, P. 2004. Computational models of the basal ganglia: from robots to membranes. *Trends Neurosci* **27**, 453–459.

Gurney, K. N. & Wright, M. J. 1992. A self-organising neural network model of image velocity encoding. *Biol Cybernetics* **68**, 173–181.

Haykin, S. 1999. *Neural Networks: A Comprehensive Foundation*. Prentice Hall.

Hebb, D. 1949. *The Organization of Behaviour*. John Wiley.

Hinton, G. E. & Shallice, T. 1991. Lesioning an attractor network – investigations of acquired dyslexia. *Psychol Rev* **98**, 74–95.

Hodgkin, A. L. & Huxley, A. F. 1952. A quantitative description of membrane current and its application to conduction and excitation in nerve. *J Physiol* **117**, 500–544.

Hopfield, J. J. 1982. Neural networks and physical systems with emergent collective computational abilities. *Proc Natl Acad Sci USA* **79**, 2554–2558.

Hubel, D. H. & Wiesel, T. N. 1968. Receptive fields and functional architecture of monkey striate cortex. *J Physiol* **195**, 215–243.

Jacobs, R. A., Jordan, M. I. & Barto, A. G. 1991. Task decomposition through competition in a modular connectionist architecture. *Cogn Sci* **15**, 219–250.

Jordan, M. I. 1986. *Serial Order: a Parallel Distributed Approach*. University of California, Institute for Cognitive Science.

Koch, C. 1999. *The Biophysics of Computation: Information Processing in Single Neurons*. Oxford University Press.

Koch, C., Mo, C. & Softky, W. 2002. Single-cell models. In *The Handbook of Brain Theory and Neural Networks* (ed. M. Arbib). MIT Press.

Kohonen, T. 1984. *Self-organization and Associative Memory*. Springer Verlag.

Lippmann, R. 1987. An introduction to computing with neural nets. *ASSP Magazine, IEEE* **4**, 4–22.
MacLeod, C. M. 1991. Half a century of research on the Stroop effect – an integrative review. *Psychol Bull* **109**, 163–203.
Makhoul, J., El-Jaroudi, A. & Schwartz, R. 1989. Formation of disconnected decision regions with a single hidden layer. In *International Joint Conference on Neural Networks, Vol. 1*, pp. 455–460.
Marr, D. 1982. *Vision: A Computational Investigation into Human Representation and Processing of Visual Information*. W.H. Freeeman and Co.
Marr, D. & Poggio, T. 1976. *From Understanding Computation to Understanding Neural Circuitry*. MIT AI Laboratory.
McClelland, J. L., McNaughton, B. L. & O' Reilly, R. C. 1995. Why there are complementary learning systems in the hippocampus and neocortex: insights from the successes and failures of connectionist models of learning and memory. *Psychol Rev* **102**, 419–457.
McClelland, J. L. & Rumelhart, D. E. 1985. Distributed memory and the representation of general and specific information. *J Exp Psychol Gen* **114**, 159–188.
McCloskey, M. 1991. Networks and theories – the place of connectionism in cognitive science. *Psychol Sci* **2**, 387–395.
McCulloch, W. S. & Pitts, W. 1943. A logical calculus of the ideas immanent in nervous activity. *Bull Math Biophysics* **7**, 115–133.
Minsky, M. & Papert, S. 1969. *Perceptrons*. MIT Press.
Movshon, J. A., Adelson, E. H., Gizzi, M. S. & Newsom, W. T. 1985. The analysis of moving visual patterns. In *Pattern Recognition Mechanisms* (ed. C. Chagas, R. Gattas & C. G. Gross), pp. 117–151. Springer-Verlag.
Parker, D. B. 1982. *Learning-logic*. Office of Technology Licensing, Stanford University.
Price, D. & Willshaw, D. 2000. *Mechanisms of Cortical Development*. Oxford University Press.
Rieke, F., Warland, D., de Ruyter van Steveninck, R. & Bialek, W. 1996. *Spikes*. MIT Press.
Rosenblatt, F. 1958. The perceptron: a probabilistic model for information storage and organization in the brain. *Psychol Rev* **65**, 386–408.
Rosin, P. L. & Fierens, F. 1995. Improving neural net generalisation. In *Proceedings of IGARSS'95*. Firenze, Italy.
Rumelhart, D. E., McClelland, J. L. & The PDP Research Group. 1986. *Parallel Distributed Processing: Explorations in the Microstructure of Cognition*. MIT Press.
Rumelhart, D. E. & Todd, P. 1993. Learning and connectionist representations. In *Attention and Performance XIV* (ed. D. Meyer & S. Kornblum), pp. 3–31. MIT Press.
Shang, Y. & Wah, B. W. 1996. Global optimization for neural network training. *IEEE Computer* **29**, 45–54.
Shepherd, G. M. & Koch, C. 1998. Introduction to synaptic circuits. In *The Synaptic Organization of the Brain* (ed. G. M. Shepherd), pp. 1–36. Oxford University Press.
Smolensky, P. 1988. On the proper treatment of connectionism. *Behav Brain Sci* **11**, 1–23.
Stone, M. 1974. Cross-validatory choice and assesment of statistical predictions. *J R Statist Soc* **B36**, 111–133.
Stroop, J. R. 1935. Studies of interference in serial verbal reactions. *J Exp Psychol* **18**, 643–662.

Sutton, R. S. 1988. Learning to predict by the method of temporal differences. *Machine Learn* **3**, 9–44.

Sutton, R. S. & Barto, A. G. 1998. *Reinforcement: An Introduction*. MIT Press.

Swindale, N. V. 1980. A model for the formation of ocular dominance stripes. *Proc R Soc B* **208**, 243–264.

Touretzky, D. S. 1995. Connectionist and symbolic representations. In *The Handbook of Brain Theory and Neural Networks, 1st Edn.* (ed. M. Arbib), pp. 243–247. MIT Press.

Tveter, D. R. 1996. Backpropagator's review. www.dontveter.com/bpr/activate.html

Vallortigara, G. & Bressan, P. 1991. Occlusion and the perception of coherent motion. *Vision Res* **31**, 1967–1978.

von der Malsburg C. 1973. Self-organization of orientation sensitive cells in the striate cortex. *Kybernetik* **14**, 85–100.

Wallach, H. 1976. On perceived identity 1: The direction of motion of straight lines. In *On Perception* (ed. H. Wallach). Quadrangle.

Werbos, P. 1974. *Beyond Regression: New Tools for Prediction and Analysis in the Behavioural Sciences*. Harvard University.

Widrow, B. & Hoff Jr, M. E. 1960. Adaptive switching circuits. In *IRE WESCON Convention Record*, pp. 96–104.

Widrow, B. & Stearns, S. D. 1985. *Adaptive Signal Processing*. Prentice Hall.

Wieland, A. & Leighton, R. 1987. Geometric analysis of neural network capabilities. In *1st IEEE International Conference on Neural Networks, Vol. III*, pp. 385–392. San Diego, California.

Williams, R. J. 1987. *Reinforcement Learning Connectionist Systems*. Northeastern University, Boston.

Willshaw, D. J. & von der Malsburg, C. 1976. How patterned neural connections can be set up by self-organization. *Proc R Soc B* **194**, 431–445.

Zipser, D. 1992. Identification models of the nervous system. *Neuroscience* **47**, 853–862.

Zipser, D. & Andersen, R. A. 1988. A back-propagation programmed network that simulates response properties of a subset of posterior parietal neurons. *Nature* **331**, 679–684.

2

Sensory ecology and perceptual allocation: new prospects for neural networks

Steven M. Phelps

2.1 Introduction

All of animal behaviour can be considered a series of choices – at any given moment an animal must decide whether to mate, eat, sleep, fight or simply rest. Such decisions require estimates of the immediate environment, and despite the diversity of those estimates, they are all carried out by sensory systems and the neural functions contingent on them. This functional diversity is central to the concept of sensory drive (Endler, 1992; Figure 2.1), which notes that animal mating, foraging and other activities are evolutionarily coupled through their shared dependence on sensory systems and local environments. In light of the many demands made of a sensory system, what does it mean to design one well?

It is often useful to consider how an ideal receiver would perform on a given task. Aside from the potentially conflicting demands posed by different aspects of one's environment, there are additional reasons to think that such an approach may not be complete. The climb to a global optimum can be a tortuous one, complicated by genetic drift, allelic diversity and phylogenetic history. Analytic models often focus on defining the best possible performance and neglect the existence of alternative local optima, or the ability to arrive at such optima through evolutionary processes. In sexual selection, researchers have suggested pleiotropy in sensory systems may be one key feature that shapes the direction of evolution (Kirkpatrick & Ryan, 1991). Pleiotropy could emerge when building a complex structure from a limited number of genes, or from the multiple functions fulfilled by a common structure. Female guppies, for example, prefer orange males, but they also prefer orange food items (Rodd *et al.*, 2002). We need means of understanding how such coincident preferences emerge, and what their consequences are for behavioural evolution. In this review I advocate a dual approach, combining first principles of sensory ecology with neural network models to gain a more balanced and nuanced view of sensory design.

Neural network models have been used to investigate the origins of hidden preferences – attributes of nervous systems that inadvertently bias their interactions with the outside world. Early studies used model visual systems, simple 'feedforward' networks, to investigate how biases toward signals that were symmetric or exaggerated could emerge as

Modelling Perception with Artificial Neural Networks, ed. C. R. Tosh and G. D. Ruxton. Published by Cambridge University Press.

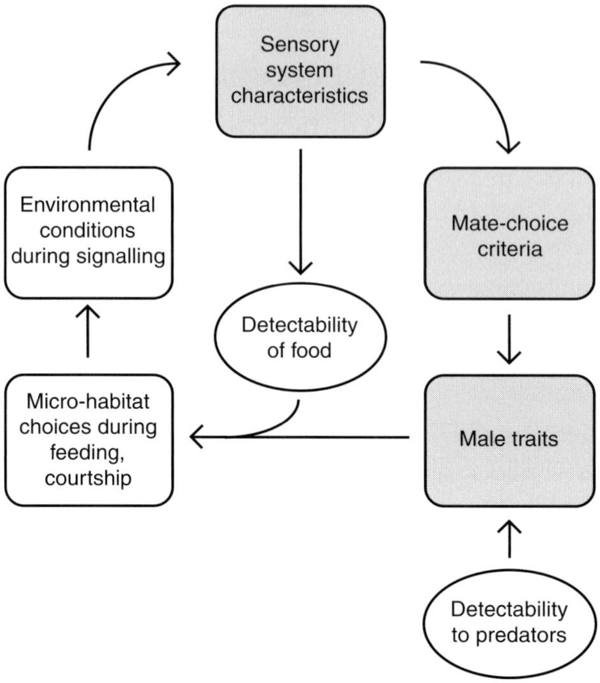

Figure 2.1. Sensory drive depicts the coupling of multiple ecological factors with the design of sensory systems and its influence on interaction between organisms. Interactions corresponding to 'sensory exploitation' are depicted in grey. Based on Endler (1992).

a by-product of selection on simple recognition tasks (Enquist & Arak, 1992, 1993; Johnstone, 1994). Additional advances came from linking such studies to empirical data from particular species (Phelps & Ryan, 1998; Phelps *et al.*, 2001), permitting an assessment of the external validity of the models. These studies provided a broader view of sexual selection, a field that has historically focused on the strategic design of signals (e.g. Zahavi, 1975). One insight from the neural network models was that preferences could emerge for stimuli as a by-product of selection in other contexts. Potential indicators of male condition were not necessarily favoured for their ability to reveal male status, but rather for their conformity to the pre-existing perceptual biases of the receiver (Basolo, 1990), a phenomenon known as 'sensory exploitation' (Ryan 1990; Ryan *et al.*, 1990b). Such complexity is not easily explained. Because neural network models simulate complex decision making, and arrive at those decisions through evolutionary processes, they provide a logical complement to more abstract models that describe performance at a global optimum.

This chapter aims to briefly survey work on each of several modalities (vision, audition, touch) and levels of sensory processing (peripheral, central). Within each domain, I describe work in sensory ecology, which takes broad models for how receivers ought to be designed and compares them to physiological and behavioural data. I also examine work from neuroethology, which often has the same general assumptions but is not always focused on testing whether sensory performance is optimal for a given task. Lastly, in each section I suggest novel neural network models that could further inform our understanding of perceptual allocation. By exploring what it would mean to design a sensory system optimally, one can test whether given systems conform to those expectations. No less importantly, one can also detect deviations from predictions that might direct us to the roles of other evolutionary forces. By making these questions explicit, and by outlining how extant neural network models could be modified to ask such questions, this review aims to stimulate thought and experiment on how evolution shapes nervous systems to accommodate a multitude of tasks.

Before beginning our survey, it may be useful to review a few basic concepts in sensory design for readers unfamiliar with neuroscience. The first is to remind the reader that each sensory modality corresponds to a type of stimulus energy able to change the voltage of a sensory receptor. Within a modality, there are subtle variations of energy that include wavelengths of light, frequencies of sound, and the depth and duration of touch. The investment in one modality or submodality over another is assumed to reflect its value. How does value interact with the costs of receiving and processing such information? Such questions are central to receiver design, and by extension to questions in foraging, mating and the various other choices which comprise an animal's behavioural repertoire. I begin our exploration with a discussion of empirical work on visual allocation at the periphery and how such allocation influences behavioural decisions.

2.2 Allocation at the periphery

2.2.1 Vision

To what extent can a visual system be said to be operating 'optimally', and in what ways may we estimate its efficacy? Some of the most sophisticated ecological analyses of animal sensation have focused on how the tuning and abundance of colour cones reflect evolutionarily important properties of the environment (see Osorio & Vorobyev, 2005 for recent review). A central formalism in this work describes the 'quantum catch', the number of photons (Q) absorbed by a sensory system (Wyszecki & Stiles, 1982).

$$Q = \int Q_0(\lambda) \cdot R(\lambda) \cdot T(\lambda, d) \cdot S(\lambda) d(\lambda) \tag{1}$$

In this equation, $Q_0(\lambda)$ is the distribution of coloured light in an environment, a value called the quantum flux and expressed as a function of wavelength (λ). This has been shown to vary as a function of the openness of terrestrial habitats, or depth of aquatic ones (Lythgoe & Partridge, 1991; Endler, 1992, 1993). It is shaped by both the distribution of wavelengths in sunlight, and by the reflectance and absorbance properties of organisms within the local environment. $R(\lambda)$ is the reflectance pattern for an object to be detected – whether that object is a patch of coral or the epaulet of a red-winged blackbird. The product of these two represents the stimulus energy emerging from the target. $T(\lambda, d)$ represents the transmittance of light in the environment at a distance d. Lastly, $S(\lambda)$ is the spectral sensitivity of the photopigment of interest. The quantum catch of multiple kinds of cones can be calculated independently. These can be summed across cone classes to generate a measure of luminance, or differences between cones can be used to estimate chroma. Based on such calculations one can visually sample local environments and calculate how well equipped the visual system is for detecting objects varying in brightness and hue. As a perceptual allocation problem, the aim is to understand how the organism should tune its eye to extract the most useful information.

Extracting chromatic information requires multiple cones tuned to different wavelengths. The simplest such system is dichromatic vision, found in most mammals, in which animals possess a pair of cones each tuned to a different wavelength. In contrast, stomatopods may possess eight or more narrowly tuned receptor types, providing the potential for fine discrimination of individual differences in body spot colouration (Cheroske & Cronin, 2005). The eyes of *Lycaena* butterflies express four cone types; wing colour and receptor tuning vary by species, and sexes vary in the relative abundance of cone types across the eye. Females use long-wavelength cones to detect host plants and, in accord, this cone can be found in the dorsal portion of the female eye (Bernard & Remington, 1991). Assessments of colour perception by bees have utilised knowledge of cone absorbance and colour opponency (combinations of cones used to perceive colour) to predict performance in foraging tasks. These studies demonstrate that chromatic contrast predicts rates of flower detection (Lunau *et al.*, 1996; Spaethe *et al.*, 2001).

Among fishes, the diversification of visual signalling is particularly interesting in the cichlid fishes of Lake Victoria, a group well known for the recent and explosive radiation. Closely related sympatric species differ substantially in sexually dimorphic colouration; environmental degradation has caused increased turbidity and a breakdown of visual barriers to hybridisation (Seehausen *et al.*, 1997). Cones differ substantially in their relative abundance across species (Carleton & Kocher, 2001) and there is evidence of selection on opsin loci associated with variation in colour and habitat (Terai *et al.*, 2002; Spady *et al.*, 2005). Interestingly, there is also substantial phenotypic plasticity in receptor allocation. Halstenberg *et al.* (2005) demonstrate a diurnal rhythm in opsin expression that is entrained by light. Similarly, in killifish, Fuller *et al.* (2003, 2004, 2005) find both heritable and environmental variation in opsin distribution across killifish lineages occupying clear and tannin-stained waters.

Most mammals are dichromats, and so the evolution of trichromatic vision within primates is remarkable. The emergence of trichromacy has evolved in two distinct ways. First, in old world apes, the medium/long wavelength opsin locus has duplicated and diverged, generating red and green responsive cones (Surridge *et al.*, 2003). Visual samples of primate habitats combined with quantum flux calculations reveal that trichromacy substantially improves the detectability of both new leaves and ripe fruits (Sumner & Mollon, 2000; Surridge *et al.*, 2003). New leaves tend to reflect more red wavelengths, and are both more nutritious and less tough (Dominy & Lucas, 2004). In both old and new world primates, trichromats are more likely to eat red-shifted leaves (Lucas *et al.*, 2003). In new world primates, there is a surprising condition in which trichromacy has emerged not by duplication of a locus, but by allelic divergence within a locus. Some new world monkeys have as many as five different medium/long wavelength alleles at a single locus (Jacobs & Deegan, 2005). Using visual models and images of natural tamarin foods, Osorio *et al.* (2004) suggest both frequency-dependent selection and heterozygote (trichromatic) advantage in the evolution of such polymorphism.

2.2.1.1 Modifying a visual neural network

The feedforward network seems a promising tool for modelling complex scenarios of visual allocation. Enquist & Arak (1993) used a simple feedforward network to discriminate between stimuli representing long-tailed conspecific and short-tailed heterospecifics, and observed emergent preferences for still longer tails. (A network similar to theirs is presented in Figure 2.2a.) As one might predict, co-evolutionary simulations result in the evolution of still longer tails. What variations on this architecture might be used to investigate sensory drive more broadly?

The traditional feedforward network used in these studies has an input layer corresponding to a retina that detects illumination, a hidden layer that extracts features from the pattern on the retina, and a single output neuron that conveys whether a target pattern has been detected. Genetic algorithms code network architectures as lists of weights, and select those best at performing the desired task (Mitchell, 1996). Instead of this original Enquist & Arak (1993) formulation, one might allow each pixel to have a distribution of reflectance intensities defined by the function $R(\lambda)$. The image falling on the retina is derived from the product of the reflectance intensities, the environmental transmittance $T(\lambda)$ and the ambient light $Q_0(\lambda)$ (Figure 2.2b). Figure 2.2c depicts distributions of reflected wavelengths ($Q_0(\lambda)\ R(\lambda)$) for each of several potential views. The investigator could allow the sensitivity of the receiver $S(\lambda)$ to wavelengths to vary genetically, perhaps assigning each spot in the retina one or more of several available cone classes. Such a model could also incorporate some formulation of colour theory, defining the networks as dichromats or trichromats and using existing models to extract hue and luminance (Wyszecki & Stiles, 1982).

A second modification might allow networks to make multiple decisions – using a pair of output neurons for mating and feeding, for example (Figure 2.2a). Would the processing of food-related cues compel signallers to evolve exaggerated versions of these stimuli? Such a finding would be consistent with sensory exploitation and sensory drive. It would also be strikingly reminiscent of the ethological notions of 'ritualisation' and 'emancipation' whereby cues from one domain become exaggerated and freed from their original source, as suggested for the elaboration of pheasant trains (Bradbury & Vehrencamp, 1998). These considerations begin to take us beyond peripheral allocation, into how peripheral coding impinges on the decisions that comprise an animal's behavioural repertoire.

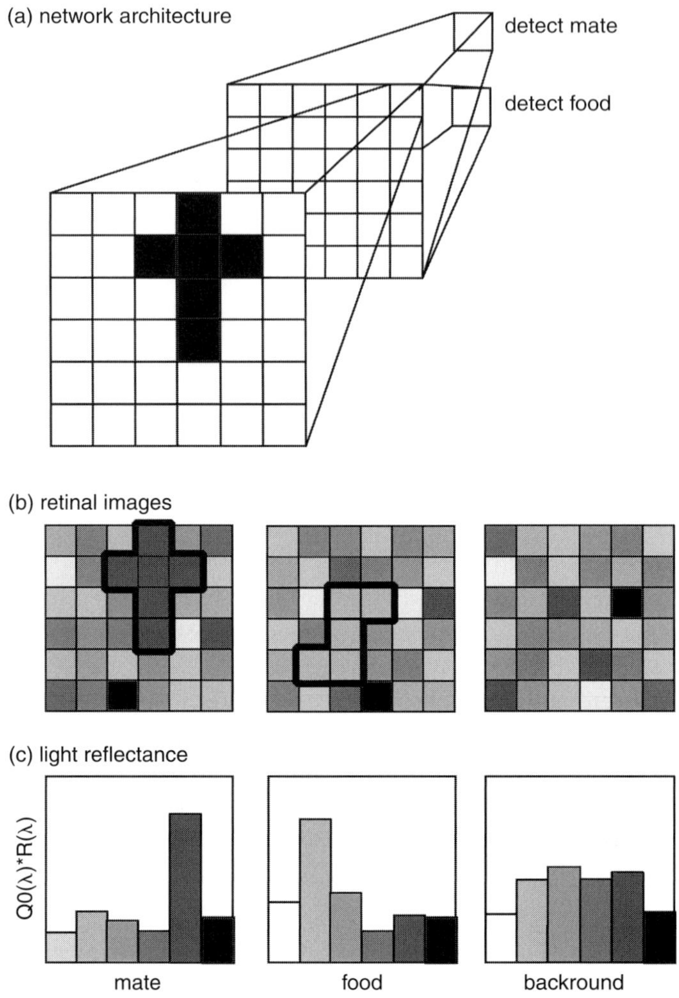

Figure 2.2. Use of a feedforward network to explicitly investigate sensory drive. The proposed architecture at top (a) is similar to that used by Enquist & Arak (1993). (b) The task is to require recognition of either a conspecific (left, outlined 't') or a food item (middle, outlined 's') using distinct output neurons, but not to respond to a stimulus consisting of only background illumination (right). In this simplified scenario, white cells correspond to short wavelengths of light, dark cells to longer wavelengths. (c) A hypothetical distribution of reflected light from targets and background, posing the neural network task in terms of sensory ecology. The horizontal axis corresponds to wavelength, the vertical axis to abundance of reflected light.

In short, such neural network models can readily be cast in terms of conventional sensory ecology. They provide valuable supplements to more general models because they facilitate the investigation of complex interactions between phenotypes and settings. Because parameters of interaction can be precisely controlled, they can be used to uncouple the components of sensory drive and to observe how these components interact to produce the higher-level patterns that characterise behavioural evolution.

2.2.3 Audition

Neuroethologists recording from the periphery of the auditory system find that its frequency sensitivity roughly matches the distribution of frequencies in species-specific acoustic signals (Figure 2.3). Tuning favouring mating signals has been found in insects (Meyer & Elsner, 1996), fish (Sisneros *et al.*, 2004), frogs (Frishkopf *et al.*, 1968) and songbirds (Konishi, 1969; Dooling *et al.*, 1971). Similarly, both bats and barn owls contain 'acoustic fovea' matching the echolocation calls used for navigation and hunting (Bruns & Schmieszek, 1980; Vater, 1982; Koppl *et al.*, 1993). Capranica & Moffat (1983) point out that such tuning can be considered a 'matched filter', a receiver design that in many conditions serves as an optimal signal detector (Dusenberry, 1992).

The tuning curves from different species are obtained in a diverse number of ways. In amphibians, for example, these may include recording directly from the 8th nerve or from more central midbrain neurons; sounds may be broadcast as 'free field' stimuli or coupled to the auditory apparatus of the organism. Thresholds, moreover, are estimates of the

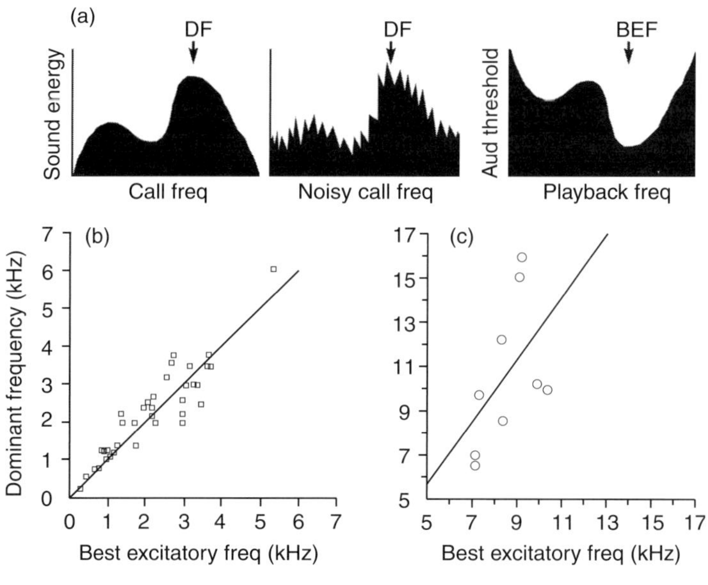

Figure 2.3. Matched filters in auditory coding. (a) Conceptual example showing the relationship between the dominant frequency (DF) for a noise-free call (left), for a noisy call (middle) and the corresponding best excitatory frequency (BEF) in an auditory organ acting as a matched filter for the species-specific call (right). After Capranica & Moffat (1983). (b) Plot of DF and BEF for 36 species of anurans, from Gerhardt & Schwartz (2001). (c) Similar plot of nine species of grasshoppers, from Meyer & Elsner (1996).

minimum stimulation required to evoke some fixed response from a population of neurons. They are an incomplete representation of the abundance and nature of neurons coding information about a given frequency. Because of these ambiguities, researchers in the neuroethology of audition have focused on the relationship between the frequency to which the receiver exhibits the lowest threshold (the 'best excitatory frequency') and the dominant frequency of an advertisement call. The match between signal and receiver is surprisingly strong (Figure 2.3). In a definitive review of matched filters in anurans, Gerhardt & Schwartz (2001) demonstrate that there is a very strong correlation between the dominant frequency of a call and the best excitatory frequency estimated from neurophysiological data; similar findings have been reported in orthoptera (Meyer & Elsner, 1996; see Gerhardt & Huber, 2002 for discussion). These data demonstrate a remarkable congruence between receiver perception and signal design, even in the absence of data concerning sources of background noise and signal degradation imposed by a given environment.

Following the 'quantum flux' model for visual observers, one can imagine a similar formalism for the sound energy (E) captured by a receiver.

$$E = \int C(f) \cdot T(f, d) \cdot S(f)\, df \qquad (2)$$

In this equation, $C(f)$ represents the spectral content of the signal as a function of frequency (f), $T(f, d)$ corresponds to the transmission of energy at a particular frequency (f) and distance (d), and $S(f)$ corresponds to the sensitivity of the receiver to various frequencies. One can describe the information capacity of the frequency domain by measuring the difference between the total sound energy perceived by a receiver when a signal is present and when it is not. Defining the energy in the target plus background as $C_{T+B}(f)$ and the background as $C_B(f)$, one can describe the information content (H_T) of a signal received at a given distance as follows.

$$H_T = \log_2\left[\int C_{T+B}(f) \cdot T(f, d) \cdot S(f)\, df\right] - \log_2\left[\int C_B(f) \cdot T(f, d) \cdot S(f)\, df\right] \qquad (3)$$

As in vision, the efficacy of communication is contingent on receiver, signal and environment. In contrast to studies of vision, there is little work that combines all three measures into a common assessment. Most studies have focused on the interaction of signal and environment, the two simplest parameters to measure. The acoustic adaptation hypothesis (Morton, 1975; Wiley & Richards, 1978, 1982), for example, predicts that signals evolve to minimise degradation due to environmental differences between habitats. In Eq. 3, this is analogous to designing signals to maximise $C_{T+B}(f)\ T(f, d)$. A related idea, that of an 'acoustic window' (Waser & Waser, 1977; Waser & Brown, 1984) posits that signals should use frequencies least evident in the background, roughly equivalent to minimising $C_B\ (f)$. These ideas have certainly met with considerable success. Even after considering such factors, however, there is a great deal of variation in signal structure that is not readily explained by habitat. Ryan & Brenowitz (1985), for example, report that although natural variation in vocalisation frequency is significantly predicted by habitat, body size is a far better predictor.

To consider how one might integrate perceptual allocation into considerations of transmission, environmental noise and call structure, it is worth considering the communication system of the cricket frog, *Acris crepitans*. Cricket frogs are small temperate frogs inhabiting the south and eastern USA. Studies by Ryan, Wilczynski and colleagues have focused on two populations in central Texas, *Acris crepitans crepitans*, which lives in wet pine forest, and *Acris crepitans blanchardi*, which lives in open habitats. Perhaps surprisingly, it does not seem that each call is specialised to minimise degradation within its own habitat. Instead, calls from the more acoustically challenging pine forest transmit with greater fidelity in both pine and open habitats (Ryan *et al.*, 1990a). To examine population differences in receivers, Wilczynski *et al.* (1992) measured the auditory tuning of frogs and found that the average tuning of the basilar papilla is similar to the dominant frequency in each call (see also Capranica *et al.*, 1973; Ryan *et al.*, 1992). To examine the interaction of receiver tuning, call structure and transmission fidelity, Sun *et al.* (2000) recorded calls at varying distances from a speaker. They removed background noise from recordings, passed the calls through an acoustic filter approximating the auditory tuning curve for each population, and estimated how well receivers would be able to detect a call at short and long distances. Their results describe how the transmission environment can alter the ideal receiver tuning depending on the distance at which a signal is perceived. What is striking, however, is that receivers from the more challenging pine forest are much better in both habitats. Similarly, Witte *et al.* (2005) find that tuning curves from pine forest animals are better at filtering equivalent intensities of background noise from either habitat. Combined with earlier data, it seems both signal and receiver from the pine habitat are more effective.

Although the cricket frog work uniquely assesses the interaction of auditory allocation, transmission environment and signal structure, Witte *et al.* (2005) note the importance of one missing variable: potential habitat differences in absolute background noise level. As background gets more intense, the relative difference between a sound sample containing the signal and one that does not gets smaller (Eq. 3). An appropriate follow-up might measure calls without filtering out background noise, and use Eq. 3 to compare the sound energy received with and without the call present. If the information is no more valuable in the pine forest than in the open forest, and the pine forest proves to be louder, the total amount of information transmitted in the two systems may be equivalent. Indeed, if the costs of communication are higher in pine forest, one may find that the net information transmitted has gone down despite the efficiencies of signal and receiver. This hints at the need to assess the costs of communication to both receiver and signaller, features rarely treated empirically (see Bradbury & Vehrencamp, 1998, 2000).

I have emphasised the optimal design of receivers, and how they might interact with signallers. As mentioned in the introduction, however, matching of signal and receiver do not explain the entirety of receiver design. In auditory communication, there are a number of examples of poor matches between signallers and receivers (e.g. Schul & Patterson, 2003). In the case of cricket frogs, though the pine forest animals have a peak frequency tuned to within 80 Hz of the dominant call frequency, the mismatch in open habitat animals is 740 Hz. Does this reflect other uses of the auditory system? Certainly it is possible. Auditory systems are likely means of detecting nearby predators, and the presence of frequency responses outside the range of mating signals is often interpreted as serving this function (e.g. frogs, Frishkopf *et al.*, 1968; birds, Konishi, 1969; insects, Schul & Patterson, 2003). Receiver compromises between detecting mates and predators are in principle amenable to analytic treatments. Are there other, less readily explained patterns of peripheral tuning? In another anuran example, it at first seems that the basilar papilla of túngara frogs is tuned to roughly match the dominant frequency of a call component called a chuck. Congeners possess similar tuning, however, despite lacking the capacity to chuck altogether – a finding suggesting the tuning substantially predates the evolution of its 'matched' target (Ryan *et al.*, 1990b; Wilczynski *et al.*, 2001). In this example, the matching of signal and receiver can be more easily attributed to the evolution of the signal than to the design of the sensory system. Any complete understanding of sensory function and perceptual allocation must include the possibility that receivers depart from optimal performance due to the contingencies of history and genetic architecture. I now describe a neural network model of call recognition that enabled the exploration of such forces. As with the visual networks discussed above, these auditory networks model higher-level decision processes as well as peripheral mechanisms of stimulus filtering.

2.2.4 A simple auditory model

We began with the intent of understanding how historical and contemporary selection on species recognition mechanisms could produce patterns of mate choice preference. This approach was motivated both by prior work emphasising how selection for simple recognition mechanisms could generate hidden preferences (Enquist & Arak, 1993), and by data suggesting evolutionary history was a significant contributor to the responses of túngara frogs (Ryan & Rand, 1995). Unlike prior neural network studies, however, we hoped to anchor the evolutionary simulations to a specific model system so that we could compare our results to data from behavioural studies.

We chose a simple recurrent network similar to an Elman net (Elman, 1990), consisting of a frequency-specific set of input neurons, two reciprocally connected hidden layers (called feature detector and context layers, by analogy with an Elman net) and a single output layer (Phelps & Ryan, 1998; Figure 2.4). The reciprocal connections between hidden layers enabled neurons to make responses to current frequency inputs contingent upon prior patterns of stimulation. This struck us as the simplest possible abstraction of neural processing that could recognise a túngara frog call in a biologically realistic manner. We settled on this general model because the precise mechanisms underlying túngara frog call recognition are not known.

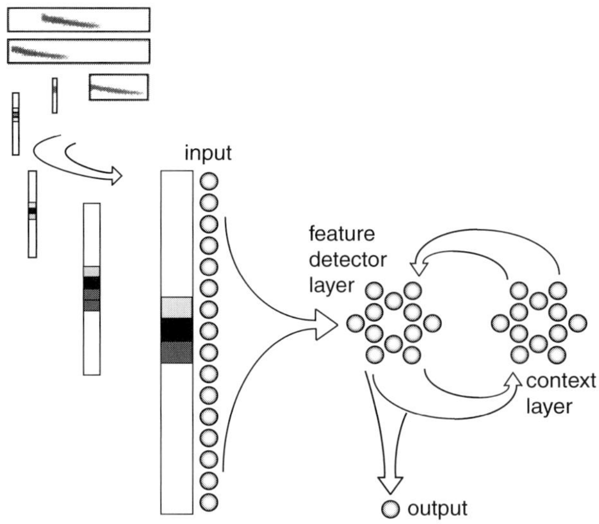

Figure 2.4. Recurrent neural network used to model call recognition in túngara frogs. The network has a recurrent loop between the feature detector and context layers that allows the detection of time-varying stimuli. The stimulus is a matrix of Fourier transform coefficients from a túngara call presented to the input layer one time interval per step. From Phelps & Ryan (1998).

We next performed fast Fourier transforms to generate a time-by-frequency breakdown of the túngara call suitable for presenting to the networks. We trained networks using a genetic algorithm in which networks were represented as binary strings corresponding to a concatenated list of neuron weights and biases. A more complete discussion of the network architecture and the training algorithm are provided in Phelps & Ryan (1998), and Phelps (2001). During training, the calls were randomly placed within a time window large enough to house any of the potential test stimuli. We made a matching noise signal by randomly assigning the energies in a given time window to a new frequency. Additional noise was added to both stimuli by adding a small, fixed probability of being assigned a value at random. Networks were selected to distinguish between calls and noise according to a fitness function defined as

$$W = \left[\sum (C_i - N_i)^2/n\right]^{1/2} + 0.01 \tag{4}$$

In this equation, W is fitness, C_i is the network response to a given call i, N_i its response to a noise and n is the number of calls presented. The small constant 0.01 provides an external component of fitness that minimises the chance networks will get caught at early local optima. We found that not only could neural networks evolve to discriminate this call from noise, but the generalisations they made to novel call stimuli were excellent predictors of the responses made by females in phonotaxis tests (34 stimuli, $R^2 = 0.88$, $p < 0.001$; Phelps & Ryan, 1998). A surprisingly large amount of variation in female responses could be explained solely on the basis of selection for conspecific recognition.

Because of data suggesting that historical forces had shaped responses of female túngara frogs, we next set out to manipulate histories of neural network models and observe the consequences on the emergent patterns of preference. To do so, we selected networks to recognise a call corresponding to the reconstructed ancestor of the túngara frog clade (Figure 2.5). Once networks were reliably able to do so, responses to this call were no longer explicitly selected, and networks were selected to recognise the next node along the trajectory leading to the call of the túngara frog. The influence of past recognition mechanisms was immediately evident in the increasing ease with which networks evolved to recognise each new target call (Phelps, 2001). To control for the cumulative variety of target calls, we also generated a control history in which the trajectory was rotated in a multidimensional call space defined by a principal component analysis of variation within the clade. This 'mirrored' history possessed an equal number of steps as the 'mimetic' history. Also like the mimetic history, the mirrored history converged on the call of the túngara frog. We found that the two history types were equally able to recognise the call of the túngara frog, but differed substantially in how they generalised to other novel calls (Phelps & Ryan, 2000). Assessing the pattern of responses across such novel calls, we found the mimetic history was significantly better than the mirrored history at predicting female responses, a finding consistent with the hypothesis that females harboured biases that were vestiges of their evolutionary histories (Phelps & Ryan, 2000).

To further assess vestigial preferences, we constructed a series of stimuli that varied in only a single call parameter – the time-to-half frequency, a measure that varied significantly between mimetic and mirrored histories. The resulting stimuli were different from training stimuli, yet networks exhibited clear asymmetries in their patterns of preference, favouring calls resembling those of ancestors to those that did not (Phelps *et al.*, 2001; Figure 2.5b). To examine females for comparable markers of vestigial preferences, we constructed a series of stimuli that varied along a dimension that ranged from the extant túngara call to the reconstructed call of an ancestor in one direction, to equidistant calls that did not resemble an ancestor in another direction. We tested such calls in both one-choice and two-choice phonotaxis experiments. We found that females, like the neural networks, exhibited strong preferences for calls resembling those of ancestors (Figure 2.5c).

Despite their simplicity, the neural network models were able to demonstrate how history and species recognition could interact to produce a complex pattern of responses in an extant species. One can imagine supplementing the extant studies with the sort of modifications described for feedforward networks – assessments and manipulations of the ambient noise or call transmission, or the performance of multiple tasks. I suggest, however, a novel set of parameters often omitted from sensory ecology, that of perceptual cost.

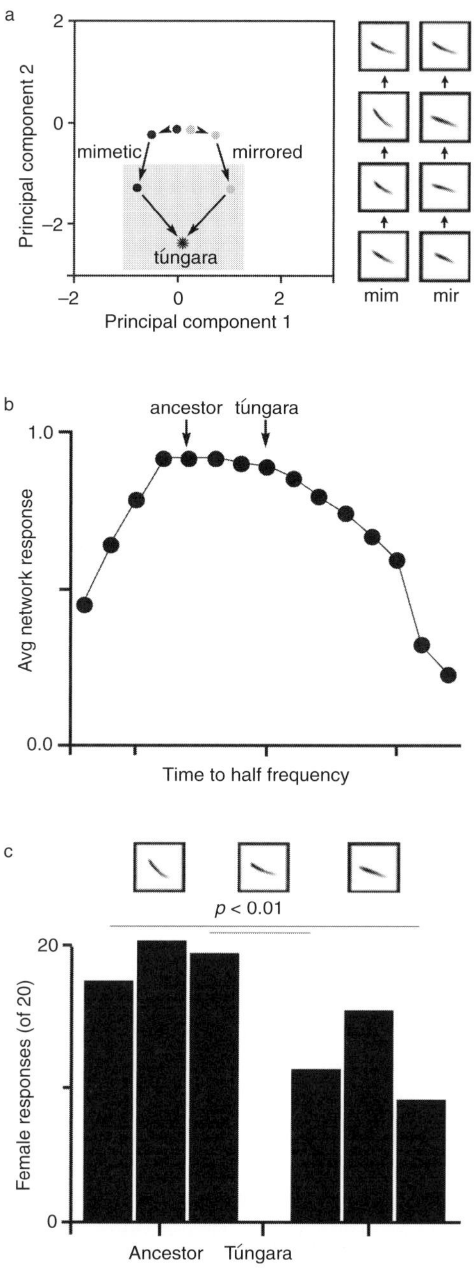

Figure 2.5. Vestigial preferences in neural networks and túngara frogs. (a) Networks were given one of two histories matched for the diversity of the calls in training set. The mimetic history consisted of a series of representing reconstructed ancestral nodes in the path leading to túngara frogs; the mirrored history is a control made by rotating this trajectory in a call space defined by PCA and synthesising the resulting calls. Sonograms on right depict frequency on the y-axis and time on the x-axis for corresponding histories. (b) Mimetic history networks retain an ability to recognise calls with short time-to-half frequencies, resembling ancestors. Mirrored histories exhibit biases in the opposite direction (not shown). (c) Number of females (of 20) approaching calls synthesised to resemble a reconstructed ancestor (left of túngara), or control calls of comparable similarity to túngara frog calls (right of túngara). Real females respond significantly more often to ancestor-like calls than to controls that do not resemble an ancestor. Figure modified from Phelps *et al.* (2001).

2.2.5 Incorporating costs

Because perceptual resources are limited, organisms must decide how much they will invest in information processing, and how to allocate those resources across alternative tasks. Consider two extreme examples: the energy budget attributed to the human brain is 20% of basal metabolic rate (Clarke & Sokoloff, 1999); and for a mormyrid electric fish it is 60% of basal metabolic rate (Nilsson, 1996). Within mammalian nervous systems, Ames (2000) estimates ~5–15% of this budget is attributable to maintaining neuron function, ~30–50% to maintaining and recycling the contents of synapses and ~50–60% to maintaining ion gradients needed for initiation and propagation of voltage changes.

Fortunately, each of these major categories has an obvious and quantifiable analogue in neural network models. Neuron maintenance costs can be assigned by multiplying the maintenance cost per neuron by the number of neurons in the network ($c_m{*}n$) or, if neuron number is not allowed to vary, the number of neurons active in the task. The costs of maintaining synapses can be assigned by summing the values of weights across the network ($\sum\sum w_{ij}{*}c_w$, where w_{ij} is the weight between a pair of neurons *i* and *j*, and c_w is the cost per unit of synapse). If one keeps track of the output of each neuron in each time step, one could assign costs based on the activity of each neuron as well ($\sum\sum c_s(a)$, where $c_s(a)$ is the function describing how signally costs accrue as a function of activity, summed across neurons and time-steps). In fact, it seems as if costs of perceptual allocation could be modelled using extant neural networks simply by altering fitness functions. Using the example given for the túngara frog network, the modified fitness function would simply be

$$\begin{aligned} W &= \left[\sum (C_i - N_i)^2/n\right]^{1/2} - costs + 0.01 \\ &= \left[\sum (C_i - N_i)^2/n\right]^{1/2} - \left(c_m * n + \sum\sum w_{ij} * c_w + \sum\sum c_S(a)\right) + 0.01 \qquad (5) \end{aligned}$$

The function $c_s(a)$ describing how costs increase with firing rate merits special attention, because it lies at the heart of work on the design of efficient coding schemes. Laughlin *et al.* (1998) point out that the relationship between energy use and information coding (bit-rate) is not linear. Rather, energy use is an accelerating function of bit-rate – at high bit-rates, doubling the amount of information encoded more than doubles the cost. As a result, it is often more efficient to parse the coding of information between two neurons rather than sustain the high costs of coding the same information in a single neuron. Indeed, this is a general feature of intensity coding (Gardener & Martin, 2000). In the auditory system of anurans, for example, many of the frequency thresholds reported in Figure 2.3b correspond to the minimum intensity needed to elicit some basal level of population-level activity. An examination of the firing thresholds for individual neurons reveals that the intensity of a given frequency is coded by a distributed set of neurons that vary in their individual thresholds (e.g. Konishi, 1969; Capranica & Moffat, 1983). If efficient coding is a major concern of sensory design, neural networks could prove invaluable tools for its exploration.

2.3 Central allocation

Sensory information does not, of course, end at the periphery. That information must ultimately make it to brain regions that will combine sensory and affective information, and translate them into a decision about how to behave. A comprehensive review of higher levels of sensory processing or its intersection with downstream decision mechanisms is well beyond our current aims. Even a broad survey of classic literature, however, reveals a remarkably congruent picture of perceptual allocation.

Sensory receptors coming from a peripheral organ converge on relay sites, each of which transforms the information into more useful forms. These relays and the higher representations they feed into exhibit a topographical representation of sensation. That is, adjacent neurons represent similar domains of sensory space. Notably, the configuration of a map does not faithfully reproduce the structure of the sensed world, but is distorted in favour of certain highly sensitive and highly important areas. The primary visual cortex, for example, consists of a series of columns of neurons that arrange visual information by eye, position and orientation. The representation of the fovea, however, is far in excess of the size of the corresponding visual field. Similarly, in moustache bats the primary auditory cortex exhibits an inflated representation of 60–62 kHz sounds, a range corresponding to the echolocation calls used to navigate and hunt (reviewed in Covey, 2005). I now discuss one well-studied representation, that of somatosensation, or touch.

2.3.1 Touch

The mammalian somatosensory cortex has been a long-standing subject for neurophysiologists. Marshall *et al.* (1941) recorded activity of neurons in somatosensory cortex in response to touching different regions of an animal's body surface. Penfield and colleagues demonstrated that such maps were causally related to perception by asking epileptic patients, whose cortices were being probed to uncover the foci of seizures, what they felt in response to electrical stimulation at different sites (Rasmussen & Penfield, 1947). The resulting maps depict a homunculus distorted by the density of sensory innervations in distinct body parts (indeed, we now know there are multiple such maps; Kaas *et al.*, 1979).

It is now accepted that topographic maps are not static representations of the periphery, but are gradually modified by experience. Jenkins *et al.* (1990) trained monkeys to perform a task that involved touching a disc with their fingertips; over the ensuing months the representation of fingertips grew at the expense of the adjacent phalanges. String players display an enlarged representation of left but not right fingers, and such changes are correlated with the ages at which musicians began playing (Elbert *et al.*, 1995). Even the nature of the topography can be modified by experience. Adjacent fingers are next to one another, for example, but somatosensory maps contain sharp divisions between neurons of one finger and those of another. A notable exception occurs when patients suffer from congenital syndactyly, in which fingers are fused. Surgically freeing the digits causes the individuation of the representations of the formerly joined fingers (Mogilner *et al.*, 1993).

Topographical representations and their distortions reflect the underlying ability of an organism to resolve physical differences between stimuli. This relationship is borne out by studies of species diversity in the organisation of somatosensory cortex. Welker and colleagues, for example, demonstrated that coatimundis, which have relatively sparse representation of the front paw in the somatosensory cortex, have less digit dexterity than the related and more elaborately represented raccoon (Welker & Seidenstein, 1959; Welker & Campos, 1963). More dramatic, perhaps, is a series of studies on the star-nosed mole, *Condylura cristata*, an insectivore which possesses an elaborate star-shaped mechanosensory organ on the tip of its nose. The organ appears to be an elaboration of a basal mole pattern in which the rostrum is specialised for touch (Catania, 2005). The high surface area of the star is combined with a fovea-like sensitivity of the central appendages (Catania & Kaas, 1997). This peculiar arrangement facilitates capture of small prey on the order of 230 ms, a handling time that makes even tiny items surprisingly profitable (Catania & Remple, 2005). This high value for mechanosensory input is accompanied by an enlarged representation of somatosensory cortex, and a disproportionate amount of this region allocated to the central appendages of the star (Catania & Kaas 1997; Catania & Remple, 2005).

2.3.2 Neural network models of topographic maps

It may seem at first that such complexity is beyond the scope of immediate modelling efforts in animal behaviour. Although the rules by which the nervous system assembles maps of sensory space are an active area of investigation (for recent review of somatosensory cortex, see Feldman & Brecht, 2005), there are a number of surprisingly simple algorithms appropriate for modelling the emergence of such maps. Such rules tend to focus on changing weights based on patterns of correlation between neurons ('Hebbian' rules, after Hebb, 1949), or on competitive interactions between neurons (e.g. Kohonen, 1982; Kohonen & Hari, 1999). Hebbian mechanisms have long been the focus of physiological work on map assembly (e.g. Bear *et al.*, 1987). Competitive learning methods, in contrast, are the focus of more computational work and less physiological work; their mechanisms are biologically feasible, however, and seem likely to be a component of natural map formation. Because one type of competitive-learning map, Kohonen's self-organising map (Kohonen, 1982; Kohonen & Hari, 1999) is readily available in a number of software packages (e.g. Matlab), I will briefly describe how it could be used to investigate perceptual allocation. The same questions could be asked of Hebbian maps as well.

Kohonen's self-organising map (SOM) commonly begins with an array of neurons which randomly weight each dimension of an input vector. The dimensions of this vector can be considered the activities of a set of input neurons. In each round of training, the map neuron which is most active in response to an input pattern, the 'winner', increases

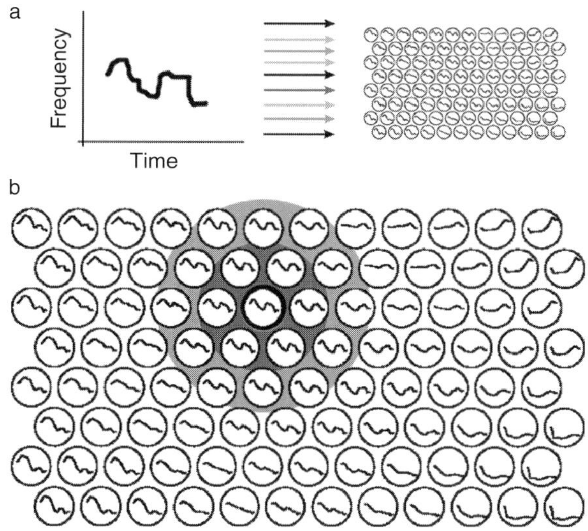

Figure 2.6. A self-organising feature map trained by Kohonen's winner-take-all algorithm. Graphs depict frequency contours over short time frames (~100 ms) for Finnish phonemes. (a) Points along the contour served as inputs to a map. (b) In a trained map, adjacent neurons map similar space. The phoneme in (a) is most similar to the 'winning' " neuron at the centre of the concentric grey circles. In a training phase, the winning neuron would cause the neighbouring neurons to update their weights to be more similar to the winning weights. This neighbourhood function drives the map's ability to topographically map complex stimuli. Figure modified from Kohonen (2003).

each of its weights in proportion to the activity of each input (Figure 2.6), thereby increasing its response to the same input on future presentations. The winning neuron influences adjacent neurons to perform an analogous update, and the magnitude of the modification declines with the distance from the most active neuron. This simple procedure produces a two-dimensional map of a complex, multi-dimensional input, in which the resulting topography reflects underlying similarities between input patterns. In one example, maps trained on short time-spectra of Finnish language sounds produce an ordered map of phoneme structure (Figure 2.6; Kohonen, 2003).

Studies of these maps reveal interesting patterns. First, as in Hebbian learning, the maps tend to be biased toward the most common types of input patterns. Interestingly, the magnification of the representation is found to correspond to the frequency of a pattern raised to a power (Haykin, 1994). That power typically varies from 1/3 to 1, indicating that as a rule, the number of neurons that represent a feature is an increasing but decelerating function of its frequency of occurrence (Kohonen & Hari, 1999). How can one relate such parameters to natural patterns of allocation? I have mentioned that innervation density at the periphery tends to predict cortical area and behavioural discrimination. What is the quantitative nature of such relationships and how are they modified by mechanisms of map formation? How do changes in peripheral allocation alter the ability to represent information at higher levels? Such questions seem central to linking sensory drive to the compatible traditions of neuroethology.

In order to ask such questions, one must first identify a metric that describes how well the map preserves information relevant to identification of ecologically relevant patterns. This is not a trivial task, but assume for the moment that has been done. How could one incorporate it into evolutionary models? Using a genetic algorithm, one can encode the relevant learning parameters, the size of the map, and perhaps the distribution of responses at the periphery (the 'innervation density'). Based on a set of inputs, one could train networks with these parameters to produce a map of the input space. Using a fitness function analogous to the one described for peripheral allocation, one could select networks to form efficient representations that preserved meaningful distinctions from within that input space. How would the mechanisms of map formation – Hebbian, competitive or otherwise – interact with the demands of efficient coding to produce higher level allocations? Are there regularities one can predict between the costs of representation and the magnification of peripheral inputs? Can such considerations help us understand why some animals have multiple maps for complex representations, while others have lost all but the lowest level maps (e.g. the lack of secondary auditory and visual cortex in the brains of shrews; Catania, 2005)? Such questions seem highly relevant to receiver design, but are unlikely to merit serious treatment by biologists focused on understanding human brain function.

2.4 Common themes, uncommon opportunities

This review has touched on a broad range of examples from both empirical and modelling studies. Running through this diversity are common threads worth articulating. The first is that the principles of sensory ecology have proved remarkably predictive of natural variation. This power has the limits of any theory that describes how animals ought to be designed – selection can only act locally, and the topology of the fitness landscape is likely to be influenced by the sub-optimal contingencies of genetics and history. A second thread is that treatments of receiver design need to explicitly address the costs of perceptual resources. Such costs are implicit in analytic models but are rarely dealt with directly. Beyond these themes are a number of more specific commonalities

In principle, meaningful information could be extracted at either early or late stages of processing. Nevertheless, the periphery does attend selectively to certain kinds of information. This is evident not only in the peripheral tuning of sensory systems, but in electrophysiological studies that demonstrate that biologically relevant stimuli are often more faithfully represented than arbitrary stimuli (e.g. Machens *et al.*, 2001). Given that the allocation of neurons to higher-order representations is shaped by the distribution of resources at the periphery, the earlier in the processing stream one can isolate information likely to be meaningful, the greater the efficiency of emergent representations. As knowledge of peripheral sensory structures becomes more advanced, more explicit and more elaborate evolutionary models for such processing seem on the horizon.

The efficacy of a nervous system must ultimately be measured by its ability to make decisions that contribute to an organism's fitness. In sensory ecology, as in the communications theory upon which much of the field is based, this is measured as an attempt to discriminate meaningful from unmeaningful stimuli. Dusenberry (1992) notes that there are multiple such measures, each having a different formulation, but all maximising the ratio of the likelihood the stimulus is present to the likelihood it is not. In signal detection theory, this decision corresponds to a threshold (which might be further weighted by the values of various kinds of decisions, as by Green & Swets, 1966). In information theory, this would likely take the form of the logarithm of this ratio, as in Eq. 3. In particular applications in sensory ecology similar measures may be used without explicit reference to such theory. In an elegant study on dichromatic marine fishes, for example, Cummings (2004) measures the distribution of colours in the environment and in food-laden corals; using quantum flux calculations, she finds that for multiple species the receptor tuning is ideally suited to discriminate a coral target from background lighting. The same conceptual framework can describe the fitness functions of most evolutionary neural network studies. In our túngara frog example, we selected a network to maximise its output in response to a túngara call, and minimise its response to background noise. This is synonymous with maximising how much information the output of the network conveys regarding the presence of a conspecific call. Similarly, in an elegant study on the coevolution of model signallers and receivers, Hurd *et al.* (1995) demonstrate that visual signals and feedforward neural networks coevolve to maximise signal discriminability. Darwin's 'principle of antithesis' and the ethological 'sign stimulus' both emphasise the tendency of signals to evolve to extreme, easily detected forms (Darwin, 1872; Tinbergen, 1951). The ability to discriminate among biologically meaningful categories can be formalised in many ways, but the common crux is that detection is valuable, and receivers are active investors in information acquisition.

Thus far discussion of perceptual allocation has focused on the ability to detect a target within a background noise – certainly a critical task, and the focus of much study in both sensory ecology and network evolution. It is worth noting, however, that not only can there be multiple targets, as exemplified by the numerous mate-choice and feeding examples discussed, but that there are tasks not readily described as target detection. A number of studies have, for example, investigated the tuning of visual receptors for wavelengths abundant in the natural environment, a task referred to as 'luminance detection' (Wyszecki & Stiles, 1982). The need to navigate in one's surroundings, for example, requires the ability to detect abundant and variable stimulus energies. Such background sampling is often at odds with a simple 'matched filter' strategy. A complete examination of perceptual allocation will need to address how background and target detection are accomplished simultaneously. The allocation of resources to potentially conflicting tasks is interesting in its own right, and more so when coupled to the interacting participants in sensory drive.

Another limitation of our first-order description of perceptual allocation is that not all tasks will have equal value. The optimal allocation to a task must balance the cost of each additional neuron with the value expected from the information gained. The most effective representation does not simply maximise the resolution of frequent stimuli – it maximises the expected value of that information, which is the product of its frequency and value. Returning to the example of somatosensory cortex, over a lifetime the glans of the penis is not likely to receive much more frequent use than a corresponding area of skin on the adjacent body surface – and yet, for obvious reasons, its uses contribute much more to fitness, and its innervation density reflects that value. It is interesting to note that the importance of value on perceptual allocation is currently an exciting topic in sensory neuroscience. For example, playing tones to rats while stimulating ascending projections from the ventral tegmentum, a region encoding positive value, enhances the representation of those frequencies in auditory cortex (Bao *et al.*, 2001; see also Kilgard & Merzenich, 1998). Similarly, Suga and colleagues have used bats to show that making tones more meaningful by linking them to negative reinforcement causes an expansion of their representation in early auditory centres (Gao & Suga, 2000). The view from behavioural ecology suggests this is likely to be a rather general process.

Sensory ecology and neural network models have been successful in isolation. The underlying unity in their concern with sensory design suggests how neural network studies could gain from the clarity provided from the mathematical methods of sensory ecology; sensory ecology, in turn, could stand to be sullied by the complexities of evolutionary processes captured in genetic algorithms and elaborate receivers. These complementary approaches make it possible to titrate the compromises faced by model organisms – to alter costs, histories and tasks to observe the complex interplay of ecological and evolutionary forces. The hybrid promises not extinction but diversity and vigour. By exploring the strengths and weaknesses of sensory ecology through neural network models, we are likely to gain a deeper understanding of its power, its limits, and its ability to direct novel experiments in a diversity of taxa.

Acknowledgements

I would like to thank Dr Alex Ophir for his help in preparing the manuscript, and Drs Colin Tosh and Graeme Ruxton for their thoughtful reviews. Dr Rebecca Fuller suggested many of the innovations in the visual network example developed here. Lastly, I must acknowledge Dr Michael Ryan and the late and much missed Dr Stan Rand. Both have been invaluable mentors and collaborators.

References

Ames, A. 2000. CNS energy metabolism as related to function. *Brain Res Rev* **34**, 42–68.

Bao, S. W., Chan, W. T. & Merzenich, M. M. 2001. Cortical remodelling induced by activity of ventral tegmental dopamine neurons. *Nature* **412**, 79–83.

Basolo, A. 1990. Female preference predates the evolution of the sword in swordtails. *Science* **250**, 808–810.

Bear, M. F., Cooper, L. N. & Ebner, F. F. 1987. A physiological basis for a theory of synapse modification. *Science* **237**, 42–48.

Bernard, G. D. & Remington, C. L. 1991. Colour-vision in *Lycaena* butterflies – spectral tuning of receptor arrays in relation to behavioural ecology. *Proc Natl Acad Sci* **88**, 2783–2787.

Bradbury, J. W. & Vehrencamp, S. L. 1998. *Principles of Animal Communication.* Sinauer Associates.

Bradbury, J. W. & Vehrencamp, S. L. 2000. Economic models of animal communications. *Animal Behaviour* **59**, 259–268.

Bruns, V. & Schmieszek, E. 1980. Cochlear innervation in the greater horseshoe bat – demonstration of an acoustic fovea. *Hearing Res* **3**, 27–43.

Capranica, R. R., Frishkopf, L. S. & Nevo E. 1973. Encoding of geographic dialects in the auditory system of the cricket frog. *Science* **182**, 1272–1275.

Capranica, R. R. & Moffat, A. J. M. 1983. Neurobehavioral correlates of sound communication in anurans. In *Advances in Vertebrate Neuroethology* (eds J. P. Ewert, R. R. Capranica & D. J. Ingle), pp. 701–730. Plenum Press.

Carleton, K. L. & Kocher, T. D. 2001. Cone opsin genes of African cichlid fishes: Tuning spectral sensitivity by differential gene expression. *Mol Biol Evol* **18**, 1540–1550.

Catania, K. C. 2005. Evolution of sensory specializations in insectivores. *Anat Rec* **287A**, 1038–1050.

Catania, K. C. & Kaas, J. H. 1997. Somatosensory fovea in the star-nosed mole: behavioural use of the star in relation to innervation patterns and cortical representation. *J Comp Neurol* **387**, 215–233.

Catania, K. C. & Remple, F. E. 2005. Asymptotic prey profitability drives star-nosed moles to the foraging speed limit. *Nature* **433**, 519–522.

Cheroske, A. G. & Cronin, T. W. 2005. Variation in stomatopod (*Gonodactylus smithii*) colour signal design associated with organismal condition and depth. *Brain Behav Evol* **66**, 99–113.

Clarke, D. D. & Sokoloff L. 1999. Circulation and energy metabolism of the brain. In *Basic Neurochemistry: Molecular, Cellular and Medical Aspects* (eds G. J. Siegel, B. W. Agranoff, R. W. Albers, S. K. Fisher & M. D. Uhler), pp. 637–669. Lippincott-Raven.

Covey, E. 2005. Neurobiological specializations in echolocating bats. *Anat Rec* **287A**, 1103–1116.

Cummings, M. E. 2004. Modelling divergence in luminance and chromatic detection performance across measured divergence in surfperch (Embiotocidae) habitats. *Vision Res* **44**, 1127–1145.

Darwin, C. 1872. *The Expression of Emotions in Man and Animals.* (3rd Edn, 1998, ed. P. Ekman.) Oxford University Press.

Dooling, R. J., Mulligan, J. A. & Miller, J. D. 1971. Auditory sensitivity and song spectrum in the common canary (*Serinus canaries*). *J Acoustic Soc Am* **50**, 700–709.

Dominy, N. J. & Lucas, P. W. 2004. Significance of colour, calories, and climate to the visual ecology of catarrhines. *Am J Primatol* **62**, 189–207.

Dusenberry, D. B. 1992. *Sensory Ecology: How Organisms Acquire and Respond to Information.* W. H. Freeman and Company.

Elbert, T., Pantev, C., Wienbruch, C., Rockstroh, B. & Taub, E. 1995. Increased cortical representation of the fingers of the left hand in string players *Science* **270**, 305–307.

Elman, J. L. 1990. Finding structure in time. *Cogn Sci* **14**, 179–211.

Endler, J. A. 1992. Signals, signal conditions, and the direction of evolution. *Am Naturalist* **139**, S125–S153.
Endler, J. A. 1993. The colour of light in forests and its implications. *Ecol Monogr* **63**, 1–27.
Enquist, M. & Arak, A. 1992. Symmetry, beauty and evolution. *Nature* **372**, 169–172.
Enquist, M. & Arak, A, 1993. Selection of exaggerated male traits by female aesthetic senses. *Nature* **361**, 446–448.
Feldman, D. E. & Brecht, M. 2005. Map plasticity in somatosensory cortex. *Science* **310**, 810–815.
Frishkopf, L. S., Capranica, R. R. & Goldstein, M. H. 1968. Neural coding in bullfrogs auditory system – a teleological approach. *Proc Inst Electric Electron Eng* **56**, 969–983.
Fuller, R. C., Carleton, K. L., Fadool, J. M., Spady, T. C. & Travis, J. 2004. Population variation in opsin expression in the bluefin killifish, *Lucania goodei*: a real-time PCR study. *J Comp Physiol A* **190**, 147–154.
Fuller, R. C., Carleton, K. L., Fadool, J. M., Spady, T. C. & Travis, J. 2005. Genetic and environmental variation in the visual properties of bluefin killifish, *Lucania goodie*. *J Evol Biol* **18**, 516–523.
Fuller, R. C., Fleishman, L. J., Leal, M., Travis, J. & Loew, E. 2003. Intraspecific variation in retinal cone distribution in the bluefin killifish, *Lucania goodie*. *J Comp Physiol A* **189**, 609–616.
Gao, E. Q. & Suga, N. 2000. Experience-dependent plasticity in the auditory cortex and the inferior colliculus of bats: role of the corticofugal system. *Proc Natl Acad Sci* **97**, 8081–8086.
Gardener, E. P. & Martin, J. H. 2000. Coding of sensory information. In *Principles of Neural Science, 4th Edn* (eds. E. R. Kandel, J. H. Schwartz & T. M. Jessel). McGraw-Hill.
Gerhardt, H. C. & Huber, F. 2002. *Acoustic Communication in Insects and Anurans: Common Problems and Diverse Solutions.* University of Chicago Press.
Gerhardt, H. C. & Schwartz, J. J. 2001. Auditory tuning and frequency preferences in anurans. In *Advances in Anuran Communication* (ed. M. J. Ryan), pp. 73–85. Smithsonian Press.
Green, D. M. & Swets, J. A. 1966. *Signal Detection Theory and Psychophysics.* Wiley.
Halstenberg, S., Lindgren, K. M., Samagh, S. P. S. *et al.* 2005. Diurnal rhythm of cone opsin expression in the teleost fish Haplochromis burtoni. *Vis Neurosci* **22**, 135–141.
Haykin, S. 1994. *Neural Networks: A Comprehensive Foundation.* Macmillan Press.
Hebb, D. O. 1949. *The Organization of Behaviour: A Neuropsychological Theory*. Wiley.
Hurd, P. L., Wachtmeister, C.-A. & Enquist, M. 1995. Darwin's principle of thesis and antithesis revisited: a role for perceptual biases in the evolution of intraspecific signals. *Proc R Soc B* **259**, 201–205.
Jacobs, G. H. & Deegan, J. F. 2005. Polymorphic New World monkeys with more than three M/L cone types. *J Optical Soc Am A* **22**, 2072–2080.
Jenkins, W. M., Merzenich, M. M., Ochs, M. T., Allard, T. & Guicrobles, E. 1990. Functional reorganization of primary somatosensory cortex in adult owl monkeys after behaviorally controlled tactile stimulation. *J. Neurophysiol* **63**, 82–104.
Johnstone, R. A. 1994. Female preferences for symmetric males as a by-product of selection for mate recognition. *Nature* **372**, 172–175.
Kaas, J. H., Nelson, R. J., Sur, M., Lin, C. S. & Merzenich, M. M. 1979. Multiple representations of the body within the primary somatosensory cortex of primates. *Science* **204**, 521–523.

Kilgard, M. P. & Merzenich, M. M. 1998. Cortical map reorganization enabled by nucleus basalis activity. *Science* **279**, 1714–1718.

Kirkpatrick, M. & Ryan, M. J. 1991. The evolution of mating preferences and the paradox of the lek. *Nature* **350**, 33–38.

Kohonen, T. 1982. Analysis of a simple self-organizing process. *Biol Cybernet* **44**, 135–140.

Kohonen, T. 2003. Self-organized maps of sensory events. *Phil Trans R Soc A* **361**, 1177–1186.

Kohonen, T. & Hari, R. 1999. Where the abstract feature maps of the brain might come from. *Trends Neurosci* **22**, 135–139.

Konishi, M. 1969. Hearing, single-unit analysis, and vocalizations in songbirds. *Science* **166**, 1178–1181.

Koppl, C., Gleich, O. & Manley, G. A. 1993. An auditory fovea in the barn owl cochlea. *J Comp Physiol A* **171**, 695–704.

Laughlin, S. B., van Steveninck, R. R. D. & Anderson, J. C. 1998. The metabolic cost of neural information. *Nat Neurosci* **1**, 36–41.

Lucas, P. W., Dominy, N. J., Riba-Hernandez, P. *et al.* 2003. Evolution and function of routine trichromatic vision in primates. *Evolution* **57**, 2636–2643.

Lunau, K., Wacht, S. & Chittka, L. 1996. Colour choices of naive bumble bees and their implications for colour perception. *J Comp Physiol A* **178**, 477–489.

Lythgoe, J. N. & Partridge, J. C. 1991. The modelling of optimal visual pigments of dichromatic teleosts in green coastal waters. *Vision Res* **31**, 361–371.

Machens, C. K., Stemmler, M. B., Prinz, P. *et al* . 2001. Representation of acoustic communication signals by insect auditory receptor neurons. *J Neurosci* **21**, 3215–3227.

Marshall, J. C., Woolsey, C.N. & Bard, P. 1941. Observations on the cortical somatic sensory mechanisms of cat and monkey. *J Neurophysiol* **4**, 1–24.

Meyer, J. & Elsner, N. 1996. How well are frequency sensitivities of grasshopper ears tuned to species-specific song spectra? *J Exp Biol* **199**, 1631–1642.

Mitchell, M. 1996. *An Introduction to Genetic Algorithms*. MIT Press.

Mogilner, A., Grossman, J. A. L., Ribary, U. *et al.* 1993. Somatosensory cortical plasticity in adult humans revealed by magnetoencephalography. *Proc Natl Acad Sci* **90**, 3593–3597.

Morton, E. 1975. Ecological sources of selection on avian sounds. *Am Naturalist* **109**, 17–34.

Nilsson, G. E. 1996. Brain and body oxygen requirements of Gnathonemus petersii, a fish with an exceptionally large brain. *J Exp Biol* **199**, 603–607.

Osorio, D., Smith, A. C., Vorobyev, M. & Buchanan-Smith, H. M. 2004. Detection of fruit and the selection of primate visual pigments for colour vision. *Am Naturalist* **164**, 696–708.

Osorio, D. & Vorobyev, M. 2005. Photoreceptor spectral sensitivities in terrestrial animals: adaptations for luminance and colour vision. *Proc R Soc B* **272**, 1745–1752.

Phelps, S. M. 2001. History's lessons: a neural network approach to the evolution of animal communication. In *Advances in Anuran Communication* (ed. M. J. Ryan), pp. 167–180. Smithsonian Press.

Phelps, S. M. & Ryan, M. J. 1998. Neural networks predict the response biases of female túngara frogs. *Proc R Soc B* **265**, 279–285.

Phelps, S. M. & Ryan, M. J. 2000. History influences signal recognition: neural network models of túngara frogs. *Proc R Soc B* **267**, 1633–1639.

Phelps, S. M., Ryan, M. J. & Rand, A. S. 2001. Vestigial preference functions in neural networks and túngara frogs. *Proc Natl Acad Sci* **98**, 13161–13166.

Rasmussen, T. & Penfield, W. 1947. The human sensorimotor cortex as studied by electrical stimulation. *Federation Proc* **6**, 184–185.

Rodd, F. H., Hughes, K. A., Grether, G. F. & Baril, C. T. 2002. A possible non-sexual origin of mate preference: are male guppies mimicking fruit? *Proc R Soc B* **269**, 475–481.

Ryan, M. J. 1990. Sexual selection, sensory systems and sensory exploitation. *Oxford Surv Evol Biol* **7**, 157–195.

Ryan, M. J. & Brenowitz, E. A. 1985. The role of body size, phylogeny and ambient noise in the evolution of bird song. *Am Naturalist* **126**, 87–100.

Ryan, M. J., Cocroft, R. B. & Wilczynski, W. 1990a. The role of environmental selection in intraspecific divergence of mate recognition signals in the cricket frog, Acris crepitans. *Evolution* **44**, 1869–1872.

Ryan, M. J., Fox, J. H., Wilczynski, W. & Rand, A. S. 1990b. Sexual selection for sensory exploitation in the frog *Physalaemus pustulosus*. *Nature* **343**, 66–67.

Ryan, M. J., Perrill, S. A. & Wilczynski, W. 1992. Auditory tuning and call frequency predict population-based mating preferences in the cricket frog, Acris crepitans *Am Naturalist* **139**, 1370–1383.

Ryan, M. J. & Rand, A. S. 1995. Female responses to ancestral advertisement calls in túngara frogs. *Science* **269**, 390–392.

Schul, J. & Patterson, A. C. 2003. What determines the tuning of hearing organs and the frequency of calls? A comparative study in the katydid genus Neoconocephalus (Orthoptera, Tettigoniidae). *J Exp Biol* **206**, 141–152.

Seehausen, O., van Alphen, J. J. M. & Witte, F. 1997. Cichlid fish diversity threatened by eutrophication that curbs sexual selection. *Science* **277**, 1808–1811.

Sisneros, J. A., Forlano, P. M., Deitcher, D. L. & Bass, A. H. 2004. Steroid-dependent auditory plasticity leads to adaptive coupling of sender and receiver. *Science* **305**, 404–407.

Spady, T. C., Seehausen, O., Loew, E. R. *et al.* 2005. Adaptive molecular evolution in the opsin genes of rapidly speciating cichlid species. *Mol Biol Evol* **22**, 1412–1422.

Spaethe, J., Tautz, J. & Chittka, L. 2001. Visual constraints in foraging bumblebees: flower size and colour affect search time and flight behaviour. *Proc Natl Acad Sci* **98**, 3898–3903.

Sumner, P. & Mollon, J. D. 2000. Catarrhine photopigments are optimized for detecting targets against a foliage background. *J Exp Biol* **203**, 1963–1986.

Sun, L. X., Wilczynski, W., Rand, A. S. & Ryan, M. J. 2000. Trade-off in short- and long-distance communication in tungara (*Physalaemus pustulosus*) and cricket (*Acris crepitans*) frogs. *Behav Ecol* **11**, 102–109.

Surridge, A. K., Osorio, D. & Mundy, N. I. 2003. Evolution and selection of trichromatic vision in primates. *Trends Ecol Evol* **18**, 198–205.

Terai, Y., Mayer, W. E., Klein, J., Tichy, H. & Okada, N. 2002. The effect of selection on a long wavelength-sensitive (LWS) opsin gene of Lake Victoria cichlid fishes. *Proc Natl Acad Sci* **99**, 15501–15506.

Tinbergen, N. 1951. *The Study of Instinct*. Oxford University Press.

Vater, M. 1982. Single unit responses in cochlear nucleus of horseshoe bats to sinusoidal frequency and amplitude modulated signals. *J Comp Physiol* **149**, 369–388.

Waser, P. M. & Brown, C. H. 1984. Is there a sound window for primate communication? *Behav Ecol Sociobiol* **15**, 73–76.

Waser, P. M. & Waser, M. S. 1977. Experimental studies of primate vocalization – specializations for long-distance propagation. *Z Tierpsychol* **43**, 239–263.

Welker, W.I. & Campos, G. B. 1963. Physiological significance of sulci in somatic sensory cerebral cortex in mammals of family Procyonidae. *J Comp Neurol* **120**, 19–31.

Welker, W. I. & Seidenstein, S. 1959. Somatic sensory representation in the cerebral cortex of the racoon (Procyon lotor). *J Comp Neurol* **111**, 469–501.

Wilczynski, W., Rand, A. S. & Ryan, M. J. 2001. Evolution of calls and auditory tuning in the *Physalaemus pustulosus* species group. *Brain, Behav Evol* **58**, 137–151.

Wilczynski, W., Keddy-Hector, A. C. & Ryan, M. J. 1992. Call patterns and basilar papilla tuning in cricket frogs .1. Differences among populations and between sexes. *Brain Behav Evol* **39**, 229–237.

Wiley, R. H. & Richards, D. G. 1978. Physical constraints on acoustic communication in the atmosphere: implications for the evolution of animal vocalization. *Behav Ecol Sociobiol* **3**, 69–94.

Wiley, R. H. & Richards, D. G. 1982. Adaptations for acoustic communication in birds: sound transmission and signal detection. In *Acoustic Communication in Birds, Vol. I* (eds. D. Kroodsma, E. H. Miller & H. Ouellet), pp. 131–181. Academic Press.

Witte, K., Farris, H. E., Ryan, M. J. & Wilczynski, W. 2005. How cricket frog females deal with a noisy world: habitat-related differences in auditory tuning. *Behav Ecol* **16**, 571–579.

Wyszecki, G. & Stiles, W. S. 1982. *Colour Science: Concepts and Methods, Quantitative Data and Formulae*. Wiley.

Zahavi, A. 1975. Mate selection: a selection for a handicap. *J Theor Biol* **53**, 205–214.

Part II

The use of artificial neural networks to elucidate the nature of perceptual processes in animals

3

Correlation versus gradient type motion detectors: the pros and cons

Alexander Borst

3.1 Introduction

In motion vision, two distinct models have been proposed to account for direction-selectivity: the Reichardt detector and the gradient detector (Figure 3.1). In the Reichardt detector (also called 'Hassenstein–Reichardt' detector or correlation-type motion detector), the luminance levels of two neighbouring image locations are multiplied after being filtered asymmetrically (Figure 3.1, left). This operation is performed twice in a mirror-symmetrical fashion, before the outputs of both multipliers are subtracted from each other (Hassenstein & Reichardt, 1956; Reichardt, 1961, 1987; Borst & Egelhaaf, 1989). The spatial or temporal average of such local motion detector signals is proportional to the image velocity within a range set by the detector time-constant (Egelhaaf & Reichardt, 1987). However, it is one of the hallmarks of this model that the output of the individual velocity detectors depends, in addition to stimulus velocity, in a characteristic way on the spatial structure of the moving pattern: in response to drifting gratings, for example, the local Reichardt detector output consists of two components: a sustained (DC) component which indicates by its sign the direction of the moving stimulus, and an AC component, which follows the local intensity modulation and, thus, carries no directional information at all. Since the local intensity modulations are phase-shifted with respect to each other, the AC components in the local signals become cancelled by spatial integration of many adjacent detectors. Unlike the AC component, the DC component survives spatial or temporal averaging (integration). The global output signal, therefore, is purely directional. It is predicted to exhibit a distinct optimum as a function of stimulus velocity for each pattern wavelength. The ratio of velocity and spatial wavelength at this optimum corresponds to a certain temporal frequency, which is the number of spatial pattern periods passing one particular image location each second. Despite a different internal structure, the so-called 'Energy model' (Adelson & Bergen, 1985) is identical to the Reichardt detector with respect to the output signal (van Santen & Sperling, 1985) and consequently with respect to all of the above predictions (for review see Borst & Egelhaaf, 1993). It has been shown to account for various phenomena of vertebrate vision

Modelling Perception with Artificial Neural Networks, ed. C. R. Tosh and G. D. Ruxton. Published by Cambridge University Press.

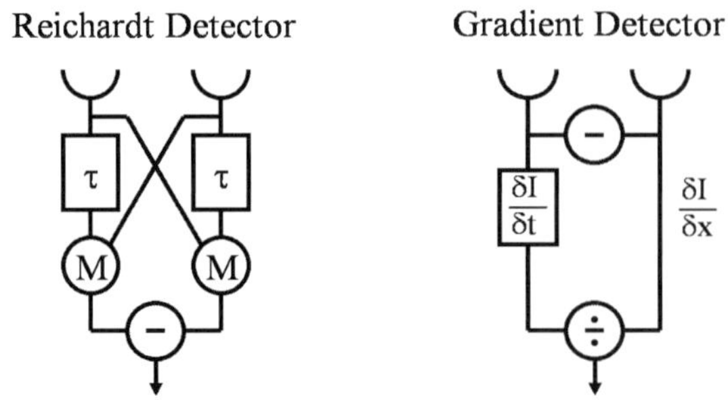

Figure 3.1. Two competing mechanisms proposed to underlie direction selectivity in fly motion detection. Reichardt detector (left). It consists of two mirror-symmetrical subunits. In each subunit, the luminance values as measured in two adjacent image locations become multiplied (M) with each other after one of them is delayed by a low-pass filter with time-constant τ. The resulting output signals of the multipliers become finally subtracted. Gradient detector (right). The temporal luminance gradient as measured after one photoreceptor ($\delta I/\delta t$, to the left) is divided by the spatial luminance gradient ($\delta I/\delta x$).

including humans, too (Borst & Egelhaaf, 1989), but shall, for the reasons outlined above, not be treated as a separate model here.

A prominent alternative model for motion detection is called the 'gradient detector' (Limb & Murphy, 1975; Fennema & Thompson, 1979; Hildreth & Koch, 1987; Srinivasan, 1990). The gradient detector computes the velocity signal by dividing the temporal derivative of local luminance $\partial I(x,t)/\partial t$ by its spatial derivative $\partial I(x,t)/\partial x$ (Figure 3.1, right). Different from the Reichardt detector, the gradient detector provides a signal that is proportional to image velocity at each point and does not depend on pattern properties. In particular, no modulations are expected in the local signals as long as the velocity is constant, and the velocity dependence of the global signal should not vary with the spatial wavelength of the pattern. The main problem with that computation is that the uncertainty in the computed velocity varies widely. If the spatial derivative $\partial I(x,t)/\partial x$ is small, the noise in the temporal derivative is amplified, and for points where the spatial derivative is zero the velocity is completely undefined. Along these lines, Potters & Bialek (1994) proposed that an ideal motion scheme would be based on the gradient detector only in the high signal-to-noise regime, whereas at low signal-to-noise ratios, a Reichardt detector would be superior (Potters & Bialek, 1994).

Many studies on the mechanisms underlying direction selectivity have been performed in the fly visual system. This system, therefore, can be said to be the best understood in terms of mechanism underlying direction selectivity. In summary, all the available evidence supports the Reichardt detector and speaks against the gradient detector. However, neither the actual neurons performing the computations nor the biophysical mechanisms underlying such operations as low-pass filtering or multiplication have been elucidated so far (for review see Borst & Haag, 2002). Rather, the studies comprise input–output experiments at the behavioural level as well as extra- and intracellular recordings from motion-sensitive neurons (Fermi & Reichardt, 1963; Götz, 1964; Buchner, 1976; Single & Borst, 1998; Haag *et al.*, 2004). In particular, the pattern dependence of both the local and the global signal of the motion detection system was recently tested at various mean luminances and, hence, at different signal-to-noise levels (Haag *et al.*, 2004): under all conditions, the pattern dependence was evident, showing modulations at the temporal frequency of the pattern in case of local signals, and characteristic shifts of the optimal velocity with different spatial wavelengths in case of the global signal. This strong experimental support of the Reichardt detector poses the question of the functional advantage this processing scheme offers over the gradient detector. In particular, it suggests that flies, for some reason, have picked a sub-optimal solution in evolution since optimal motion detectors, according to Potters & Bialek (1994), switch over from Reichardt to gradient detectors with increasing signal-to-noise, and flies' motion detectors do not. Briefly, the argument presented in the following will be that the major advantage of Reichardt detectors, besides their superior noise suppression at low luminance levels, relies in their intrinsic adaptive properties allowing for an automatic gain adjustment to the statistics of the velocity stimulus and, thus, for a maximum information transmission under all conditions.

3.2 Materials and methods

In all simulations, an array of 20 detectors covering two spatial periods of the grating was simulated. The output of all detectors was added giving rise to the output signal. The filter in the Reichardt detector was a first-order low-pass with a time-constant τ_L of 50 ms. In the gradient detector, the spatial and temporal derivatives were approximated by the difference between two adjacent input stages of the detector and the difference between two subsequent input signals in time, respectively. In order to avoid division by zero, a small epsilon was added to the absolute value of the spatial derivative in the denominator, and the sign of the spatial gradient was taken care of in the nominator. Calculations were done at 1 ms temporal resolution.

3.3 Results and discussion

A classical functional difference between the two detector models is that the steady-state velocity dependence does not follow the velocity in a linear way in Reichardt detectors, but does so in gradient detectors (Figure 3.2). Moreover, the response of Reichardt detectors is dependent on the pattern wavelength and contrast, and completely independent of pattern properties in gradient detectors. It should be noted that all these statements refer to the steady-state response of the spatial average over an array of detectors. When the output signal of an individual detector is considered, further differences appear: when stimulated by a periodic grating moving at a constant velocity, the output of a single Reichardt detector is not constant but rather modulated by the temporal frequency of the pattern. Due to phase offsets between the individual detectors, these modulations cancel in the spatial average (Single & Borst, 1998; Haag *et al.*, 2004). In contrast, even individual gradient detectors

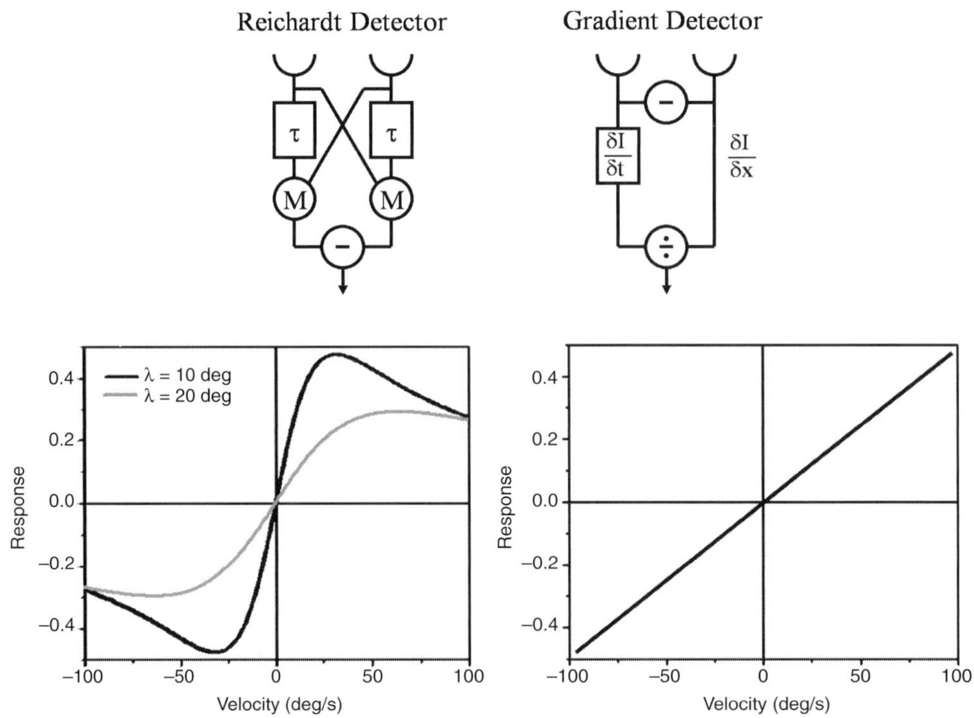

Figure 3.2. Steady-state velocity dependence of Reichardt (left) and gradient (right) detectors in response to moving sine gratings (spatial wavelength λ as indicated). The Reichardt detector shows a peaked velocity dependence. For velocities higher than the optimal velocity, the response gradually returns to zero. Furthermore, the optimal velocity is different for different pattern wavelengths. In contrast, the response of the gradient detector follows the pattern velocity in a linear way and is independent of the pattern wavelength.

report a smooth signal proportional to local pattern velocity at every point in time, without the need of spatial averaging. In general, thus, Reichardt detectors do not qualify as reporters of local retinal velocity because (a) their output is proportional to image velocity only within a certain range and only after spatial averaging and (b) their output depends, in addition to image velocity, on the spatial structure of the moving pattern.

The above distinctions would suffice to expect that any motion vision system evolved in biology should implement a gradient detector and not a Reichardt detector. However, these statements apply only to ideal noise-free circumstances. And reality is noisy, in particular vision due to the Poisson nature of photon emission. When comparing the response properties of both detector models under noisy conditions, the major disadvantage of gradient detectors becomes apparent: its superb performance is paid for by an exquisite sensitivity to input noise. The simulations shown in Figure 3.3 exemplify this point. Here, a sine grating was moved with a white noise velocity profile, low-pass filtered with a cut-off frequency at 10 Hz. The luminance at each point in time and at each image

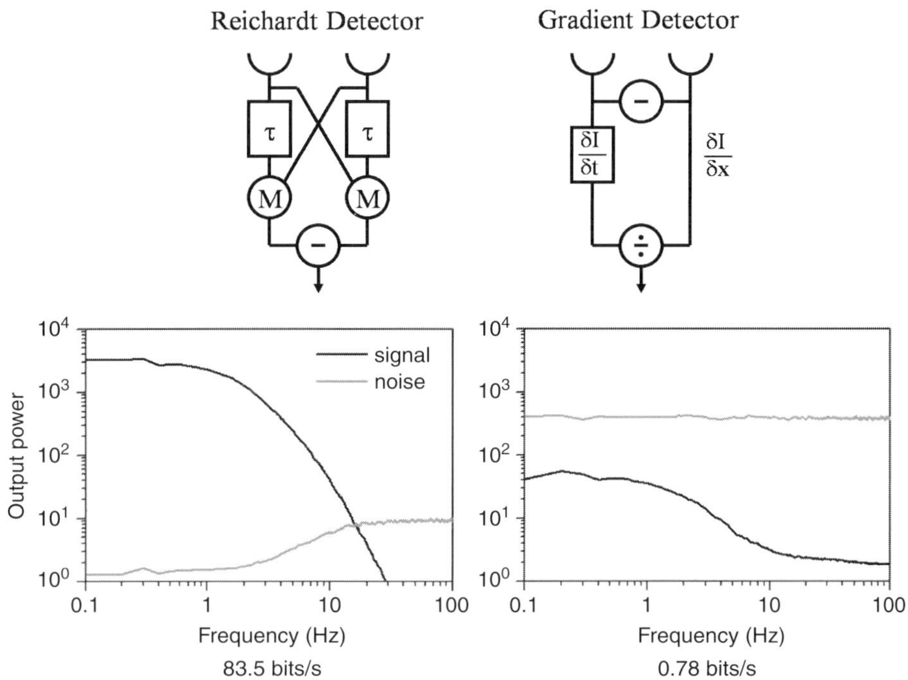

Figure 3.3. Signal-to-noise ratios of Reichardt and gradient detector responses at low luminance levels. A sine grating of 100% contrast was moved with a Gaussian velocity fluctuation at a luminance level corrupted by photon noise corresponding to 1 cd/m^2. The velocity waveform was created by a white noise signal with an autocorrelation time-constant of 100 ms and a standard deviation σ of 5 Hz. The resulting signal and noise spectra at the output of the detector arrays are shown for both types of detectors. Reichardt detectors perform much better than gradient detectors under such conditions, resulting in high information rates for Reichardt detectors and negligible ones for gradient detectors.

location was determined by the local luminance of the pattern plus a Poisson distributed photon noise, corresponding to a luminance level of 1 cd/m^2. Such low luminance levels would occur during dusk or dawn, or in a room with dim lights. The same velocity profile was used to move the pattern many times, and the response of the detector array was calculated for each trial. From the resulting response array, the signal and noise power spectra were calculated and are shown in Figure 3.3. It is clearly seen, that the Reichardt detectors show a much superior performance under these conditions: Here, the signal power is orders of magnitude larger than the noise power, in particular in the low frequency range up to 1 Hz, with the cross-over occurring between 10 to 20 Hz. In contrast, the signal of the gradient detector is drowned in a sea of noise at all frequencies. As a result, the information rate, calculated from the signal-to-noise power ratios, of the Reichardt detector is at about 80 bits/s under these conditions, whereas the gradient detectors signal almost 0 bits/s. An analytical investigation of this point indeed reveals how the internal processing structure of Reichardt detectors explicitly supports noise suppression in the detector, maximising information transmission by the 'water-filling principle' (Lei & Borst, 2006).

At this point, one would argue that each detector model has its pros and cons: while the Reichardt detector is ideal for operation under noisy conditions, the gradient detector shows superior and uncorrupted velocity dependence even locally. Therefore, the proposal made by Potters & Bialek (1994) made a lot of sense: the ideal motion detector should be of Reichardt type under low luminance conditions where noise is prominent, and switch over to a gradient detector under high luminance conditions when signal-to-noise-levels at the photoreceptor input are high. However, as already outlined in the introduction, the available experimental evidence does not support such a switch over, as plausible as it seems.

The functional considerations dealt with so far have left out an important point in image processing, and this pertains to its adaptive properties. First experiments addressing this point were performed, again on the fly visual system using the motion-sensitive neuron H1, by Brenner *et al.* (2000), later confirmed and extended by Fairhall *et al.* (2001) and Borst (2003). In all three studies, flies were stimulated with a rigidly moving spatial pattern. The velocity profile had zero mean and a band-limited white-noise spectrum. From repeated stimulations using identical velocity profiles, stimulus-response functions were constructed using the first principal component of the spike-triggered stimulus ensemble extracted from all the responses. Such experiments were performed using identical time-courses but different velocity amplitudes. When comparing the resulting stimulus-response functions for the different conditions, the response gain was observed to vary depending on the stimulus used: for high-velocity amplitudes (large standard deviations of the velocity distribution), the stimulus-response function was seen to have a small slope or low gain, while for small velocity amplitudes (small standard deviations of the velocity distribution), the stimulus-response function was seen to have a large slope or high gain. Such an adaptive gain seemed to be very useful since it allows the system to dynamically adjust its gain to the prevailing stimulus statistics, making use of its full dynamic range under all circumstances.

The surprising outcome of model simulations on the Reichardt detector was that such detectors do exactly this, namely to adjust their transfer function automatically to the width of the velocity distribution (Borst *et al.*, 2005). The even bigger surprise was that this adaptation occurs in Reichardt detectors without any parameter adjustment: As an intrinsic property of the nonlinear processing structure together with a temporal filter, no internal parameter needs to be changed in order to exhibit this characteristic. As was shown analytically, the effect relies crucially on the temporal dynamics of the stimulus: while the gain adaptation is always the more pronounced the larger the amplitude of the velocity fluctuations is, another crucial parameter is the ratio of the autocorrelation time of the velocity fluctuations and the time-constant in the motion detector filter. With the velocity changing slowly over time (slowly with respect to the detector filter time-constant), gain adaptation is only marginal; with the velocity changing quickly (again with respect to the detector filter time-constant), gain adaptation is pronounced. These

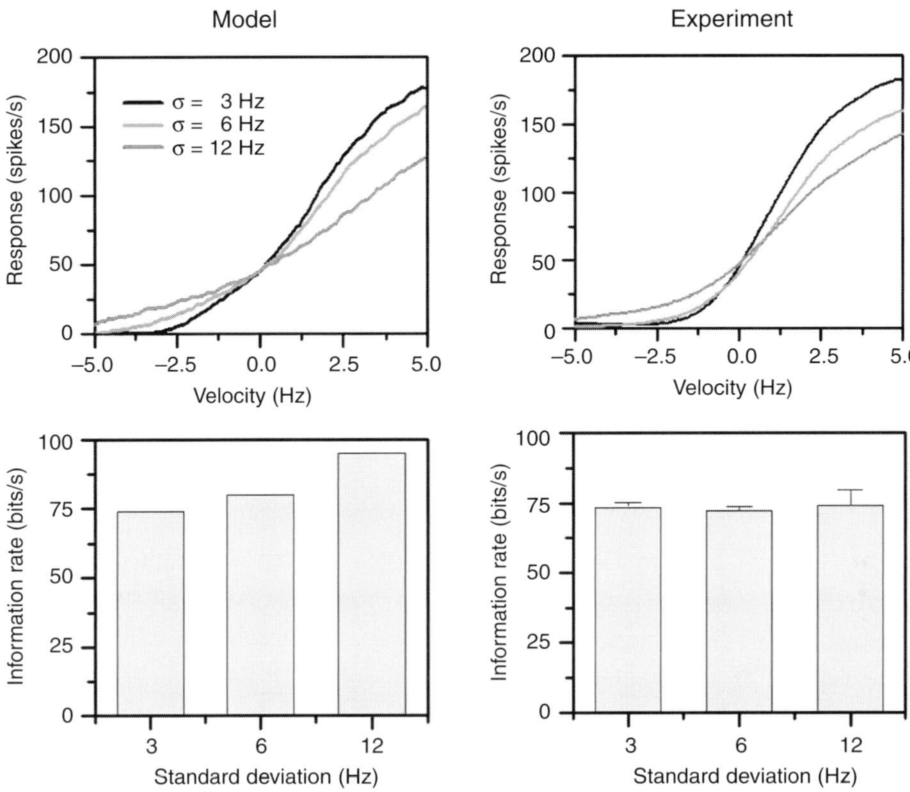

Figure 3.4. Dynamic gain control in Reichardt detectors. Using a white-noise velocity stimulus with three different amplitudes (indicated in temporal frequencies corresponding to the number of periods passing one image location per second), the Reichardt detector can be seen to adapt its velocity gain to the stimulus statistics (upper left). Due to this adaptive gain control, the information rate rises only by a small amount with increasing stimulus amplitude (lower left). The same phenomenon is observed in the fly motion sensitive neuron H1 (right panels, data from Brenner *et al.*, 2000). Note that in this simulation, in order to compare the model and experimental data in a one-to-one fashion and in order to use identical software to evaluate responses from both sources, the output of the Reichardt detector array was used to drive a leaky integrate-and-fire neuron, resulting in a spike output of the model, too. Information was calculated from the spike output according to the method as outlined in de Ruyter van Steveninck *et al.* (1997) and Borst (2003).

considerations could lead to the assumption that the gain is set by image acceleration, but as calculations show, this is an oversimplification (Borst *et al.*, 2005).

In Figure 3.4, model simulations of an array of Reichardt detectors are shown side by side with experimental results obtained in the H1 neuron. Here, instead of the method reported in Brenner *et al.* (2000), Fairhall *et al.* (2001) and Borst (2003), the immediate responses of both the model and the fly neuron are displayed. In order to

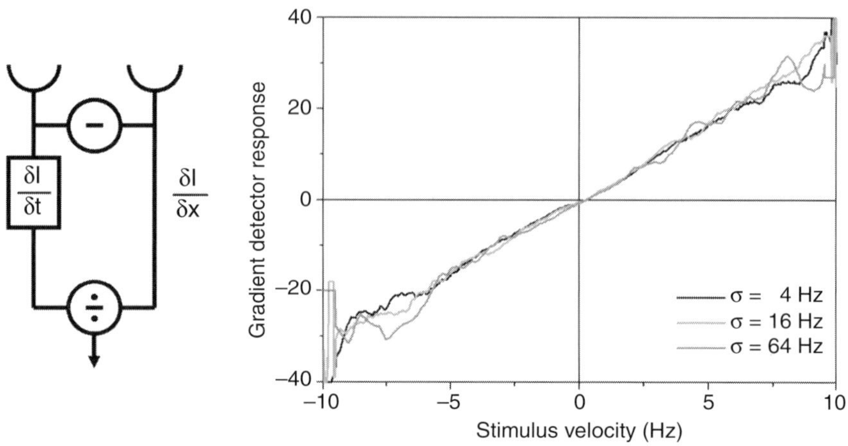

Figure 3.5. Gradient detectors do not show adaptive gain control: When stimulated by white-noise velocity fluctuations with three different amplitudes, identical stimulus-response curves are obtained.

have comparable output signals, the model response was fed into a leaky integrate-and-fire neuron. This also allowed for an analysis of the information rate using the direct method (de Ruyter van Steveninck *et al.*, 1997) without the prerequisite of Gaussian signal and noise distributions. As can be seen, the adaptive gain makes the signals largely independent of the stimulus variance. Nevertheless, the compensation is not complete in the model responses: to some degree, the information rate still grows with stimulus entropy. This again can be understood when calculating the exact propagation of photon noise through the detector (Lei & Borst, 2006) and considering the dependence of the gain adaptation on stimulus dynamics, in addition to stimulus amplitude.

In contrast, gradient detectors do not exhibit such an inherent adaptive velocity gain. As is shown in Figure 3.5, when stimulated with Gaussian velocity distributions of three different amplitudes, the velocity-response curves of an array of gradient detectors are identical for the different conditions: the same linear relationship between velocity and response holds, no matter how large the velocity signals are distributed. While this makes the output of gradient detectors unambiguous with respect to the actual velocity input, it is most unfavourable when the noisiness of the visual input is considered (see Figure 3.2). This decisive difference between the two models of motion detection is further exemplified when looking at the distributions of their output signals under different input signal distributions (Figure 3.6): while Reichardt detectors have a rather similar output signal distribution under various conditions (Figure 3.6, left), output signals of gradient detectors simply follow the input signal distribution (Figure 3.6, right).

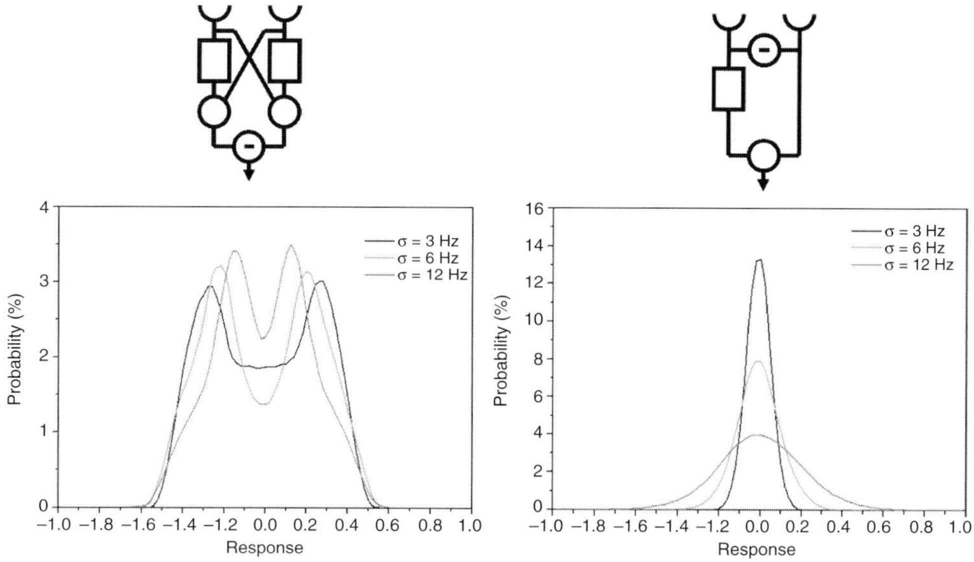

Figure 3.6. Response histograms of Reichardt and gradient detectors stimulated by white-noise velocity fluctuations with three different amplitudes. Whereas Reichardt detectors exhibit similar response distributions, gradient detector response distributions simply follow the corresponding stimulus distributions.

3.4 Conclusion

The work summarised above has given a possible explanation for the experimental finding that fly motion vision is based on Reichardt detectors over a wide range of signal-to-noise ratios: The inherent adaptive velocity gain control of Reichardt detectors allows this algorithm to cover a wide range of velocities resulting in a nearly optimal information transmission for different stimulus amplitudes. Nevertheless, the question remains of how a flying animal can deal with the resulting ambiguity of the course control signals with respect to the absolute retinal velocities. Here, it is important to realise that animals work under closed-loop conditions where any steering action is automatically fed back onto the sensory input: when flying straight requires only the zeroing of the rotational input, the absolute values of the error signal might be of less importance. However, a clear answer to this question will require a thorough and quantitative investigation of the transformation of the sensory signal into the motor output, including the biophysics and aerodynamics of wing motion (Dickinson *et al.*, 1999), as well as the typical innate flight behaviour that flies display under free-flight conditions (van Hateren *et al.*, 2005).

References

Adelson, E. H. & Bergen, J. R. 1985. Spatiotemporal energy models for the perception of motion. *J Opt Soc Am A* **2**, 284–299.

Borst, A. & Egelhaaf, M. 1989. Principles of visual motion detection. *Trends Neurosci* **12**, 297–306.

Borst, A. & Egelhaaf, M. 1993. Detecting visual motion: theory and models. In *Visual Motion and its Role in the Stabilization of Gaze* (ed. F. A. & J. Wallman), pp. 3–27. Elsevier.

Borst, A. & Haag, J. 2002. Neural networks in the cockpit of the fly. *J Comp Physiol* **188**, 419–437.

Borst, A. 2003. Noise, not stimulus entropy, determines neural information rate. *J Computat Neurosci* **14**, 23–31.

Borst, A., Flanagin, V. & Sompolinsky, H. 2005. Adaptation without parameter change: dynamic gain control in motion detection. *PNAS* **102**, 6172–6176.

Brenner, N., Bialek, W. & de Ruyter, R. R. 2000. Adaptive rescaling maximizes information transmission. *Neuron* **26**, 695–702.

Buchner, E. 1976. Elementary movement detectors in an insect visual system. *Biol Cybern.* **24**, 85–10.

de Ruyter van Steveninck, R. R., Lewen, G. D., Strong, S. P. *et al.* 1997. Reproducibility and variability in neural spike trains. *Science* **275**, 1805–1808.

Dickinson, M. H., Lehmann, F.-O. & Sane, S. P. 1999. Wing rotation and the aerodynamic basis of insect flight. *Science* **284**, 1954–1960.

Egelhaaf, M. & Reichardt, W. 1987. Dynamic response properties of movement detectors: theoretical analysis and electrophysiological investigation in the visual system of the fly. *Biol Cybern* **56**, 69–87.

Fairhall, A. L., Lewen, G. D., Bialek, W. & de Ruyter, R. R. 2001. Efficiency and ambiguity in an adaptive neural code. *Nature* **412**, 787–792.

Fermi, G. & Reichardt, W. 1963. Optomotorische Reaktionen der Fliege Musca domestica. Abhängigkeit der Reaktion von der Wellenlänge, der Geschwindigkeit, dem Kontrast und der mittleren Leuchtdichte bewegter periodischer Muster. *Kybernetik* **2**, 15–28.

Fennema, C. L. & Thompson, W. B. 1979. Velocity determination in scenes containing several moving objects. *Comp Graph Im Process* **9**, 301–315.

Götz, K. G. 1964. Optomotorische Untersuchungen des visuellen Systems einiger Augenmutanten der Fruchtfliege Drosophila. *Kybernetik* **2**, 77–92.

Hassenstein, B. & Reichardt, W. 1956. Systemtheoretische Analyse der Zeit-Reihenfolgen- und Vorzeichenauswertung bei der Bewegungsperzeption des Rüsselkäfers Chlorophanus. *Z Naturforsch* **11b**, 513–524.

Haag, J., Denk, W. & Borst, A. 2004. Fly motion vision is based on Reichardt detectors regardless of the signal-to-noise ratio. *PNAS* **101**, 16333–16338.

Hildreth, E. & Koch, C. 1987. The analysis of motion: from computational theory to neural mechanisms. *Ann Rev Neurosci* **10**, 477–533.

Limb, J. O. & Murphy, J. A. 1975. Estimating the velocity of moving images in television signals. *Comp Graph Im Process* **4**, 311–327.

Lei, S. & Borst, A. 2006. Propagation of photon noise and information transfer in visual motion detection. *J Computat Neurosci 20*, 167–178.

Potters, M. & Bialek, W. 1994. Statistical mechanics and visual signal processing. *J Physiol France* **4**, 1755–1775.

Reichardt, W. 1961. Autocorrelation, a principle for the evaluation of sensory information by the central nervous system. In *Sensory Communication*, (ed. W. A. Rosenblith), pp. 303–317. MIT Press and John Wiley & Sons.

Reichardt, W. 1987. Evaluation of optical motion information by movement detectors. *J Comp Physiol* **A 161**, 533–547.

Srinivasan, M. V. 1990. Generalized gradient schemes for measurement of image motion. *Biol Cybern* **63**, 421–443.

Single, S. & Borst, A. 1998. Dendritic integration and its role in computing image velocity. *Science* **281**, 1848–1850.

van Santen, J. P. H. & Sperling, G. 1985. Elaborated Reichardt detectors. *J Opt Soc Am A* **2**, 300–320.

van Hateren, J. H., Kern, R., Schwerdtfeger, G. & Egelhaaf, M. 2005. Function and coding in the blowfly H1 neuron during naturalistic optic flow. *J Neurosci* **25**, 4343–4352.

4

Spatial constancy and the brain: insights from neural networks

Robert L. White III and Lawrence H. Snyder

4.1 Introduction

As you read this article, your eyes make a sequence of rapid movements (saccades) to direct your gaze to the words on this page. Although saccades move the eyes at speeds of up to 400° per second, you do not perceive that the visual world is moving rapidly. The characters, which are stationary on the page, do not appear to move, even though their projection onto the retina changes. The ability of the visual system to account for self-motion is called space constancy (Holst & Mittelstaedt, 1950; Helmholtz, 1962; Stark & Bridgeman, 1983; Bridgeman, 1995; Deubel *et al.*, 1998) and is an important component of goal-directed behaviour: in order to interact with objects in our surroundings, we must account for motion of the eyes, head and body for accurate visually guided movements (e.g. saccades, reaching).

Space constancy is particularly important when remembering the locations of objects that become occluded from view. In this case, compensation for self-motion is critical for maintaining an accurate representation of an object's location while it is being remembered. Both humans and monkeys are capable of performing such compensation, which is also known as spatial updating.

To update accurately, the brain must synthesise information regarding movements of various types (eye-in-head, head-on-body, body-in-world) with remembered visuospatial information. However, multiple possible solutions exist. It has been suggested that spatial locations could be stored with respect to an allocentric, world-fixed coordinate system that does not vary with the motion of the observer. An object that is stationary in the world would be stationary in such a world-fixed coordinate frame. However, not all objects are stationary in the world, and often their motion is linked to movements of the observer. For example, when tracking a moving object, features of the moving object remain stationary with respect to the retina. Similarly, objects fixed with respect to the head or body move on the retina, but remain stationary in a head-fixed or body-fixed reference frame. Thus, one would need a multitude of redundant static representations of space to encode the variety of ways in which objects move in the world.

Modelling Perception with Artificial Neural Networks, ed. C. R. Tosh and G. D. Ruxton. Published by Cambridge University Press.

There is evidence that the primate brain contains multiple spatial representations in different coordinate frames (e.g. Duhamel *et al.*, 1997; Snyder *et al.*, 1998; Galletti *et al.*, 1993; Olson, 2003; Martinez-Trujillo *et al.*, 2004; Park *et al.*, 2006), however a large number of representations are sensoritopic – that is, represented with respect to the spatial characteristics of the input. Visual spatial representations are often gaze-centred, auditory spatial representations are often head-centred, etc. How might a sensoritopic neural representation update spatial locations in response to shifts in gaze? In the lateral intraparietal area (LIP), a cortical area in monkeys known to be involved in the planning of saccadic eye movements (Andersen *et al.*, 1992), neurons have gaze-centred receptive fields (Colby *et al.*, 1995), but the responsiveness (gain) to targets presented in the receptive field can be linearly modulated by changes in gaze position. This modulation is called a gaze position gain field (Andersen & Mountcastle, 1983).

Neural network modelling has been a useful tool in elucidating the mechanisms of spatial updating. Zipser & Andersen (1988) demonstrated gaze position gain fields in hidden layer units that combine retinal and gaze position information in a feedforward neural network. Xing & Andersen (2000) extended these results, and showed that a recurrent neural network could dynamically update a stored spatial location based on changes in a position signal. In this network, gain fields for position were observed in the hidden layer. It has been proposed that gain fields might serve as a general mechanism for updating in response to changes in gaze, since gain fields have been observed in many of the areas where spatial signals are updated (Andersen *et al.*, 1990).

Gaze position signals are not the only means by which to update a remembered location in a gaze-centred coordinate system. Droulez & Berthoz (1991) demonstrated that world-fixed targets could be updated entirely within a gaze-centred coordinate system, with the use of a gaze velocity signal. The use of velocity signals may be particularly relevant for updating with vestibular cues (signals from the inner ear that sense rotation and translation of the head). For example, humans and monkeys are able to update accurately in response to whole-body rotations, and neurons in LIP appear to reflect this updating (Baker *et al.*, 2002). In the monkey, gain fields for body position have not been observed in LIP (Snyder *et al.*, 1998). Perhaps vestibular velocity signals provide the necessary inputs for updating in this case without a necessity for a gain field mechanism.

In order to explore how spatial memories are stored and manipulated in complex systems, we compared the responses of macaque monkeys and individual LIP neurons with the responses of recurrent neural networks and individual network nodes. We trained two monkeys to memorise the location of a visual target in either a world-fixed or gaze-fixed frame of reference. The reference frame was indicated by the colour of the fixation point, as well as an instructional sequence at the beginning of each trial. After the target disappeared, each animal's gaze was shifted horizontally to a new location by a smooth pursuit eye movement, a saccade, or a whole-body rotation. In whole-body rotations, the fixation point moved with the same angular velocity as the monkey so that gaze remained aligned with the body. Following the gaze shift, the fixation point disappeared, cuing the animal to make a memory-guided saccade to the remembered location of the target.

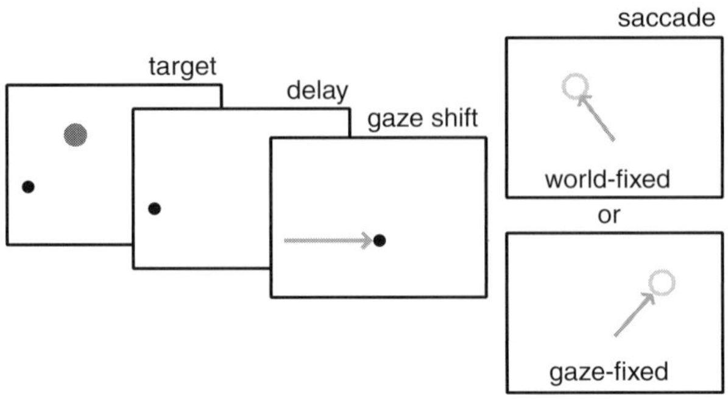

Figure 4.1. Schematic of the task. Animals began each trial by looking at a fixation point. The target was briefly flashed, and could appear at one of many different locations. It disappeared, and gaze was shifted to either the left or right, either by whole-body rotation (VOR cancellation), smooth pursuit or a saccade. The delay period concluded with the disappearance of the fixation point. The animal was rewarded for moving to either the world-fixed or gaze-fixed location of the original target. The neural network model tracked only the horizontal component of target location. The network was given a transient visual target and a sustained reference frame cue. A slow gaze shift followed, after which the network was queried for a saccade vector to either the world-fixed or gaze-fixed location of the original target.

Figure 4.1 depicts a schematic of the task. Subsequent analyses in this report focus on the whole-body rotation case.

Next, we created and trained a recurrent neural network model to perform an analogue of the animals' task. The network consisted of an input layer, a fully recurrent hidden layer and an output layer. The input nodes had activations corresponding to the retinal input, gaze position and gaze velocity, and the reference frame to be used. The responses of the output nodes were decoded in order to determine the saccadic goal location in gaze-centred coordinates. The model only traced one dimension of visual space (analogous to the horizontal component of the monkeys' task). We first established that our network was a reasonable model of the animals' behaviour, and then used the network model to investigate issues that are difficult to investigate in the animal. We were principally interested in the inputs used to drive updating and whether or not the mechanism for updating involves gain fields.

4.2 Methods

We implemented a three-layered recurrent neural network model (White & Snyder, 2004), trained it on a flexible updating paradigm, and compared the output and internal activity of the model to the behaviour and neuronal activity obtained from monkeys performing an analogous task (Baker *et al.*, 2003). Detailed methods are presented in the Appendix.

The model's input layer was designed to represent a visual map in one dimension of space and to provide information about gaze position and gaze velocity. Output was in eye-centred coordinates, representing the goal location of an upcoming saccade. For flexible behaviour, we added a binary cue that indicated whether the target was world-fixed or gaze-fixed. The network had full feedforward connections between successive layers and full recurrent connections in the hidden layer. The recurrent connections provided the network with memory over time.

The task of the network was to store and, if necessary, transform a pattern of activity representing the spatial location of a target into a motor command. In order to correctly perform the task, the network had to either compensate or ignore any changes in gaze that occurred during the storage interval, depending on the instruction provided by the two reference frame input units. The network output provides an eye-centred representation of the stored spatial location. The network was trained using a 'back-propagation through time' algorithm (Rumelhart *et al.*, 1986).

In an analogous task, two rhesus monkeys performed memory-guided saccades to world-fixed and gaze-fixed target locations following whole body rotations (vestibulo-ocular reflex [VOR] cancellation), smooth pursuit eye movements or saccades that occurred during the memory period (Baker *et al.*, 2003). Fixation point offset at the conclusion of the memory period cued monkeys to make a saccade to the remembered target location. Changes in eye position as well as the neuronal activity of single neurons in cortical area LIP were recorded.

4.3 Results

Animals made accurate memory-guided saccades to target locations in the appropriate reference frame. Figure 4.2a (left panel) compares the average memory-guided saccade horizontal amplitude to the ideal amplitude. Animals were equally accurate in both reference frame conditions across a range of target locations, although animals tended to undershoot the most peripheral targets in both conditions. When we examined the precision of memory-guided saccade endpoints, we found a difference between the performance on world-fixed and gaze-fixed trials. Figure 4.2b (left panel) shows the standard deviation of saccade endpoints as a function of ideal saccade amplitude. World-fixed saccades had a larger standard deviation than gaze-fixed saccades.

We found that our recurrent neural network model, once trained, had a similar behavioural performance to the animals. We used the centre of mass of the output activations to 'read out' the saccadic goal location. When noise was injected into the inputs to produce variability in the model's output, the saccade readout was largely accurate (Figure 4.2a, right panel), but precision was worse for world-fixed trials compared with gaze-fixed trials (Figure 4.2b, right panel). This emergent property of the model's behaviour was not explicitly designed or trained into the architecture, and therefore indicates that our computational model captures at least one feature of spatial updating that occurs in biology.

We next examined the responses of single neurons in LIP during the flexible updating task. We found that many cells updated their activity in response to gaze shifts during

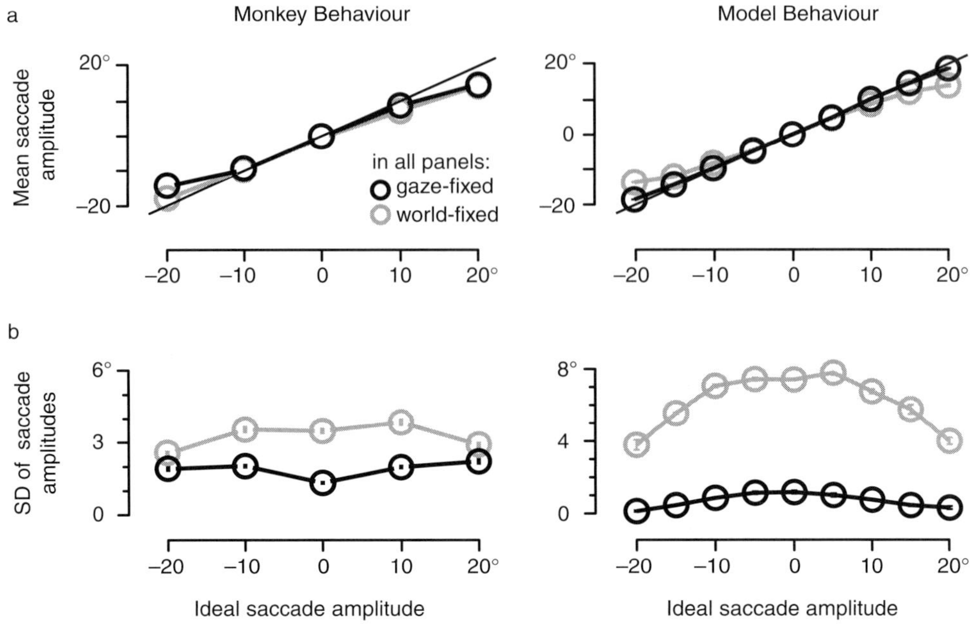

Figure 4.2. (a) Accuracy of network output (left) and animal behaviour (right). Left panel: mean coded output (± standard error) from three networks of 35 hidden units at the last time step versus the ideal saccade amplitude for world-fixed (grey) and subject fixed (black) trials. Right panel: mean horizontal saccade amplitudes versus ideal saccade amplitudes from two monkeys (Baker *et al.*, 2003). The straight black line in each panel indicates perfect performance (actual = ideal). (b) Precision of network output and animal saccades, measured as the standard deviation (SD) of the saccade amplitudes, for the same data as shown in (a).

world-fixed trials, but held their activity constant during gaze-fixed trials. This pattern of activity is consistent with dynamic updating of spatial memories in a gaze-centred coordinate system. Responses from a single cell are shown in Figure 4.3a. Consider first the response to target presentation prior to the gaze shift (left-hand panel). A target appearing inside the cell's RF produced a transient burst of activity followed by sustained elevation in activity that continued after the target had disappeared from view (solid traces). A target appearing outside the cell's RF produced no change in activity, or even a slight decrease to below baseline (dotted traces).

Next we considered the effect of a gaze shift. In the middle panel, whole-body rotations (±20°; VOR cancellation paradigm) brought the world-fixed location of the target either into (solid traces) or out of (dotted traces) the cell's RF. A short time later (right-hand panel), the animal was rewarded for an accurate memory-guided saccade that treated the target as either world-fixed (grey traces) or gaze-fixed (black traces). A world-fixed target whose remembered location moved from inside to outside the RF resulted in a decrease in firing rate (solid grey trace). Conversely, when the location moved from outside to inside

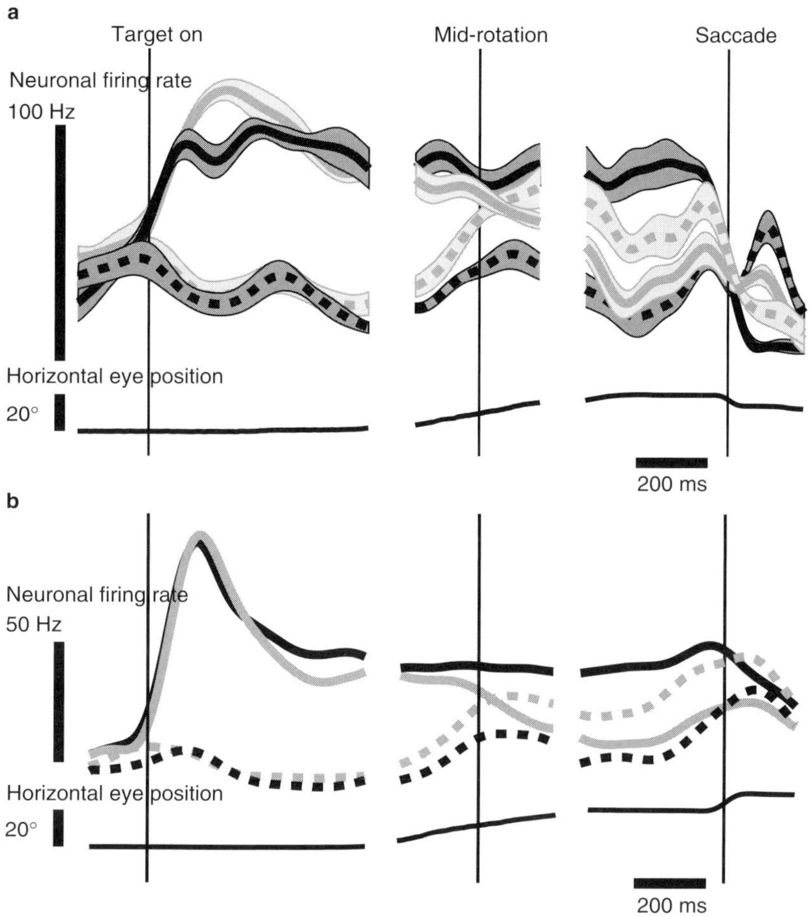

Figure 4.3. (a) Smoothed instantaneous firing rates for a single LIP neuron. Mean firing rates (thick lines) ± SEM (thin lines) are shown for targets presenting inside the neurons RF (solid lines) and outside the RF (dashed lines). On world-fixed trials (grey), a subsequent 20° whole body rotation to the left or right brought the goal location out of or into the neuron's RF, resulting in a decrement or increment in the neuron's activity, respectively. On gaze-fixed trials (black), target locations and rotations were identical, but the goal location remained constant (either inside or outside the RF). Little modulation of activity was observed for this neuron during the delay on gaze-fixed trials. Data are aligned on target presentation (left), middle of the rotation (centre) and the memory-guided saccade (right). Seven trials were performed for each condition. On the bottom, the average horizontal eye position is shown for one of the conditions. (b) Smoothed instantaneous firing rates for the average activity from 42 LIP neurons during whole-body rotations. Same format as (a), except standard error traces are omitted for clarity.

the RF, the firing rate increased (dotted grey trace). Activity was unaffected by either type of gaze shift when the target was gaze-fixed (solid and dotted black traces).

Figure 4.3b shows the normalised average response from 42 LIP cells in one monkey during whole-body rotations of the same type as Figure 4.3a. Like the example cell, it shows a pattern consistent with dynamic updating in a gaze-centred coordinate system. The population average activity decreases when a target's remembered location leaves the RF on world-fixed trials (solid grey trace), but does not change on gaze-fixed trials (solid black trace). Similarly, the population activity increases when a target's remembered location enters the RF (dashed grey trace), but does not change during the gaze shift (dashed black trace).

It appears that on average, LIP cells change their firing rate (update) in response to shifts of gaze for world-fixed targets but not gaze-fixed targets. However, this updating is incomplete: in Figure 4.3b, the difference between in-RF and out-RF activity is smaller on world-fixed trials that gaze-fixed trials. We quantify this below.

For an analogue of LIP cells in our model, we examined the responses of hidden (middle) layer units in the recurrent neural network. We found that many of the hidden units were active during the delay period and dynamically updated their activity on world-fixed trials, much like the LIP cells in the monkey. Activity from a single model unit is shown in Figure 4.4a. The model unit responds in much the same way as the example cell from LIP (Figure 4.3a). On gaze-fixed trials, in-RF and out-RF activity remains constant over the memory period, whereas, on world-fixed trials, the in-RF activity decreases and the out-RF activity increases following the gaze shift.

Figure 4.4b shows the population average of all 35 hidden units from the neural network. The patterns of neural activity are remarkably similar to those seen in the example cell and in the responses of LIP neurons. Like LIP, it also appears that updating in the model is incomplete: the difference between in-RF and out-RF activity at the end of the trial is greater for gaze-fixed targets than for world-fixed targets.

We next wished to examine the relative amounts of gaze-fixed and world-fixed information conveyed by individual cells. In other words, we asked if individual cells were 'specialised' for one reference frame or the other, or 'flexible', like the example cell (Figure 4.2a). To address this question, we calculated the relative amounts of gaze-fixed and world-fixed information conveyed by each unit using the receiver operating characteristic (ROC) of an ideal observer. We used the ROC to discriminate whether a target was inside or outside of a cell's RF, and calculated the cell's performance as the area under the ROC curve. If individual cells were specialised for either the world-fixed or gaze-fixed transformation, then we would expect individual cells to have ROC values that were greater than chance for one type of transformation, and at chance or systematically worse than chance for the other. In contrast, if cells were capable of contributing to both transformations, their corresponding ROC values would be greater than chance for both transformations.

Figure 4.5a displays the world-fixed and gaze-fixed ROC discrimination values (the area under the ROC curve) on rotation trials for 86 LIP cells from two monkeys. Each point represents one cell. There are many points in the upper right quadrant, indicative of many individual cells that contribute to both gaze-fixed and world-fixed transformations. A substantial subset of cells lies at or below an ordinate value of 0.5 (lower right quadrant), indicating that they retain spatial information on gaze-fixed trials but either

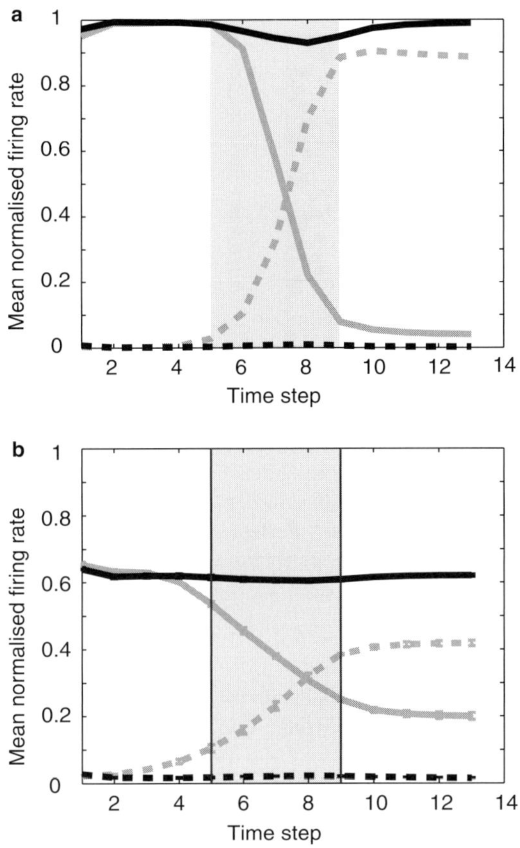

Figure 4.4. (a) Activity of one hidden layer unit during individual trials for a 20° gaze shift. When presented with a gaze-fixed cue and target inside its RF (solid black trace) or outside the RF (dashed black trace), there is little change in the unit's response over time. When a world-fixed target appears inside the unit's RF, the subsequent gaze shift moves the remembered target location outside the RF and the unit's activity decreases (solid grey trace). When a world-fixed target appears outside the unit's RF, but the gaze shift brings the remembered target location into the RF, the unit increases its activity (dashed grey trace). Grey shading indicates the time steps over which the gaze shift occurred. (b) Average response of all 35 hidden units in a single trained network. Same format as (a). Error bars represent standard errors from three networks.

contribute nothing or even provide erroneous information on world-fixed trials. Very few neurons lie in the upper left quadrant (valid information on world-fixed trials and invalid information on gaze-fixed trials). The net result is that mean discrimination was better on gaze-fixed trials compared with world-fixed trials (mean ROC: 0.83 vs. 0.58, $p < 1 \times 10^{-12}$, Wilcoxon test, $n = 86$).

We performed the same ROC analysis on the hidden units of the model (with injected noise to produce variability in the output). We found that units discriminated well in

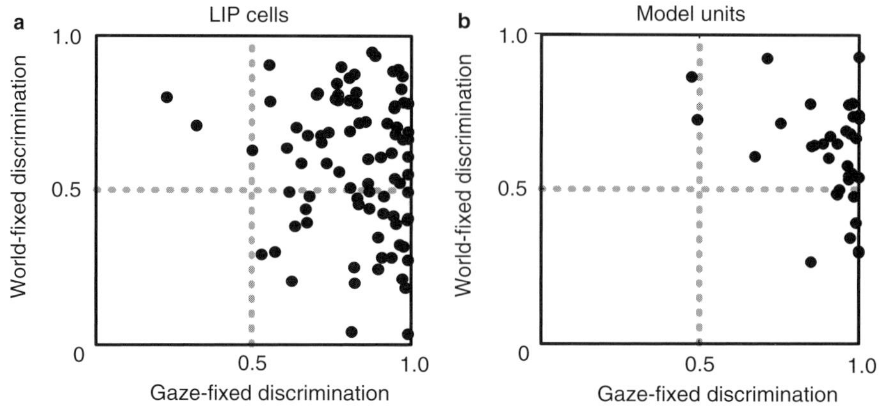

Figure 4.5. (a) Comparison of gaze-fixed vs. world-fixed discrimination for LIP neurons recorded during trials with whole-body rotation gaze shifts. For each unit, the area under the ROC curve (AUC) was calculated (see Methods). Each unit is plotted with respect to its AUC for gaze-fixed (abscissa) and world-fixed (ordinate) trials. Perfect discrimination corresponds to an AUC of 1. Dashed lines indicate the level of chance performance (AUC of 0.5). (b) Same format as (a) for units in the hidden layer of the recurrent network model (n = 35).

both types of trials (Figure 4.5b), but that mean discrimination was better on gaze-fixed trials than world-fixed trials (mean ROC: 0.90 vs. 0.61, $p < 1 \times 10^{-5}$, Wilcoxon test, $n = 35$). This pattern mimics that seen for the LIP neurons following whole body rotations (Figure 4.5a).

We next asked whether networks trained with both position and velocity signals relied preferentially on one of these inputs. To answer this question, we zeroed the gaze shift inputs of either the position or velocity input nodes of networks trained with both position and velocity inputs. This is analogous to selectively 'lesioning' either the position or velocity inputs to the network. The performance deficits that resulted indicated that these networks preferentially relied on velocity to perform spatial updating (Figure 4.6a). It was not the case that a velocity signal was required for updating, because networks trained with either velocity or position alone performed quite well. Thus, it appears that models show a preference for velocity inputs for spatial updating.

In our last model experiment, we examined the presence of gaze position gain fields in networks trained with both gaze position and velocity inputs, position alone, or velocity alone. Gaze position gain fields have been postulated to be involved in the mechanisms of spatial updating (Xing & Anderson, 2000; Balan & Ferrera, 2003). However, we postulated that hidden units of networks that relied of a gaze velocity signal would not demonstrate gaze position gain fields. When networks were trained with only gaze position signals, we found an abundance of hidden units with gain field responses. Inversely, when gaze velocity signals were used for training, no gain fields were present

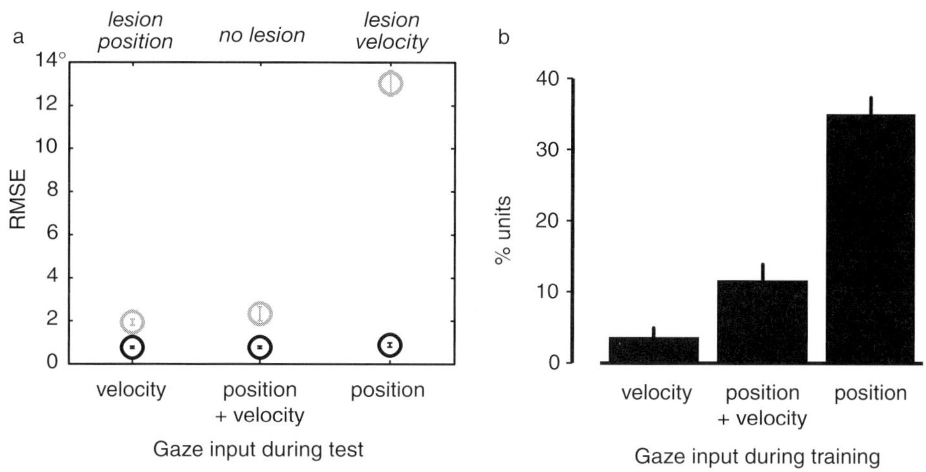

Figure 4.6. (a) Performance of the network when gaze signals are removed. RMS error for all world-fixed (grey) and gaze-fixed (black) trial types is shown for the network with position inputs removed, the intact network, and the network with velocity inputs removed. The network has a strong preference for the velocity input over the position input. Error bars represent standard errors from three networks. (b) Proportion of hidden layer units demonstrating gaze position gain fields stronger than 0.2% per degree under the different gaze input conditions. Error bars represent standard errors from three networks.

in hidden layer units (Figure 4.6b). Our results indicate that gaze position gain fields are not a necessary feature of networks that perform spatial updating.

4.4 Discussion

We have described a 3-layer recurrent neural network model that combines retinal and extra-retinal signals that is capable of encoding target locations as either fixed in the world or fixed with respect to gaze. We found the output of the network mimics the pattern of behaviour in animals trained to perform a similar task. Furthermore, the response properties of the hidden (middle) layer units in this network resemble those of neurons in LIP. By removing specific inputs to the model, we tested whether updating preferentially utilises gaze position or gaze velocity signals, and found that the network strongly preferred velocity for updating world-fixed targets. We found that gaze position gain fields were present when position signals, but not velocity signals, were available for updating.

Neural networks provide a powerful analytic tool in understanding computations performed by the brain. They provide a means of analysing and dissecting distributed coding and computation by a population of neurons. Furthermore, they provide means to test insightful manipulations that would be difficult to perform in vivo, such as the input lesion experiments described above.

One of the goals of our modelling experiments was to address whether a single population of neurons could perform both world-fixed and gaze-fixed transformations dynamically. We found that a neural network model could perform both world-fixed and gaze-fixed transformations in a flexible manner. Furthermore, the precision of memory-guided behaviour was worse on world-fixed trials compared with gaze-fixed trials, consistent with dynamic updating in a gaze-centred coding scheme (Baker *et al.*, 2003). Monkey behaviour showed an identical pattern, suggesting that a gaze-centred coding scheme could be utilised to track remembered target locations in the brain.

The LIP has a putative role in spatial working memory (Gnadt & Andersen, 1988). Neurons in this area encode spatial information using gaze-centred coordinates (Colby *et al.*, 1995). When we examined the responses of single LIP neurons during a spatial updating task, we found they were markedly similar to those in the hidden layer of the model. Both neurons and model units had gaze-centred RFs, and, on average, updated their delay-period activity during world-fixed trials and (correctly) maintained their activity during gaze-fixed trials. The marked similarity between LIP cells and model units suggests that LIP may contribute to the flexible updating in the brain necessary for correct behavioural output. However, it is quite possible that this neural activity is merely reflective of computations that occur elsewhere in the brain – a limitation of many electrophysiological recording experiments.

A second goal of the modelling experiments was to ask whether world-fixed and gaze-fixed spatial memories were held by sub-populations of units that specialised in just one of the two transformations, or by a single homogeneous population of units that flexibly performed both transformations. In the model, the world-fixed and gaze-fixed transformations were not performed by separate sub-networks, but instead by a distributed population of units that contributed to both world-fixed and gaze-fixed output. Using the same analysis, we similarly found that cells in LIP did not specialise in either the world-fixed or gaze-fixed transformation.

Spatial updating requires an estimate of the amplitude and direction of movements that have been made while a location is maintained in memory. Such an estimate could arise from sensory signals or from motor commands. By utilising a copy of a motor command, areas involved in spatial updating could quickly access gaze shift signals, without the delay associated with sensory feedback (e.g. proprioception). An input carrying such a signal would be, in essence, a duplicate of the motor command itself, and has been called either a 'corollary discharge' or 'efference copy' signal. Recent studies have identified a role for saccadic motor commands in spatial updating following shifts of eye position (Guthrie *et al.*, 1983; Sommer & Wurtz, 2002). New evidence points to FEF as a possible locus of updating in this case (White & Snyder, 2007; Opris *et al.*, 2005).

If corollary discharge signals are used for updating, what is the nature of these signals? Do they report gaze position or gaze velocity? We examined how gaze position and gaze velocity signals are used in spatial updating by testing our recurrent network model. First, we found that in networks trained with both gaze velocity and gaze position signals, removing the velocity input produced a profound deficit in performance, whereas removing the position inputs produced only a minor effect (Figure 4.6a). In the model, velocity was preferentially utilised to perform the updating computation while position only played a minor role. Second, we looked for hidden units with position-dependent gain field responses in networks trained with only one of the signals or with both signals in combination. We found that networks trained with gaze velocity did not demonstrate gaze position gain fields (Figure 4.6b). By contrast, networks trained with gaze position contained many units with gain fields.

Recent models have highlighted the importance of a gaze position gain field representation in updating for changes in gaze (Cassanello & Ferrera, 2004; Smith & Crawford, 2005). Results from our model would indicate that such a representation would rely on gaze position signals (Figure 4.6b). However, we suggest that although position signals may be used, they are not required for spatial updating. As an example, we have shown that LIP neurons update in response to whole-body rotations (Baker *et al.*, 2002) but lack gain fields for body position (Snyder *et al.*, 1998). Based on results from our recurrent network model (Figure 4.6b), LIP neurons could update in response to whole-body rotations if gaze shift signals were provided by vestibular velocity signals. Updating with velocity signals would not require that positional gain fields be present in LIP.

Based on behavioural differences, spatial updating for saccades could be a special case. In a recent study from our laboratory, a distinct pattern of updating behaviour was observed following saccades (as compared to rotations and pursuit) (Baker *et al.*, 2003). Updating for saccades might utilise signals conveying eye velocity (Droulez & Berthoz, 1991) or eye displacement (White & Snyder, 2004). However, results from our model (Figure 4.6b), predict that such a system would not demonstrate gain fields for eye position. Since eye position gain fields are observed in LIP and in FEF (Andersen *et al.*, 1990; Balan & Ferrera, 2003), we support the hypothesis that saccadic updating relies on a gain field mechanism.

4.5 Appendix: Detailed methods

4.5.1 Model architecture

We designed a neural network model containing units in three layers (input, hidden and output). The input layer is designed to represent a visual map in one dimension of space and to provide information about gaze position and gaze velocity. Output is in eye-centred coordinates and can be read out as encoding the goal location of an upcoming saccade. For flexible behaviour, we added a cue that indicated whether the target was world-fixed or gaze-fixed. The hidden layer is given recurrent connections to provide the network with memory over time. Units in each layer can take on activity values from 0 to 1 (except where stated otherwise). The network has full feedforward connections: every unit in the input layer projects to every unit in the hidden layer, and every unit in the hidden layer projects to every unit in the output layer. Each connection has an associated weight that indicates the strength of that connection. The activity of an individual unit in the hidden or output layer is determined by the logistic function $f(x) = 1/(1 - e^{-x})$ where x is the weighted sum of the inputs plus a bias. In addition to the feedforward connections, the network contains recurrent connections between every unit in the hidden layer. Thus, the input to a hidden unit at a discrete time t includes the feedforward connections from the input layer at time t plus the recurrent connections from the hidden layer at time $t - 1$. For an individual trial simulation (see Section 4.5.2), the activity of the network is calculated for 13 consecutive time steps. We define each step to be 100 ms, for ease of comparison with physiological data. Individual weights were initially set to random values between –0.1 and 0.1.

In the input layer, the retinal array is modelled as 25 visual units with receptive fields evenly distributed over a range, defined as –60° to 60°. Each retinal unit has a Gaussian receptive field with a $1/e^2$ width of 7°. Gaze position is linearly encoded by two units. Activity of the first position unit is scaled such that 0 activity corresponds to straight ahead (0°) and the activity range [–1, +1] corresponds to gaze position in the range [– 40°, 40°]. The second position unit has the opposite activity of the first (a simple push–pull model). A second pair of push–pull units linearly encodes velocity. Velocity units can encode a range of gaze speeds between –200° and 200° per second. Finally, two binary reference frame units with opposite polarity indicate whether the output should be world- or gaze-fixed. The representation of the cue signal in the cortex is likely to be complex, involving not only sensory input but also prior experience and current expectations. The model presented here does not address these complexities, but instead considers only how flexible visuospatial processing is accomplished. To that end, we chose a very simple representation of the cue: a binary switch. Most analyses were performed with 25 units in the hidden layer. The output layer consists of an array of 25 units encoding saccadic goal location in oculocentric coordinates. The units are evenly distributed over the range –60° to 60° and have Gaussian receptive fields identical to the visual input units.

4.5.2 Task and training

The task of the network was to store and if necessary transform a pattern of activity representing the spatial location of a target. In order to correctly perform the task, the network had to either compensate or ignore any changes in gaze that occurred during the storage interval, depending on the instruction provided by the two reference frame input units. The network output provides an eye-centred representation of the stored spatial location. A target was presented for one time step (defined as 100 ms for ease of comparison to physiological data) at the onset of each trial. A target could appear at 1 of 9 locations, evenly spaced within the central one third (40°) of the workspace. The reference frame cue, which determined whether gaze perturbations should be incorporated into the output or ignored, was present for the entire duration of the trial. Gaze position at the start of the trial was at one of four locations (–15°, –5°, 5° or 15°). Memory-period gaze perturbations were 10° or 20° per second to the left or right, were initiated 300 ms after the target disappeared and lasted for 500 ms. The network output (saccade amplitude or target location) was read out from the network 0–400 ms after the end of the gaze perturbation.

In an analogous task, two rhesus monkeys performed memory-guided saccades to world-fixed and gaze-fixed target locations following whole body rotations (VOR cancellation), smooth pursuit eye movements or saccades. Slow gaze perturbations (rotations, pursuit) were 10°/s to the left or right from fixation points at –20°, –10°, 0°, 10° or 20° and lasted for either 300 ms or 600 ms. Monkeys were cued 400–1200 ms following the end of the gaze perturbation to make a saccade to the remembered target location.

The network was often less accurate when the saccadic goal lay on the boundaries of the workspace (defined by the output). To minimise these edge effects, trial types requiring saccades greater than 20° in either reference frame were excluded. This left 96 remaining combinations of target location, eye position and gaze perturbation vectors that constituted the training set.

The network was trained using the 'back-propagation through time' algorithm (Rumelhart *et al.*, 1986; Williams & Zipser, 1995). The algorithm optimises the connection weights between units to minimise the error produced at the output. Briefly, for a given output unit, the output error (ε_o) is computed as the difference between the actual and desired activities in the output layer. The virtual error for a unit in the hidden layer (ε_h) is determined by propagating the output error back through the logistic function and the hidden-to-output weight (W_{ho}) that connects the two units:

$$\delta_o = f'(A_o)\varepsilon_o$$

$$\varepsilon_h = \sum_o W_{ho}\delta_o$$

A hidden-to-output weight is adjusted based on the δ value from the output unit, the activity of the unit in the hidden layer (A_h) and a learning rate η:

$$\Delta W_{ho} = \eta A_h \delta_o$$

The virtual error is backpropagated through the hidden-to-hidden weight matrix to calculate the virtual error at each earlier time step:

$$\delta_{h(t)} = f'(A_{h(t)})\varepsilon_{h(t)}$$

$$\varepsilon_{h(t-1)} = \sum_h W_{hh}\delta_{h(t)}$$

The input-to-hidden and the hidden-to-hidden weights are updated based on the sum of the errors from the first time step (t_0) to the teach time (t_n):

$$\Delta W_{ih} = \eta \sum_{t=t_0}^{t_n} A_i \delta_{h(t)}$$

$$\Delta W_{hh} = \eta \sum_{t=t_0+1}^{t_n} A_{h(t-1)} \delta_{h(t)}$$

An additional constraint was enforced during training: the hidden-to-output weights were constrained to values greater than –0.1. Forcing positive weights to the output makes the hidden units more likely to develop response fields similar to those of the output units (Mitchell & Zipser, 2001). However, removal of this constraint did not change our overall conclusions.

From our experience training monkeys on an identical task (Baker *et al.*, 2003), we suspected that training the network in stages might facilitate learning, rather than presenting the complete task and training set from the very start. This suspicion was confirmed by preliminary studies. Networks presented with the entire training set from the start of training converged slowly or converged to local minima (unpublished observations). Therefore, we chose a graduated training regime analogous to that used for the monkeys.

The network was initially trained using a simple saccade task: no gaze perturbation and no memory period. Once this was mastered, a short (100 ms) memory period was introduced, and then gradually lengthened to 1200 ms. Weights were adjusted after every cycle of the complete training set until an error threshold was reached. The threshold was set at a high value to avoid over-training the network on the memory saccade task. Again, this is analogous to our experience in training monkeys; we do not dwell too long on any one stage, else achieving the next stage becomes more difficult (unpublished observations). With the network, each time the duration of the memory period was increased, smaller and fewer weight updates needed to be performed to reach the threshold. Hence, the learning rate started at $\eta = 0.05$ and was decreased inversely with the number of steps.

Once it could perform memory saccades, the network was introduced to the full task. Training had an equal probability of occurring on any one of the four post-perturbation time steps or on the time step directly preceding the onset of gaze perturbation. The latter case promotes stability in the network output and enforces the condition that world-fixed and gaze-fixed trials should produce the same output if no perturbation has occurred. Training was not performed *during* the gaze perturbation to avoid enforcing a particular time course of output during the perturbation period. A post hoc comparison revealed slightly improved performance when training also occurred during the perturbation, but no other differences. Weights were adjusted after each complete cycle through the training set. Training proceeded for 5000 cycles at a learning rate of $\eta = 0.001$. Subsequent training for 7500 cycles occurred with the learning rate halved every 2500 cycles. This process of simulated annealing helps avoid both local minima (by using an initially high learning rate) and limit cycles (by moving to a low learning rate) and generally helps optimise both the speed and the accuracy of training (Mitchell, 1996).

4.5.3 Recording procedures

Subjects and behavioural control methods were the same as described in a previous report (Baker *et al.*, 2003). Briefly, monkeys were seated in a Lexan box. Eye movements were monitored using the scleral search coil technique. Visual stimuli were projected onto a 100 × 80 cm screen placed 58 cm from the animal. The room was otherwise completely dark, as confirmed by a dark-adapted human observer. All aspects of the experiment were computer-controlled (custom software). Eye position was logged every 2 ms. Visual stimulus presentation times were accurate to within one video refresh (17 ms).

Electrophysiological recordings were made with tungsten microelectrodes (FHC; 0.2–2.0 MΩ) Extracellular potentials were amplified (FHC) and filtered (band pass 400–5000 Hz; Krohn-Hite). Single units were isolated with a dual time-amplitude window discriminator and spike times logged to a personal computer with 1 μs precision (custom software).

All surgical and behavioural procedures conformed to National Institutes of Health guidelines and were approved by the Washington University Institutional Animal Care and Use Committee.

4.5.4 Display of neural activity

For display purposes only, single unit and population average traces were smoothed using a 181-point digital low-pass filter with a transition band spanning 2 to 15 Hz (–3 dB point = 9 Hz). For population activity, the peak firing rate of each unit was normalised to 100 Hz before averaging.

4.5.5 ROC analysis

A receiver operating characteristic (ROC) was calculated for each unit in the hidden layer as described previously (Britten *et al.*, 1992). The receiver discriminated between two conditions (whether a target was inside or outside of the RF) based on the activity of the unit. For a given criterion level of activity, the proportion of trials on which the inside-RF response exceeded the criterion was plotted against the proportion of trials on which the outside-RF response exceeded the criterion. Points were calculated for 3334 criterion values uniformly distributed over the range of possible activity values ([0, 1]). The collection of points forms a stair-step function on the unit square from (0, 0) to (1, 1).

The area under this curve was calculated and was assigned as the 'ROC value' of each neuron or model unit. The ROC value describes how well an ideal observer could correctly discriminate between the inside-RF and outside-RF conditions based on the responses of the unit.

The ROC values for individual model units were calculated from the activations in the last time bin. Using any of the three post-rotation time bins, or averaging across all three, yielded similar conclusions. We introduced a fixed amount of noise to all of the inputs ($\sigma_n = 0.25$) and measured the activity of each unit for 15 repetitions of selected trial types. The ROC values from LIP neurons were calculated using the number of spikes in the interval from 400 ms to 100 ms before the cue to initiate the memory-guided saccade. In both model and monkey experiments, trial types were selected to be those in which the target was flashed either inside or outside the RF, followed by a 20° gaze perturbation that brought the target outside or inside the RF, respectively.

References

Andersen, R. A., Bracewell, R. M., Barash, S., Gnadt, J. W. & Fogassi, L. 1990. Eye position effects on visual, memory, and saccade-related activity in areas LIP and 7a of macaque. *J Neurosci* **10**, 1176–1196.

Andersen, R. A., Brotchie, P. R. & Mazzoni, P. 1992. Evidence for the lateral intraparietal area as the parietal eye field. *Curr Opin Neurobiol* **2**, 840–846.

Andersen, R. A. & Mountcastle, V. B. 1983. The influence of the angle of gaze upon the excitability of the light-sensitive neurons of the posterior parietal cortex. *J Neurosci* **3**, 532–548.

Baker, J. T., Harper, T. M. & Snyder, L. H. 2003. Spatial memory following shifts of gaze. I. Saccades to memorized world-fixed and gaze-fixed targets. *J Neurophysiol* **89**, 2564–2576.

Baker, J. T., White, R. L. & Snyder, L. H. 2002. Reference frames and spatial memory operations: area LIP and saccade behavior. *Soc Neurosci Abstr* **57.16**.

Balan, P. F. & Ferrera, V. P. 2003. Effects of gaze shifts on maintenance of spatial memory in macaque frontal eye field. *J Neurosci* 23, 5446–5454.

Bridgeman, B. 1995. A review of the role of efference copy in sensory and oculomotor control systems. *Ann Biomed Eng* **23**, 409–422.

Britten, K. H., Shadlen, M. N., Newsome, W. T. & Movshon, J. A. 1992. The analysis of visual motion: a comparison of neuronal and psychophysical performance. *J Neurosci* **12**, 4745–4765.

Cassanello, C. R. & Ferrera, V. P. 2004. Vector subtraction using gain fields in the frontal eye fields of macaque monkeys. *Soc Neurosci Abstr* **186.11**.

Colby, C. L., Duhamel, J. R. & Goldberg, M. E. 1995. Oculocentric spatial representation in parietal cortex. *Cereb Cortex* **5**, 470–481.

Deubel, H., Bridgeman, B. & Schneider, W. X. 1998. Immediate post-saccadic information mediates space constancy. *Vision Res* **38**, 3147–3159.

Droulez, J. & Berthoz, A. 1991. A neural network model of sensoritopic maps with predictive short-term memory properties. *Proc Natl Acad Sci USA* **88**, 9653–9657.

Duhamel, J. R., Bremmer, F., BenHamed, S. & Graf, W. 1997. Spatial invariance of visual receptive fields in parietal cortex neurons. *Nature* **389**, 845–848.

Galletti, C., Battaglini, P. P. & Fattori, P. 1993. Parietal neurons encoding spatial locations in craniotopic coordinates. *Exp Brain Res* **96**, 221–229.

Gnadt, J. W. & Andersen, R. A. 1988. Memory related motor planning activity in posterior parietal cortex of macaque. *Exp Brain Res* **70**, 216–220.

Guthrie, B. L., Porter, J. D. & Sparks, D. L. 1983. Corollary discharge provides accurate eye position information to the oculomotor system. *Science* **221**, 1193–1195.

Helmholtz, H. V. 1962. *A Treatise on Physiological Optics*. Dover.

Holst, v. E. & Mittelstaedt, H. 1950. The reafferent principle: reciprocal effects between central nervous system and periphery. *Naturwissenschaften* **37**, 464–476.

Martinez-Trujillo, J. C., Medendorp, W. P., Wang, H. & Crawford, J. D. 2004. Frames of reference for eye-head gaze commands in primate supplementary eye fields. *Neuron* **44**, 1057–1066.

Mitchell, J. & Zipser, D. 2001. A model of visual-spatial memory across saccades. Vision Res 41, 1575–1592.

Mitchell, M. 1996. *An Introduction to Genetic Algorithms (Complex Adaptive Systems)*. MIT Press.

Olson, C. R. 2003. Brain representation of object-centered space in monkeys and humans. *Annu Rev Neurosci* **26**, 331–354.

Opris, I., Barborica, A. & Ferrera, V. P. 2005. Effects of electrical microstimulation in monkey frontal eye field on saccades to remembered targets. *Vision Res* **45**, 3414–3429.

Park, J., Schlag- Rey, M. & Schlag, J. 2006. Frames of reference for saccadic command, tested by saccade collision in the supplementary eye field. *J Neurophysiol* **95**, 159–170.

Rumelhart, D. E., Hinton, G. E. & Williams, R. 1986. Learning internal representations by error propagation. In *Parallel Distributed Processing: Explorations in the Microstructure of Cognition* (ed. D. E. Rumelhart & J. L. McClelland), pp. 316–362. MIT Press.

Smith, M. A. & Crawford, J. D. 2005. Distributed population mechanism for the 3-D oculomotor reference frame transformation. *J Neurophysiol* **93**, 1742–1761.

Snyder, L. H., Grieve, K. L., Brotchie, P. & Andersen, R. A. 1998. Separate body- and world-referenced representations of visual space in parietal cortex. *Nature* **394**, 887–891.

Sommer, M. A. & Wurtz, R. H. 2002. A pathway in primate brain for internal monitoring of movements. *Science* **296**, 1480–1482.

Stark, L. & Bridgeman, B. 1983. Role of corollary discharge in space constancy. *Percept Psychophys* **34**, 371–380.

White, R. L., 3rd & Snyder, L. H. 2004. A neural network model of flexible spatial updating. *J Neurophysiol* **91**, 1608–1619.

White, R. L. & Snyder, L. H. 2007. Subthreshold microstimulation in frontal eye fields updates spatial memories. *Exp Brain Res.* **181**, 477–92.

Williams, R. J. & Zipser, D. 1995. Gradient-based learning algorithms for recurrent neural networks. In *Back-Propagation: Theory, Architecture and Applications* (ed. Y. Chauvin & D. E. Rumelhart), pp. 433–486. Lawrence Erlbaum.

Xing, J. & Andersen, R. A. 2000. Memory activity of LIP neurons for sequential eye movements simulated with neural networks. *J Neurophysiol* **84**, 651–665.

Zipser, D. & Andersen, R. A. 1988. A back-propagation programmed network that simulates response properties of a subset of posterior parietal neurons. *Nature* **331**, 679–684.

5

The interplay of Pavlovian and instrumental processes in devaluation experiments: a computational embodied neuroscience model tested with a simulated rat

Francesco Mannella, Marco Mirolli and Gianluca Baldassarre

5.1 Introduction

The flexibility and capacity of adaptation of organisms greatly depends on their learning capabilities. For this reason, animal psychology has devoted great efforts to the study of learning processes. In particular, in the last century a huge body of empirical data has been collected around the two main experimental paradigms of 'classical conditioning' (Pavlov, 1927; Lieberman, 1993) and 'instrumental conditioning' (Thorndike, 1911; Skinner, 1938; Balleine *et al.*, 2003; Domjan, 2006).

Classical conditioning (also called 'Pavlovian conditioning') refers to an experimental paradigm in which a certain basic behaviour such as salivation or approaching (the 'unconditioned response' – UR), which is linked to a biologically salient stimulus such as food ingestion (the 'unconditioned stimulus' – US), becomes associated to a neutral stimulus like the sound of a bell (the 'conditioned stimulus' – CS), after the neutral stimulus is repeatedly presented before the appearance of the salient stimulus. Such acquired associations are referred to as 'CS-US' or 'CS-UR' associations (Pavlov, 1927; Lieberman, 1993).

Instrumental conditioning (also called 'operant conditioning') refers to an experimental paradigm in which an animal, given a certain stimulus/context such as a lever in a cage (the 'stimulus' – S), learns to produce a particular action such as pressing the lever (the 'response' – R), which produces a certain outcome such as the opening of the cage (the 'action outcome' – O), if this outcome is consistently accompanied by a reward such as the access to food. In this case, the acquired associations are referred to as either 'S-R' associations, when the reactive nature of the acquired behaviour is stressed, or 'A-O' associations, when the goal-directed nature of behaviour is stressed (Thorndike, 1911; Skinner, 1938; Domjan, 2006, see below).

This empirical work has been paralleled by the development, within the machine-learning literature, of 'reinforcement learning algorithms' (Sutton & Barto, 1981, 1998), that is algorithms directed to provide machines with the capacity of learning new

Modelling Perception with Artificial Neural Networks, ed. C. R. Tosh and G. D. Ruxton. Published by Cambridge University Press.

behaviours on the basis of rewarding stimuli (i.e. signals from the external environment that inform the machine about the achievement of desired goals). Interestingly, reinforcement learning algorithms have gained increasing interest within the empirical literature on animal learning as they represent theoretical models that can potentially furnish coherent explanations of organisms' learning processes. Indeed, a class of such models, the so-called *temporal-difference learning* algorithms (TD-learning), are currently considered as the best available theoretical accounts of several key empirical findings (Dayan & Balleine, 2002; Schultz, 2002; Houk *et al.*, 1995).

Notwithstanding their success, standard reinforcement learning models have several limitations from a biological point of view. In particular, three of the main drawbacks are as follows. First, such models ignore the role of internal states (e.g. hunger vs. satiety related to a certain type of food) in modulating the effects of 'external' rewards (e.g. the receiving of such a food). Such kinds of effects are exhibited by organisms, for example, in 'devaluation' experiments in which animals tend to change their reinforced behaviours when the value of a rewarding stimulus, such as a food, is suddenly decreased through satiation or its association with poison. By ignoring the role of internal states in learning and behaviour, current reinforcement learning models cannot account for such effects.

Second, standard models do not take into account the important difference existing, within instrumentally acquired behaviours, between 'habits' and 'goal-directed actions', that is between those instrumentally acquired behaviours that are automatically triggered by *antecedent stimuli* and those that are controlled by their *consequences* (Yin & Knowlton, 2006). In fact, while classical (behaviourist) reinforcement learning theory assumed that all behaviours are elicited by some antecedent stimulus from the external environment, over the last few decades a significant body of research has demonstrated that animals are able to control their own behaviour on the bases of the *expected outcomes* of their actions. The most important way to assess whether a behaviour is driven by a stimulus or by an expected outcome is through a devaluation experiment: if a stimulus (S) elicits a response (R) only if the associated reinforcing outcome (O) has not been devalued, then it is evident that the behaviour is driven by the outcome and not by the stimulus. Indeed, sensitiveness to the manipulation of outcome value has been proposed (cf. Yin & Knowlton, 2006) as part of the operational definition of *goal-directed behaviours* driven by action-outcome (A-O) associations as distinct from habits driven by stimulus-response (S-R) associations (the other part of the definition, related to the manipulation of the contingency between actions and outcomes, will not be considered here, see Yin & Knowlton, 2006).

Third, standard models tend to conflate the notions of classical conditioning and instrumental conditioning. On the contrary, accumulating empirical evidence indicates that classical and instrumental conditioning are based on different processes that rely on distinct neural systems. Furthermore, such processes interplay in complex ways (Dayan & Balleine, 2002), as demonstrated, for example, by phenomena like 'Pavlovian-Instrumental Transfer' (PIT; where a conditioned stimulus that is predictive of reward can energise the execution of instrumentally acquired behaviours), and 'incentive learning' (where, under certain conditions, the valence of an action outcome needs to be re-learned to exert its effects on behaviour).

This paper starts to address these limitations by presenting a novel computational model which (a) is strongly rooted in the anatomy and physiology of the mammal brain, (b) is embodied in a simulated robotic rat and (c) reproduces the results of empirical devaluation experiments conducted on both normal and amygdala-lesioned rats (Balleine *et al.*, 2003)[1]. Since, as indicated above, devaluation phenomena constitute the most evident demonstrations of both the role of internal states in modulating behaviour and the distinction between habits and goal-driven behaviours, and since, as indicated below, here it is assumed that devaluation phenomena depend on the modulation of instrumental processes by Pavlovian processes, the attempt to reproduce these experiments constitutes the most appropriate way of addressing the aforementioned drawbacks of standard reinforcement learning models.

The proposed model constitutes the first working computational model furnishing a coherent picture about the neural mechanisms underlying conditioning phenomena. The model is built upon the following three fundamental assumptions:

(1) The *amygdala* constitutes the CS-US associator at the core of Pavlovian conditioning phenomena.
(2) The *cortex-dorsolateral striatum*[2] pathway, forming S-R associations, constitutes the main actor involved in instrumental conditioning.
(3) The *amygdala-nucleus accumbens* pathway 'bridges' classical conditioning processes happening in the amygdala and instrumental processes taking place in the basal ganglia, and so it allows producing the behaviours exhibited by rats in devaluation experiments.

Although the main goals of this chapter have a scientific relevance, the research agenda that covers the work presented here is believed to possess the potential to also produce useful outcomes for technology. Undoubtedly, living organisms' behaviour is characterised by a degree of autonomy and flexibility that greatly surpasses those of current robots. A way to tackle these limits is to attempt to understand the mechanisms underlying organisms' behavioural flexibility so as to use them in designing robot's controllers.

The rest of the chapter is organised as follows. Section 5.2 illustrates the general methodological approach which guided the research reported in this chapter named 'Computational Embodied Neuroscience'. Section 5.3 reports the original experiments addressed by the model. Section 5.4 describes the simulated rats and environment used to test the model. Section 5.5 contains a detailed description of the model. Section 5.6 reports the main results. Finally, Section 5.7 concludes the chapter.

[1] The amygdala, an almond-shaped group of nuclei located within each medial temporal lobe of the brain, is associated with a wide range of cognitive functions, including emotional regulation, learning, action selection, memory, attention and perception. Amygdala is involved in both aversive behaviours such as those involved in fear conditioning and taste aversion experiments (Blair *et al.*, 2005; Knight *et al.*, 2005; Maren, 2005) and appetitive behaviours (Baxter & Murray, 2002a; Cardinal *et al.*, 2002; Balleine & Killcross, 2006). The functioning of amygdala relies on both its capacity to assign emotional valence to stimuli on the basis of input information related to internal body states, and on its capacity to associate neutral stimuli, biologically salient stimuli and innate responses. Some of its main functional sub-systems are (McDonald, 1998; Pitkänen *et al.*, 2000) the 'central nucleus' (CeA), responsible for triggering innate responses and neuromodulation processes underlying learning and broad brain regulation, and the 'basolateral complex' (BLA), responsible for forming CS-US associations (Mannella *et al.*, 2008).

[2] The striatum is the input portion of the basal ganglia, a set of forebrain subcortical nuclei that are traditionally considered to be responsible for instrumental conditioning phenomena (i.e. the locus of S-R associations: Packard & Knowlton, 2002; Yin & Knowlton, 2006). In rats, the striatum can be divided in: (a) dorsolateral striatum, mainly underlying motor-execution functions; (b) dorsomedial striatum, playing a role in motor-preparation and cognitive functions; (c) nucleus accumbens (or ventral striatum), considered an important interface (Mogenson *et al.*, 1980) between the motivational processes taking place in the limbic system and the motor processes taking place within the rest of the basal ganglia and cortex.

5.2 The method used: computational embodied neuroscience

This chapter addresses issues related to animal learning (in particular, conditioning phenomena) by following an approach which can be referred to as 'Computational Embodied Neuroscience' (CEN) (cf. Prescott *et al.*, 2003, 2006 who propose a research method which shares some principles with the approach proposed here). CEN is based on the following principles:

(1) *Theoretical cumulativity.* CEN aims at achieving theoretical cumulativity in the study of brain and behaviour. Indeed, the great amount of empirical data provided by neuroscience, psychology, and the other related disciplines are seldom integrated by strong and general theoretical explanations, thus failing to produce a coherent picture of the phenomena under investigation. Moreover, various approaches based on computational models tend to develop different ad-hoc models to explain different phenomena. In this respect, CEN aims at building models (see principle 4 below) which are based on general principles and can be developed in an incremental fashion so as to account for an increasing number of phenomena, data and theories of neuroscience and psychology. Indeed, many of the principles illustrated below (principles 5–8) should aid achieving theoretical cumulativity in that they furnish specific criteria usable to select models and this should lead to focus on models with a high potential for cumulativity. The principle of cumulativity is important as it is a strong drive towards the production of comprehensive accounts and general theories of brain and behaviour, that is, against the tendency to generate many unrelated and mutually incompatible theories each accounting for only a limited set of empirical data. The goal of theoretical cumulativity is in line with the 'spirit' of both 'systems computational neuroscience' (Brody *et al.*, 2004), that aims at explaining the functioning of whole brain systems rather than specific areas or physiological/chemical mechanisms, and with the computationally informed approaches proposed within psychology (e.g. Newell, 1973), which stress the need of identifying general principles and theories on behaviour.

(2) *Evolutionary framework.* The theory of evolution (Darwin, 1859) is the fundamental theoretical framework needed to understand biological phenomena and hence also to understand brain and behaviour. This has the important implication of stressing the focus of investigation on the function for the organisms' survival and reproduction of brain mechanisms and behavioural processes, and not only on their mere mechanical functioning. This is in contrast with some neuroscientific research approaches which pose their attention only on neural mechanisms per se, without trying to understand their role in organisms' adaptation.

(3) *Complex systems framework.* The brain and the brain–body–environment set are complex systems. Complex systems are systems formed by many parts (e.g. brain neurons) which interact via local rules (e.g. activation potentials). These interactions lead to the emergence of the global behaviour of the system without (fully) relying upon centralised coordination mechanisms, but rather on self-organisation principles such as positive feedback and negative feedback (Camazine *et al.*, 2001; Baldassarre, 2008). This not only implies that brain and behaviour have to be studied with computational models (see below) but also that their understanding has to rely upon the concepts of complex-system theory.

(4) *Computational models.* The investigation of brain and behaviour conducted on the basis of empirical experiments and observations (such as those of neuroscience, psychology and ethology) should be accompanied by the instantiation of theories into formal computational models, that is computer programs that simulate the mechanisms underlying brain processes

and produce behaviour as an emergent outcome of their functioning. The rationale behind this principle is that the brain, and the brain–body–environment set, are complex systems, and theories expressed only through words or analytically-solved equation systems can give only limited, non-generative accounts of these phenomena.

(5) *Constraints from behaviour.* The computational models used to instantiate the theories have to be capable of reproducing the investigated behaviour, in line with what is proposed by 'artificial ethology' (Holland & McFarland, 2001). Furthermore, the comparison between the model and the target behaviour should be done on a detailed basis (i.e. with reference to the outcomes of specific target experiments) and possibly in quantitative terms (i.e. not just with qualitative comparisons).

(6) *Constraints from brain.* Challenging models with the request to account for specific behavioural data, is not enough as, given a behaviour, many alternative models capable of reproducing it can always be built. For this reason, a second fundamental source of constraints for models are the data on the anatomy and physiology of the brain. These data should be used in two ways. First, for choosing the assumptions that drive the design of the architecture, functioning and learning mechanisms of the models. Second, for testing the low-level predictions produced by the models (i.e. the predictions produced at the neural level). This principle comes from computational neuroscience (Sejnowski *et al.*, 1988) urging computational models to keenly account for data on brain.

(7) *Embodiment.* In line with the ideas proposed by the 'animal approach' (Meyer & Wilson, 1991) and 'artificial life' (Langton, 1987), models should be capable of reproducing the addressed behaviours within 'whole' autonomous systems acting on the basis of circular interactions with the environment mediated by the body (the sensors, the actuators and the internal body states). Indeed, the brain generates behaviour by forming a large dynamic complex system together with the body and the environment (Clark, 1997; Nolfi & Floreano, 2000; Nolfi, 2006), so a full understanding of the brain and behaviour will need to rely on models that take into consideration this fundamental fact. The principle has two implications. First, the computational models should involve the simulation of organisms' brains and bodies, and of the environment where they live, as this allows understanding of how behaviours emerge from the interactions between them. Second, models should aim at being scalable to realistic setups, that is they should be capable of functioning with realistic sensors (e.g. retinas), realistic actuators (e.g. bodies should be governed by realistic Newtonian dynamics), realistic scenarios (e.g. objects and supports with complex textures, shapes and dynamics), and noise (affecting sensors, actuators, environment, etc.).

(8) *Learning and evolution.* Much of the organisms' brain architecture (and, in higher mammals, a relevant part of behaviour) is shaped by the need to learn to adapt their behaviour to varying environmental conditions during life and to avoid the need of wholly encoding behaviour in the genome (encoding learning mechanisms is often cheaper than encoding the details of behaviour). This implies that to fully understand the brain we need to understand the learning processes which it supports, and to fully understand behaviour we need to understand how it forms on the basis of learning processes (Weng *et al.*, 2001; Zlatev & Balkenius, 2001). A similar reasoning holds for evolution, whose specific functioning mechanisms (e.g. its interplay with learning during life) have likely exerted a strong influence on how brains are currently structured, and impose strong constraints and opportunities on how behaviour develops during life. For these reasons, a full understanding of behaviour and the brain will need to pass through a full understanding of how they 'historically' developed in the course of evolution (Parisi *et al.*, 1990).

5.3 Target experiment

The target data addressed with the model are reported in Balleine *et al.* (2003) which illustrates various experiments directed to investigate the relations existing between the manipulation of the value of primary rewards (devaluation) and instrumental conditioning, and the role that the amygdala (Amg) plays in them. The present work focuses on 'Experiment 1' reported in that paper, a standard 'devaluation test'. This experiment is particularly well-suited to test four important processes that the model aims to capture: (a) the association between neutral stimuli (e.g. the sight of a lever) and biologically salient stimuli (e.g. food); (b) the dependence of the evaluation of external stimuli (e.g. food, levers, etc.) on internal states (e.g. hunger for a specific food); (c) learning and use of 'habits', that is, stereotyped and rigid behaviours acquired during prolonged periods of trial-and-error learning (these processes are captured well by standard reinforcement learning models); (d) the influence of internal states, mediated by the associations between neutral stimuli and biologically salient stimuli, on the selection and triggering of habits.

In two preliminary phases of the experiment, eight sham-lesioned rats plus eight rats whose basolateral amygdala complex (BLA) was lesioned were trained in *separate* trials to press a lever or pull a chain to obtain two different kinds of food, respectively Noyes pellets and maltodextrin. The training phase was followed by an extinction test lasting 20 min (divided in groups of 2 min) where: (a) *both* manipulanda were present in the experimental chamber; (b) half of the rats had been previously satiated with Noyes pellets while the other half with maltodextrin. The main result indicated that during the first 2 min of the test non-lesioned rats performed the action corresponding to the manipulandum of the non-satiated food with a higher frequency with respect to the other manipulandum. On the other hand, BLA-lesioned rats did not show any devaluation effect: they performed the two actions at the same rate. These experiments clearly demonstrate that BLA plays a fundamental role in the transfer of the diminished hedonic value of food to instrumentally acquired habits. As we shall see in Section 5.5, this key finding is central for clarifying the relationship existing between Pavlovian and instrumental conditioning.

5.4 The simulated organism, environment and experiments

In line with the CEN approach (Section 5.2), the model presented here was tested within an embodied system. Although we are aware that the role of the 'degree of embodiment and situatedness' of the model and simulations presented here is rather limited (e.g. the sensors and actuators are rather simplified, low-level behaviours are hardwired, etc.), nevertheless the use of a simulated organism and experimental set-up forced us to design a model potentially capable of coping with the difficulties posed by more realistic set-ups. For example, the noise of behaviour execution, and the randomly variable duration of the trials, actions' execution, and rewarding effects posed interesting challenges to the robustness of the learning algorithms of the model.

The model was tested with a simulated robotic rat ('ICEAsim') developed within the EU project ICEA on the basis of the physics 3D simulator Webots^{TM}. The model was

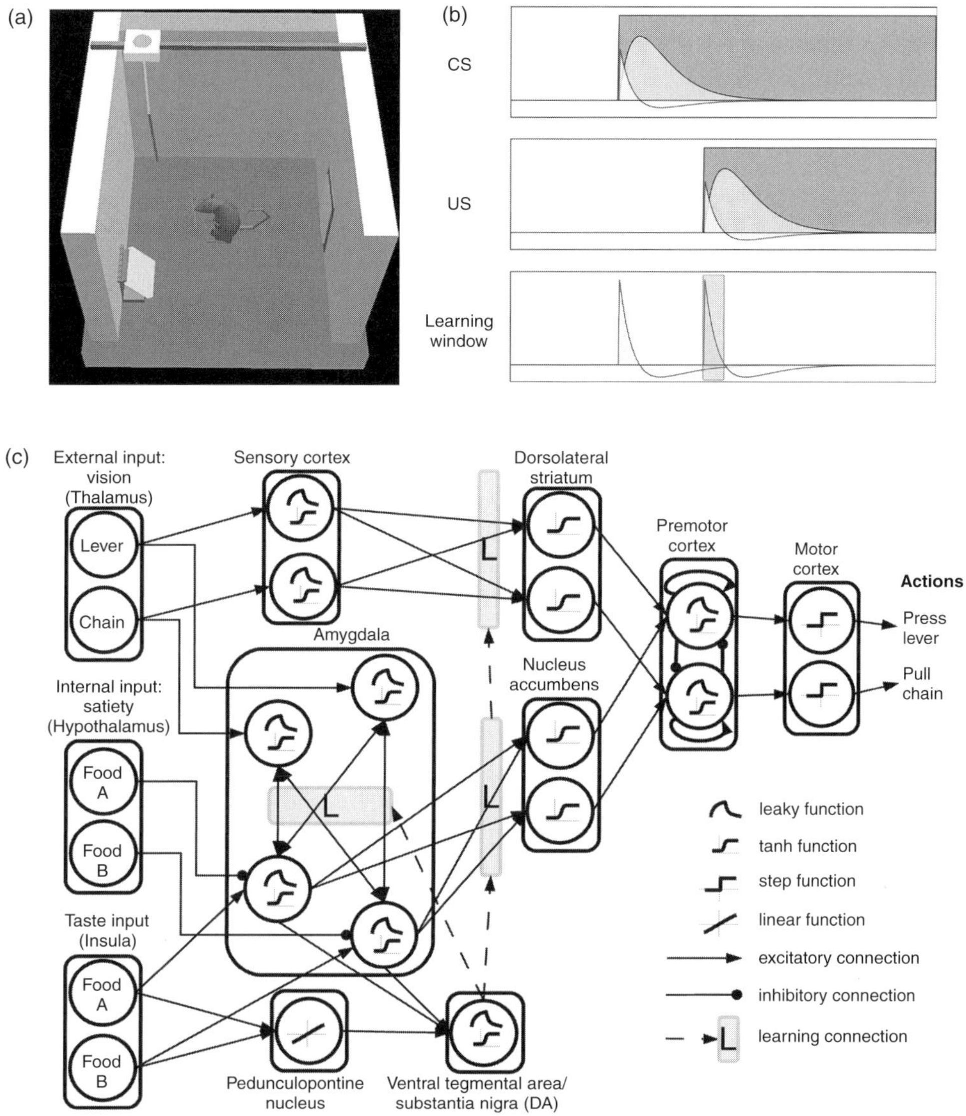

Figure 5.1. (a) A snapshot of the simulator, showing the simulated rat at the centre of the experimental chamber, the food dispenser (behind the rat), the lever (at the rat's left hand side) and the chain (at the rat's right hand side). (b) A scheme showing the key aspects of the learning process taking place within the amygdala, based on a step activation of a CS and a US unit. In the three graphs, time is reported in the x-axis. CS and US graphs: the dark grey rectangle represents the step-activation, the grey positive curve represents the onset activation of the unit, the light grey positive/negative curve represents the derivative of such onset. Learning window graph: the grey rectangle represents the window in which the derivative of the CS on-set is negative (indicating that the CS started in the past) and the derivative of the US on-set is positive (indicating that the US is starting): this is the window in which the CS-US weight is increased. (c) The architecture of the model.

written in Matlab^{TM} and was interfaced with ICEAsim through a TCP/IP connection. The robotic set-up used to test the model is shown in Figure 5.1a and is described briefly here, skipping details not central for this chapter's goals. The training and test environment is composed of a grey-walled chamber containing a yellow lever, a red chain and a grey food dispenser that turns green or blue when respectively food A or food B is delivered in it. When 'pressed' or 'pulled', the lever and chain make respectively food A or B (the rewarding stimuli) available at the dispenser.

The simulated rat is a robot equipped with two wheels and with various sensors. Among these, the experiments reported here use two cameras representing the rat's eyes (furnishing a panoramic 300° view) and the whisker sensors. The rat uses the cameras to detect the lever, the chain and the food dispenser, in particular their presence/absence (via their colour) and their (egocentric) direction. The rat uses the whiskers, activated with one if bent beyond a certain threshold and zero otherwise, to detect contacts with obstacles. The rat is also endowed with *internal* sensors related to satiety for either food A or B (these sensors assume the value of one when the rat is satiated, and zero otherwise). The rat's actuators are two motors that can independently control the speed of the two wheels.

For simplicity, the information fed to the model is only related to the presence/absence of the lever and chain in the test chamber and food A and food B in the mouth, whereas the other information is used to control four low-level hardwired behavioural routines. These routines, triggered either by the model or directly by stimuli, are as follows: (1) *obstacle avoidance routine*: this routine, triggered by the whiskers, 'overwrites' all other actions to avoid obstacles; (2, 3) *lever press routine* and *chain pull routine*: these routines, activated by the model, cause the rat to approach the lever/chain on the basis of their visually detected direction; when the lever/chain are touched they activate the food delivery in the dispenser; (4) *consummatory routine*: when the dispenser turns green or blue (this signals the presence of food in it), the rat approaches and touches it ('consumption' of the food) so causing the perception of respectively food A or food B in mouth; the routine ends after the rat touches the dispenser ten times.

The simulated devaluation experiment is divided into a training phase and two test phases. The training phase lasts 8 min and the two test phases 2 min each. Each phase is divided into trials that end either when the rat executes the correct action and consumes the food or after a 15 s time-out (the duration of the experimental phases is shorter than the duration of the original experiment's phases as the limited complexity and number of available actions of the simulated rat allowed a faster learning). In each trial the rat is set in the middle of the chamber with an orientation randomly set between the lever and the chain direction. In the trials of the training phase either the lever and food A or the chain and food B are used in an alternate fashion and the rat is always 'hungry' (the two satiation sensors are set to zero). In the two test phases, the rat is respectively satiated either with food A (the satiation sensors for food A and B are respectively set to one and zero) or with food B. In all trials of the two test phases *both* manipulanda are present and the rat is evaluated in extinction (i.e. without delivery of food). The experiment (the three phases) was run 20 times with 'unlesioned' artificial rats and 20 times with 'lesioned' rats.

5.5 The model

The model (Figure 5.1c) is formed by three major components: (a) the *amygdala* which instantiates a CS-US associator and is at the core of Pavlovian conditioning (Baxter & Murray, 2002b; Cardinal *et al.*, 2002); (b) the *cortex-dorsolateral striatum* pathway which learns, via instrumental conditioning processes, habits based on S-R associations (Yin & Knowlton, 2006); (c) the *amygdala-nucleus accumbens* pathway which 'bridges' classical conditioning processes happening in the amygdala and instrumental processes taking place in the basal ganglia (Baxter & Murray, 2002b; Cardinal *et al.*, 2002). The model's input is formed by six neurons activated by the sensors illustrated in Section 5.4: two neurons encode the presence/absence of the lever and the chain (s_{lev} and s_{cha}), two neurons encode the presence/absence of food A and food B in the rat's mouth (s_{fA} and s_{fB}), and two neurons encode the satiation for food A and food B (s_{sfA} and s_{sfB}).

5.5.1 The amygdala: a CS–US associator

The associator implements Pavlovian conditioning through the association between CSs and USs ('stimulus substitution'). In real brains this role seems to be played by the amygdala (Amg) (Baxter & Murray, 2002b; Cardinal *et al.*, 2002). There are massive reciprocal connections between Amg and several brain areas, including inferotemporal cortex (IT), insular cortex (IC), prefrontal cortex (PFC) and hippocampus (Hip) (Baxter & Murray, 2002b; Cardinal *et al.*, 2002; Price, 2003; Rolls, 2005). Furthermore, Amg receives inputs from posterior intralaminar nuclei of thalamus (PIL) (Shi & Davis, 1999). These connections underlie an interplay between processes related to perceived (or represented) external context (IT, PFC, Hip) and processes related to internal states (IC, PIL). In general, Amg can be seen as playing the function of assigning a subjective valence to external events on the basis of the animal's internal context (needs, motivations etc.), and to use this to both regulate learning processes and directly influence behaviour.

The model's associator, which is considered as an abstraction of the processes taking place in Amg, performs 'asynchronous learning/synchronous recall' associations. First, stimuli perceived in different times are associated (CSs are associated with USs): this associative learning takes place if USs cause a dopamine (DA) release (see below). When the association is established, CSs are able to synchronously re-activate the USs' representations. The associator is composed by a vector $\mathrm{amg} = (amg_{lev}, amg_{cha}, amg_{fA}, amg_{fB})'$ of four reciprocally connected leaky neurons that process the input signals as follows:

$$\tau_{\mathrm{amg}} \cdot \dot{\mathrm{amg}}_p = -\,\mathrm{amg}_p + (s_{lev}, s_{cha}, (s_{fA} - s_{sfA}), (s_{fB} - s_{sfB}))' + \mathrm{W}_{amg} \cdot \mathrm{amg} \qquad (1)$$

$$\mathrm{amg} = \varphi\left[tanh\left[\mathrm{amg}_p\right]\right]$$

where amg_p represents the activation potentials of Amg units, $\varphi[.]$ is a positive linear function ($\varphi[x] = 0$ if $x \leq 0$ and $\varphi[x] = x$ otherwise) and W_{amg} is the matrix of all-to-all lateral connection weights within Amg. Note that while external stimuli have a binary representation (0/1 for absence/presence), internal stimuli *modulate* the representation of external stimuli. In particular s_{sfB} and s_{sfB} assume a value in {0, 5} when the corresponding satiation has respectively a low or high value, and this simulates the fact that satiation for a food inhibits the hedonic representation of such food within Amg. This assumption is supported by evidence indicating that a similar computation is performed in the secondary taste areas of the prefrontal/insular cortex (Rolls, 2005) connected with Amg. This part of the model is particularly important because, as we shall see, it mediates the influence of the shifts of primary motivations on both learning and behaviour.

The associator's learning is based on the *onset* of input signals, detected as follows (see Figure 5.1b). First, 'leaky traces' (tr) of the derivatives of Amg, trunked to positive values, are computed:

$$\tau_{\text{tr}} \cdot \dot{\text{tr}} = -\,\text{tr} + C_{Amg} \cdot \varphi[\dot{\text{amg}}] \tag{2}$$

where C_{Amg} is an amplification coefficient. Second, the derivatives of tr are computed: when positive, these derivatives detect the onset of the original signals, whereas when negative they detect the fact that some time elapsed since such onset took place.

The weights between Amg's neurons are updated on the basis of the signs of $\dot{\text{tr}}$ and the dopamine (DA) signal (see below). In particular, when (and only when) the derivative of the presynaptic neuron's trace is negative and the derivative of the postsynaptic neuron's trace is positive (notice that this happens when the presynaptic neuron fires *before* the postsynaptic neuron) the related connection is strengthened (for all couples of neurons this condition is encoded in the matrix L with Boolean elements):

$$\Delta W_{amg} = \eta_{amg} \cdot \varphi[da - th_{da}] \cdot L \tag{3}$$

where η_{amg} is a learning rate coefficient, da is the dopamine signal and th_{da} is a threshold over which dopamine elicits learning.

Dopamine release (corresponding to activation in the ventral tegmental area, VTA, and in the substantia nigra pars compacta, SNpc) is triggered by Amg through the units representing the hedonic impact of food and by the primary reward signals received from the pedunculopontine tegmental nucleus (PPT) (Kobayashi & Okada, 2007):

$$\tau_{da_p} \cdot \dot{da}_p = -\,da_p + da_{baseline} + w_{amg-da} \cdot \left(amg_{fA} + amg_{fB}\right) + w_{ppt-da} \cdot ppt \tag{4}$$

$$da = \varphi\left[tanh\left[da_p\right]\right]$$

where $ppt = s_{fA} + s_{fB}$ is the PPT's primary reward signal. Dopamine drives learning in both the associator and the action selectors (see Sections 5.5.2 and 5.5.3).

5.5.2 The cortex-dorsolateral striatum pathway: an S-R associator

The action selector based on the cortex-dorsolateral striatum learns 'habits' (S-R associations) through reinforcement learning processes. In real brains this function might be implemented in the cortex-dorsolateral striatum pathway (Yin & Knowlton, 2006). In the model this component receives s_{lev} and s_{cha} as input and, on the basis of this, selects one of the two lever-press/chain-pull actions (together with NAc, see Section 5.5.3).

The component is formed by four layers of neurons corresponding to four vectors: (a) a visual sensory cortex (SC) leaky-neuron layer: sc; (b) a neuron layer corresponding to the DLS's encoding of the 'votes' for the two actions: dls; (c) a neuron layer corresponding to premotor cortex (PM), formed by reciprocally inhibiting neurons that implement a competition for selecting one of the two actions (this function might be implemented by the reciprocal thalamo-cortical connections, Dayan & Balleine, 2002): pm; (d) a layer corresponding to motor cortex (M), representing the selected action with a binary code: m.

The visual leaky-neuron layer processes the input signal in a straightforward fashion:

$$\tau_{sc} \cdot \dot{\mathrm{sc}}_p = -\mathrm{sc}_p + (s_{lev}, s_{cha})' \tag{5}$$

$$\mathrm{sc} = \varphi[tanh[\mathrm{sc}_p]]$$

The SC is fully connected with DLS. The DLS's (non-leaky) neurons collect the signals from SC that tend to represent the evidence ('votes') in favour of the selection of either one of the two actions:

$$\mathrm{dls}_p = \mathrm{W}_{(sc-dls)} \cdot \mathrm{sc} \tag{6}$$

$$\mathrm{dls} = \varphi[tanh[\mathrm{dls}_p + dls_{baseline}]]$$

The selection of actions is performed on the basis of these votes (and NAc's votes, see Section 5.5.3) through a competition taking place between the leaky neurons of PM:

$$\tau_{pm} \cdot \dot{pm}_p = -pm_p + w_{dls-nac-pm}\,(\mathrm{dls} + \mathrm{nac}) + \mathrm{W}_{pm} \cdot \mathrm{pm} + \mathrm{n} \tag{7}$$

$$\mathrm{pm} = \varphi[tanh[\mathrm{pm}_p]]$$

where $w_{dls-nac-pm}$ is a coefficient scaling the votes, W_{pm} are the PM's lateral connection weights, and n is a noise vector with components uniformly drawn in $[-n, n]$. The assumption for which the action selection takes place within PM, used here for simplicity, raises interesting theoretical problems which are discussed in Section 5.7.

When one of the PM neurons reaches an activation threshold th_A, the execution of the corresponding action is triggered via M:

$$\mathrm{m} = \psi[\mathrm{pm} - th_A] \tag{8}$$

where $\psi[x]$ is a step function ($\psi[x]=0$ if $x \leq 0$ and $\psi[x]=1$ otherwise). Once the execution of the routine corresponding to the selected action terminates, the connection weights between SC and DLS, W_{sc-dls}, are modified according to the dopamine signal (this might be null if the wrong action has been selected):

$$\Delta W_{sc-dls} = \eta_{sc-dls} \cdot \varphi[da - th_{da}] \cdot \mathrm{m} \cdot \mathrm{sc}' \tag{9}$$

where η_{sc-dls} is a learning coefficient. Note that here M activations were directly used to train both the DLS and NAc (see Section 5.7 on this strong assumption).

5.5.3 The amygdala-nucleus accumbens pathway: a bridge between Pavlovian and instrumental processes

The Amg-NAc pathway 'bridges' Pavlovian processes to instrumental processes in that it learns A-O associations between the USs encoded in Amg (which might be thought of as desired outcomes of actions, corresponding to ingested elements of food in the presence of hunger for such food, elicited by the CSs, e.g. the sight of a lever) and actions encoded in the SC-DLS-PM pathway. In real brains this function might be implemented by the neural pathway connecting the BLA nuclei of Amg to NAc (Baxter & Murray, 2002b). In the model the pathway is implemented through an all-to-all connection matrix $W_{amg-nac}$ linking the Amg's hedonic representation of food, $amg_{f\,A}$ and $amg_{f\,B}$, to the NAc's (non-leaky) neurons:

$$\mathrm{nac}_p = W_{amg-nac} \cdot (mg_{f\,A},\ amg_{f\,B})' \tag{10}$$

$$\mathrm{nac} = \varphi\left[\tanh\left[\mathrm{nac}_p + nac_{baseline}\right]\right]$$

NAc's neurons play the same function as DLS neurons, namely they represent 'votes' that bias the action competition taking place in PM. Similarly to SC-DLS connections, Amg-NAc connections $W_{amg-nac}$ are modified, after action execution, on the basis of the dopamine signal:

$$\Delta W_{amg-nac} = \eta_{amg-nac} \cdot \varphi[da - th_{da}] \cdot \mathrm{m} \cdot \left(amg_{f\,A},\ amg_{f\,B}\right) \tag{11}$$

where $\eta_{(amg-nac)}$ is the learning rate coefficient. Note that in the experiments reported in Section 5.6 the lesions of rats' BLA have been simulated by setting the Amg-NAc connections $W_{amg-nac}$ to zero.

The importance of the Amg-NAc action selector resides in the fact that its 'votes' for the various actions can be *modulated on the fly* by the system's motivational states, e.g. by satiety for either one of the two foods. In general, this mechanisms opens up the possibility for the motivational-sensitive Pavlovian system (mainly the Amg in the model) to exert a direct effect on actions without the need of relearning processes, as it will be exemplified by the devaluation experiments illustrated in the next section.

5.5.4 Parameter setting and justification

The model's parameters were set as follows. The model's equations were integrated with a 50 ms time step: this rather long value allows running fast simulations and at the same time avoiding stability problems. The decay coefficient of most leaky neurons of the model are set to a rather high value (which implies a slow dynamic) as such neurons are intended to abstract the activation of *populations* of real neurons: $\tau_{sc} = 500$ ms, $\tau_{amg} = 500$ ms, $\tau_{pm} = 500$ ms. The decay of DA is set to a rather low value (which implies a fast dynamic) to reproduce the fast dynamics of phasic dopamine bursts underlying learning (Schultz, 2002): $\tau_{da} = 50$ ms. The decay of learning traces is set to a high value in order to allow the association between stimuli having onsets separated by time intervals ranging within a few seconds, as happens in real rats: $\tau_{tr} = 1000$ ms. The trace-derivative amplification coefficient is set to a high value to suitably amplify the low value of the derivative of Amg neurons' activation: $C_{amg} = 50$. The NAc and DLS baseline coefficients, and the weights connecting them to PM, are set to suitable values so as to not overcome the action-triggering threshold in PM: $nac_{baseline} = 0.3$, $dls_{baseline} = 0.3$, $w_{dls-nac-pm} = 0.5$, $th_A = 0.6$. The baseline of DA is set below the DA threshold which triggers learning: $th_{DA} = 0.6$, $da_{baseline} = 0.3$. The Amg-DA connections are set to a value lower than the PPN-DA to have a DA signal stronger for primary rewards (US) than for secondary rewards (CS): $w_{amg-da} = 0.3$, $w_{ppn-da} = 0.6$. The noise level is set to a rather high value to allow triggering of actions in the initial exploratory phase where the signal activation of NAc and DLS is null or low: $n = 0.6$. Learning coefficients are set to relatively low values to have a progressive stable learning: $\eta_{amg} = 0.015$, $\eta_{amg-nac} = 0.02$, $\eta_{sc-dls} = 0.02$. The weights of lateral connections between PM neurons are set to values which lead to a stable and reliable competition:

$$w_{pm} = \begin{pmatrix} 1 & -.5 \\ -.5 & 1 \end{pmatrix}.$$

5.6 Results

This section describes the basic functioning of the model on the basis of Figure 5.2. The figure shows the activations of various neurons related to the lever (data related to the chain are omitted as they are qualitatively similar) during both the training and testing phases of an experiment run with a non-lesioned simulated rat. It also shows the activations of the same neurons in the two test phases of a lesioned rat.

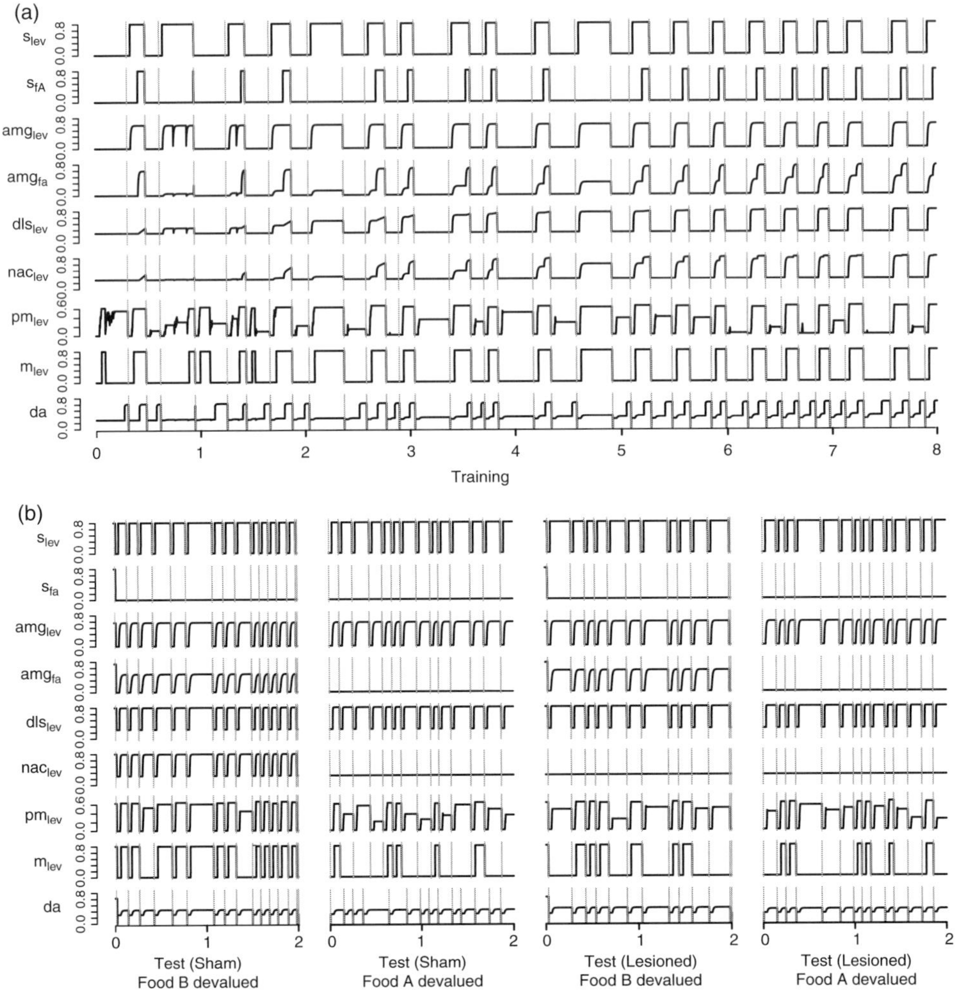

Figure 5.2. (a) Activations of some key neurons of a non-lesioned rat during the training phase. (b) Activations of the same neurons in two test phases where the same rat was satiated either with food A or B (first and second block, respectively); activations of the same neurons of a BLA-lesioned rat in two similar test phases (third and fourth block). Trials are separated by short vertical lines.

At the beginning of the training phase, the baseline activations of DLS and NAc (see dls_{lev}, nac_{lev}), together with noise, are sufficient to occasionally trigger the execution of an action (m_{lev} in the graph) by the competition taking place in PM (pm_{lev}). When the behavioural routine corresponding to the selected action is appropriate for the environment configuration (e.g. 'lever press' in the presence of lever), the dispenser becomes green and the rat approaches it and consumes the corresponding food (s_{fA}). The food consumption activates the internal hedonic representation of food in Amg (amg_{fA}) and hence the neurons in VTA/SNpc release DA in DLS (DA). This drives the learning of the cortex-dorsolateral striatum instrumental pathway. The effect of these events is that after a few learning trials the model learns the S-R habits which perform the action which is appropriate to the current context: 'sight of lever-press lever' and 'sight of chain-pull chain'. The progress of habits' learning can be seen in terms of: (a) the increase of DLS's votes for the press lever action (dls_{lev}) in the trials in which the lever is present (s_{lev}); (b) the increase of the regularity of the peaks of the food A amygdala neurons (amg_{fA}); (c) the DA release in VTA/SNpc (da).

When instrumental S-R associations begin to form, the vision of the neutral stimuli of the lever (s_{lev}, amg_{lev}) starts to be reliably followed, within a relatively small time interval, by the food perception in mouth (s_{fA}). The food perception, as mentioned above, causes a DA release (da). This contingency and the DA signal allow the Pavlovian learning taking place within Amg to 'take off' and form CS-US associations between the lever and Amg's food A representation. This is evident from the fact that after a few successful trials the Amg's food-A neuron (amg_{fA}) not only shows an activation peak when food A is delivered but it is also pre-activated by the presence of the lever: this reveals that a Pavlovian association is being acquired between the CS (lever) and the US (food).

Note how these processes show a rather interesting interaction between Pavlovian and instrumental processes. In the model, as in organisms (Lieberman, 1993), Pavlovian CS-US associations can form only if the two stimuli are separated by a time lag lasting at maximum a few seconds (in the model, this is due to the dynamics of Amg's traces, see Eq. 2 and Section 5.5.4). As with the progress of the S-R instrumental learning, e.g. involving the 'sight of lever-press lever' association, the sight of the lever (CS) is followed progressively more readily and regularly by the food (US); this allows Pavlovian processes to form the association CS-US which would not otherwise form (roughly speaking, it might be said that 'Pavlov' observes and registers a contingency suddenly appearing in the environment due to 'Skinner').

The pre-activation of the amg_{fA} neuron due to the perception of the conditioned stimulus is responsible for the early DA release (da) which anticipates the future delivery of reward. Even if this process does not play any particular function in the current model, it reproduces an important well-known phenomenon observed in real animals (Schultz, 2002), and shows how Amg can play an important role in the neuromodulation of brain, in this case the DA release.

A last important learning process, related to the main topic of the chapter, takes place in the Amg-NAc pathway. This process is at the basis of the influence of Amg on the selection of habits taking place in the SC-DLS pathway. Once the CS-US associations are formed in the Amg, CSs, such as the lever, can trigger the activation of the Amg's hedonic representation of the related food and, via this, influence DLS action selection via NAc. This process is shown by the fact that, after some training, NAc starts to activate and to vote for the correct actions (nac_{lev}). The importance of the formation of this Stimuli-Amg-NAc-PM pathway resides in the fact that it constitutes the fundamental bridge between the Pavlovian processes happening in the Amg and the instrumental processes happening in the SC-DLS pathway. We claim that this pathway plays a central role in the flexibility exhibited by real organisms. In particular, it is through this pathway that instant motivational manipulations that characterise Pavlovian conditioning are able to directly affect instrumentally learned behaviours, as in the devaluation tests which are now illustrated.

To see this, let us focus on the test with sham rats (Figure 5.2b, first two blocks). During the two test phases, in which the rat perceives both the lever and the chain, the satiety of respectively food A or B are kept at its maximum level, namely five (the other satiety level is kept at zero). Recall that the tests are performed 'in extinction', that is without food delivery (see s_{fA}). The satiety for a food causes a strong inhibition to the Amg's hedonic representation of such food. As a consequence, both the direct consumption of that food and the perception of the conditioned stimulus previously associated with it cannot elicit the related Amg's hedonic reaction. This is shown by the lack of activation of the Amg's Food-A neuron (amg_{fA}) during the second test phase when the rat is satiated with food A.

The perception of both the lever and the chain during the tests leads DLS to 'vote' for both the lever press and chain pull actions at the same time (compare the dls_{lev} activation in the two test phases). This rules out the influences of the S-R instrumental pathway on action selection: rigid habits are not capable of driving the rat to make the suitable decision as they lack information on internal states (incidentally, note that this experimental condition was precisely designed by Balleine *et al.* (2003) to stop the effects of habits that would otherwise 'mask' the motivation-sensitive Pavlovian influence on action selection). On the other hand, satiation stops only one of the two influences of the Amg-NAc pathway on action selection in that it inhibits only the amygdala representation of the unconditioned stimulus which has been satiated (compare the nac_{lev} activation in the two test phases). The fact that the Amg-NAc pathway 'votes' only for the action associated with the non-satiated food breaks the symmetry between two habits and makes the related action to win the competition in PM with a high probability (compare the pm_{lev} and m_{lev} activations in the two test phases).

The comparison between the lesioned and non-lesioned conditions (Figure 5.2b, last two blocks) reproduces the basic finding of the target experiment of Balleine *et al.* (2003) and confirms the aforementioned interpretation of the devaluation tests: as happens in real rats, a lesion to the BLA pathway linking the amygdala to the NAc prevents the devaluation of food from having any effect on the action selection process. More in particular (see Figure 5.3), during the test non-lesioned (Sham) rats perform the action associated with the non-devaluated food A 11.2 times on average whereas they perform the action associated with the devaluated food A 2.9 times on average: the difference between the two conditions is statistically significant (paired t-test, $t = 15.7003$, $df = 19$, $p < 0.001$). On the contrary, BLA-lesioned rats select actions randomly as indicated by the fact that the number of performed actions associated with the non-devaluated and the devaluated food A have an average of 6.2 and 6.5: the difference between the two conditions is not statistically significant (paired t-test, $t = -0.4346$, $df = 19$, $p > 0.05$). These results show the plausibility of the hypothesis for which the Amg-(BLA)-NAc pathway bridges the Pavlovian processes happening in the amygdala with the instrumental processes happening in the cortexbasal ganglia pathway, so allowing the current state of animals' motivational system to modulate *on the fly* their action selection mechanisms.

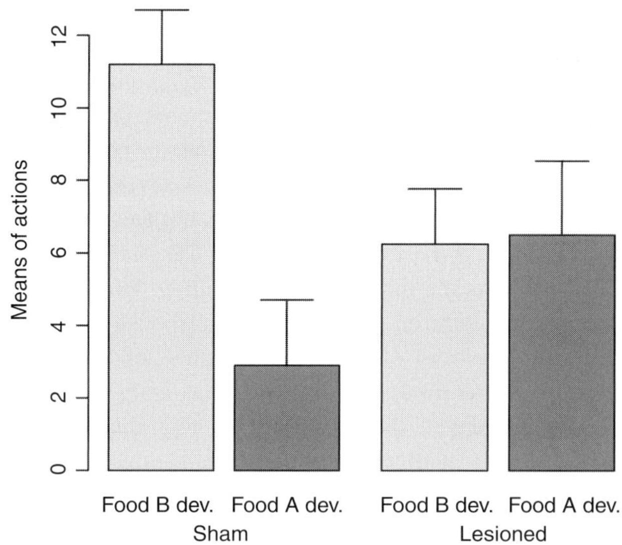

Figure 5.3. Averages and standard deviations of the number of actions selected (y-axis) by Sham and BLA lesioned rats during different tests involving devaluation of either food A or food B.

5.7 Conclusions and future work

This chapter presented an embodied model of some important relations existing between Pavlovian and instrumental conditioning. The model's architecture and functioning was constrained with relevant neuroscientific knowledge on the brain anatomy and physiology. The model was validated by successfully reproducing the primary outcomes of some instrumental conditioning devaluation tests conducted with normal and amygdala-lesioned rats. These tests are particularly important for studying the Pavlovian-instrumental interplay as they show how the sensitivity to motivational states exhibited by the Pavlovian system can transfer to instrumentally acquired behaviours.

To the best of the authors' knowledge, the model represents the first attempt to propose a comprehensive interpretation of the aforementioned phenomena, tested in an embodied model. The works most closely related to this one are those of Armony *et al.* (1997), Dayan & Balleine (2002), Morén & Balkenius (2000) and O'Reilly *et al.* (2007). The model presented here differs from these works in that it proposes an embodied model (absent in all mentioned researches), presents a fully developed model (Dayan & Balleine, 2002, presented only a 'sketched' model), and tackles the issue of the relations existing between Pavlovian and instrumental conditioning (Armony *et al.*, 1997, Morén & Balkenius, 2000 and O'Reilly *et al.*, 2007 focused only on Pavlovian conditioning).

Notwithstanding the proposed model has these several strengths, it will be improved in many directions in future work. The first limit of the work is that the model was tested with an embodied system where input signals were heavily pre-processed before being fed into the model in the form of 'localistic representations' (one neuron-one object), and where actions were specified at a rather abstract level by relying on hardwired low-level behavioural routines. In the future the whole model, or some of its parts (e.g. the amygdala component), will be tested with more challenging embodied systems where the model will be fed with realistic distributed input patterns (e.g. the activations of the retina's pixels) and will be required to issue low-level motor commands (e.g. the desired displacement and turning speed). Second, the model has several limitations with respect to available biological evidence. For example, it does not learn to inhibit the dopamine signal at the onset of the USs if these are preceded by CSs, as happens in real organisms (Schultz, 2002). This prevents the model from performing 'extinction' (i.e. to unlearn a classical conditioning association or an instrumental response if these are not followed any more by a reward) and from stopping the weights' update. In future work, this capability will be added to the model by drawing ideas from other works, for example O'Reilly *et al.* (2007). Moreover, the model cannot reproduce classical conditioning-based modulation of the *vigour* with which instrumental actions are performed (Niv *et al.*, 2007), nor is it capable of triggering innate actions on the basis of classical-conditioning (e.g. approaching a US, or approaching a CS after this has been associated to a US; Dayan & Balleine, 2002). Finally, the model assumes that the selection of actions takes place within premotor cortex. However, there is strong evidence (Redgrave *et al.*, 1999) that in real brains action selection takes place at the level of the DLS itself, and so PM activations might only reflect such selection without causing it (cf. Cisek, 2007). This possibility, however, opens up the problem of how the NAc might influence such action selection, as requested for the Pavlovian processes to exert an influence on instrumental processes. In this respect, an interesting neural pathway through which this influence might be implemented are the striatonigro-striatal connections (or 'dopaminergic spirals'; Haber *et al.*, 2000). These topics will be addressed in future work.

Notwithstanding these limitations, the proposed model represents an important step in the construction of an integrated picture on how animals' motivational systems can both drive instrumental learning and directly regulate behaviour. Constructing such a picture is of paramount importance from the scientific point of view as psychology and neuroscience have now amassed a large body of evidence and knowledge on the phenomena investigated here which would greatly benefit from theoretical systematisation. In this respect, we believe that computation modeling carried out under the principles of computational embodied neuroscience illustrated in Section 5.2 can greatly aid this process.

As mentioned in Section 5.1, although this work has mainly a scientific relevance, the research agenda of the work presented here has also a potential interest for overcoming the limited autonomy of current robots. In fact, a way to tackle these limits is to attempt to understand the mechanisms underlying organisms' behavioural flexibility so as to use them in designing robot's controllers. In this respect, notwithstanding the motivational and emotional regulation of behaviour is very important for behavioural flexibility, it has been almost completely overlooked by autonomous robotics. For this reason Parisi (2004) has advocated the need for an 'Internal Robotics' research agenda dedicated to the study of these processes. In line with this, recently machine learning and robotics communities have been devoting increasing efforts to the study of autonomous learning by trying to improve the standard reinforcement learning algorithms mentioned in Section 5.1 on the basis of ideas coming from the study of real organisms (Weng *et al.*, 2001; Zlatev & Balkenius, 2001). In this respect, the investigations on emotional regulation of learning and behaviour in animals, such as those reported here, are expected to produce important insights on possible new principles and techniques to be used to design more powerful learning algorithms exhibiting a degree of autonomy similar to that of real organisms (see Barto *et al.*, 2004, and Schembri *et al.*, 2007, for two examples of this).

Acknowledgements

This research was supported by the EU-funded projects *ICEA – Integrating Cognition, Emotion and Autonomy*, contract no. FP6-IST-027819IP, and *IM-CLeVeR – Intrinsically Motivated Cumulative Learning Versatile Robots*, contract no. FP7-ICT-IP-231722. A preliminary version of this work was published as Mannella *et al.* (2007).

References

Armony, J. L., Servan-Schreiber, D., Romanski, L. M. & LeDoux, D. J. J. E. 1997. Stimulus generalization of fear responses: effects of auditorycortexlesions in a computational model and in rats. *Cereb Cortex* **7**(2), 157–165.

Baldassarre, G. 2008. Self-organization as phase transition in decentralized groups of robots: a study based on Boltzmann entropy. In *Advances in Applied Self-Organizing Systems* (ed. M. Prokopenko), pp. 127–146. Springer-Verlag.

Balleine, B. W., Killcross, A. S. & Dickinson, A. 2003. The effect of lesions of the basolateral amygdala on instrumental conditioning. *J Neurosci* **23**(2), 666–675.

Balleine, B. W. & Killcross, S. 2006. Parallel incentive processing: an integrated view of amygdala function. *Trends Neurosci* **29**(5), 272–279.

Barto, A., Singh, S. & Chentanez, N. 2004. Intrinsically motivated learning of hierarchical collections of skills. In *International Conference on Developmental Learning (ICDL)*, LaJolla, CA.

Baxter, M. G. & Murray, E. A. 2002a. The amygdala and reward. *Nat Rev Neurosci* **3**(7), 563–573.

Baxter, M. G. & Murray, E. A. 2002b. The amygdala and reward. *Nature Rev Neurosci* **3**(7), 563–573.

Blair, H. T., Sotres-Bayon, F., Moita, M. A. P. & LeDoux, J. E. 2005. The lateral amygdala processes the value of conditioned and unconditioned aversive stimuli. *Neuroscience* **133**(2), 561–569.

Brody, C., Pouget, A., Shadlen, M. & Zador, A. (Eds.) 2004. *Abstracts of Papers Presented at the 2004 Meeting on Computational & System Neuroscience*. Cold Spring Harbor Laboratory.

Camazine, S., Deneubourg, J. L., Franks, N. R., Sneyd, J., Theraulaz, G. & Bonabeau, E. (Ed.) 2001. *Self-organization in Biological Systems*. Princeton University Press.

Cardinal, R. N., Parkinson, J. A., Hall, J. & Everitt, B. J. 2002. Emotion and motivation: the role of the amygdala, ventral striatum, and prefrontal cortex. *Neurosci Biobehav Rev* **26**(3), 321–352.

Cisek, P. 2007. Cortical mechanisms of action selection: the affordance competition hypothesis. *Phil Trans R Soc B* **362**(1485), 1585–1599.

Clark, A. 1997. *Being There: Putting Brain, Body and World Together Again*. MIT Press.

Darwin, C. 1859. *The Origin of Species*. Retrieved from http://www.literature.org/authors/darwincharles/the-origin-of-species/index.html.

Dayan, P. & Balleine, B. 2002. Reward, motivation and reinforcement learning. *Neuron* **36**, 285–298.

Domjan, M. 2006. *Principles of Learning and Behaviour*. Thomson Wadsworth.

Haber, S. N., Fudge, J. L. & McFarland, N. R. 2000. Striatonigrostriatal pathways in primates form an ascending spiral from the shell to the dorsolateral striatum. *J Neurosci* **20**(6), 2369–2382.

Holland, O. & McFarland, D. 2001. *Artificial Ethology*. Oxford University Press.

Houk, J. C., Adams, J. L. & Andrew, G. B. 1995. A model of how the basal ganglia generate and use neural signals that predict reinforcement. In *Models of Information Processing in the Basal Ganglia* (ed. J. C. Houk, J. L. Davids & D. G. Beiser), pp. 249–270. MIT Press.

Knight, D. C., Nguyen, H. T. & Bandettini, P. A. 2005. The role of the human amygdala in the production of conditioned fear responses. *Neuroimage* **26**(4), 1193–1200.

Kobayashi, Y. & Okada, K.-I. 2007. Reward prediction error computation in the pedunculopontine tegmental nucleus neurons. *Ann N Y Acad Sci* **1104**, 310–323.

Langton, C. (Ed.) 1987. *The First International Conference on the Simulation and Synthesis of Living Systems (ALifeI)*.

Lieberman, D. A. 1993. *Behavior and Cognition*. Brooks/Cole.

Mannella, F., Mirolli, M. & Baldassarre, G. 2007. The role of amygdala in devaluation: a model tested with a simulated robot. In *Proceedings of the Seventh International Conference on Epigenetic Robotics* (ed. L. Berthouze, C. G. Prince, M. Littman, H. Kozima & C. Balkenius), pp. 77–84. University of Lund.

Mannella, F., Zappacosta, S. & Baldassarre, G. 2008. A computational model of the amygdala nuclei's role in second order conditioning. In *Proceedings of the Tenth International Conference on Simulation of Adaptive Behavior: From Animals to Animals 10* (ed. M. A. J. Tani, J. Hallam & J.-A. Meyer). Springer-Verlag.

Maren, S. 2005. Building and burying fear memories in the brain. *Neuroscientist* **11**(1), 89–99.

McDonald, A. J. 1998. Cortical pathways to the mammalian amygdala. *Prog Neurobiol* **55**(3), 257–332.

Meyer, J.-A. & Wilson, S. W. (Ed.) 1991. *From Animals to Animats 1: Proceedings of the First International Conference on Simulation of Adaptive Behaviour*. MIT Press.

Mogenson, G. J., Jones, D. L. & Yim, C. Y. 1980. From motivation to action: functional interface between the limbic system and the motor system. *Prog Neurobiol* **14**(2–3), 69–97.

Morén, J. & Balkenius, C. 2000. A computational model of emotional learning in the amygdala. In *From Animals to Animats 6: Proceedings of the 6th International Conference on the Simulation of Adaptive Behaviour* (ed. J.-A. Meyer, A. Berthoz, D. Floreano, H. L. Roitblat & S. W. Wilson). MIT Press.

Newell, A. 1973. You can't play 20 questions with nature and win: projective comments on the papers of this symposium. In *Visual Information Processing* (ed. W. G. Chase), pp. 283–308. Academic Press.

Niv, Y., Daw, N. D., Joel, D. & Dayan, P. 2007. Tonic dopamine: opportunity costs and the control of response vigor. *J Psychopharmacol* **191**(3), 507–520.

Nolfi, S. 2006. Behaviour as a complex adaptive system: on the role of self-organization in the development of individual and collective behaviour. *ComplexUs* **2**(3–4), 195–203.

Nolfi, S. & Floreano, D. 2000. *Evolutionary Robotics: The Biology, Intelligence, and Technology*. MIT Press.

O'Reilly, R., Frank, M., Hazy, T. & Watz, B. 2007. PVLV: The primary value and learned value pavlovian learning algorithm. *Behav Neurosci* **121**, 31–49.

Packard, M. G. & Knowlton, B. J. 2002. Learning and memory functions of the basal ganglia. *Annu Rev Neurosci* **25**, 563–593.

Parisi, D. 2004. Internal robotics. *Connection Sci* **16**(4), 325–338.

Parisi, D., Cecconi, F. & Nolfi, S. 1990. Econets: Neural networks that learn in an environment. *Network* **1**, 149–168.

Pavlov, I. P. 1927. *Conditioned Reflexes: An Investigation of the Physiological Activity of the Cerebral Cortex*. Oxford University Press.

Pitkänen, A., Jolkkonen, E. & Kemppainen, S. 2000. Anatomic heterogeneity of the rat amygdaloid complex. *Folia Morphol. (Warsz)* **59**(1), 1–23.

Prescott, T. J., Gonzalez, F. M., Humphries, M. & Gurney, K. 2003. Towards a methodology for embodied computational neuroscience. In *Proceedings of the Symposium on Scientific Methods for the Analysis of Agent-Environment Interaction (AISB2003)*. AISB Press.

Prescott, T. J., Gonzalez, F. M. M., Gurney, K., Humphries, M. D. & Redgrave, P. 2006. A robot model of the basal ganglia: behavior and intrinsic processing. *Neural Netw* **19**(1), 31–61.

Price, J. L. 2003. Comparative aspects of amygdala connectivity. *Ann N Y Acad Sci* **985**(1), 50–58.

Redgrave, P., Prescott, T. J. & Gurney, K. 1999. The basal ganglia: a vertebrate solution to the selection problem? *J Neurosci* **89**(4), 1009–1023.

Rolls, E. T. 2005. Taste and related systems in primates including humans. *Chem Senses* **30** Suppl. 1, i76–i77.

Schembri, M., Mirolli, M. & Baldassarre, G. 2007. Evolving internal reinforcers for an intrinsically motivated reinforcement-learning robot. In *Proceedings of the 6th International Conference on Development and Learning (ICDL)* (ed. Y. Demiris, D. Mareschal, B. Scassellati & J. Weng), pp. E1–6. Imperial College London.

Schultz, W. 2002. Getting formal with dopamine and reward. *Neuron* **36**, 241–263.

Sejnowski, T. J., Koch, C. & Churchland, P. S. 1988. Computational neuroscience. *Science* **241**(4871), 1299–1306.

Shi, C. & Davis, M. 1999. Pain pathways involved in fear conditioning measured with fear potentiated startle: lesion studies. *J Neurosci* **19**(1), 420–430.

Skinner, B. F. 1938. *The Behavior of Organisms*. Appleton-Century-Crofts.

Sutton, R. S. & Barto, A. G. 1981. Toward a modern theory of adaptive networks: Expectation and prediction. *Psychol Rev* **88**, 135–140.

Sutton, R. & Barto, A. 1998. *Reinforcement Learning: An Introduction*. MIT Press.

Thorndike, E. L. 1911. *Animal Intelligence*. Transaction Publishers.

Weng, J., McClelland, J., Pentland, A. *et al.* 2001. Autonomous mental development by robots and animals. *Science* **291**, 599–600.

Yin, H. H. & Knowlton, B. J. 2006. The role of the basal ganglia in habit formation. *Nat Rev Neurosci* **7**, 464–476.

Zlatev, J. & Balkenius, C. 2001. Introduction: Why epigenetic robotics? In *Proceedings of the First International Workshop on Epigenetic Robotics: Modeling Cognitive Development in Robotic Systems* (ed. C. Balkenius, J. Zlatev, H. Kozima, K. Dautenhahn & C. Breazeal), University of Lund. pp. 1–4.

6

Evolution, (sequential) learning and generalisation in modular and nonmodular visual neural networks

Raffaele Calabretta

6.1 Introduction

In general terms, modular systems are systems that can be decomposed in functional and/or structural independent parts. In cognitive science, modularity of mind (Fodor, 1983) is the controversial cognitivist view according to which human mind is made up of specialised innate modules. In contrast, the connectionist view tends to conceive of mind as a more homogeneous system that results from development and learning during life (see Karmiloff-Smith, 2000).

What is a module? Are modules innate? What is the relationship between modularity, robustness and evolvability? And what is the role of nonmodularity? These are only a few of the open central questions in the nature–nurture debate.

In a paper published in the journal *Cognition*, Gary Marcus (2006) deals with the vexata quaestio of modularity of mind from an enlightening point of view. He does not offer a detailed definition of modularity, nor an answer to the controversial issue of what is innate and what is learned during life. More simply, he identifies and contrasts two competing 'hypothetical conceptions of modularity', which would represent distinct perspectives about modularity of mind, implicitly present in the scientific literature and different in their implications: a 'sui generis modularity' and a 'descent with modification modularity'. According to the former conception, 'each cognitive (or neural) domain would be an entity entirely unto itself'; according to the latter conception, 'current cognitive (or neural) modules are to be understood as being the product of evolutionary changes from ancestral cognitive (or neural) modules'.

Marcus argues that much of the criticism directed toward modularity per se in reality militates against an extreme, sui generis version of modularity, and not against a descent with modification version of modularity. Moreover, he claims that a range of experimental data is incompatible with the former view, and is compatible with the latter.

An analogous 'descent with modification' view has characterised research conducted in the last ten years by Calabretta and colleagues (1997, 1998a, 1998b, 2000, 2003, 2004, 2008), who proposed an 'evolutionary connectionism' approach to studying and simulating

Modelling Perception with Artificial Neural Networks, ed. C. R. Tosh and G. D. Ruxton. Published by Cambridge University Press.

the evolution of modularity (Calabretta, 2002; Calabretta & Parisi, 2005), based on the use of neural networks as a model of the brain (Rumelhart & McClelland, 1986), and of genetic algorithms as a model of darwinian biological evolution (Holland, 1992).

This approach was based on the idea that, in order to understand modularity of mind, one has to simulate its biological evolution, taking into account not only adaptations ('feature that promotes fitness and was built by selection for its current role'; Gould & Vrba, 1982), as evolutionary psychology does (e.g. Cosmides & Tooby, 1994), but also exaptations ('characters evolved for other usages (or for no function at all), and later [coopted] for their current role'; Gould & Vrba, 1982), chance and non-adaptive factors (see Gould, 1997) and natural mechanisms such as duplication of genes (Ohno, 1970). In the evolutionary connectionism approach, brain modules are not the theoretical entities postulated in 'boxes and arrows' cognitive modules (see, for example, Pinker & Prince, 1988), but are modelled by using a neural network model; computer simulations are adopted for testing evolutionary scenarios and exploring what can be innate and what can be learned (see Elman *et al.*, 1996).

Marcus (2006) stresses that 'evolution never starts from scratch, but instead by modifying the structures of organisms already in place', a phenomenon sometimes known as 'bricolage', and that 'through mechanisms such as "duplication and divergence" [. . .], evolution often creates new systems by making modified copies of old systems, whether at the level of a single gene, or by the force of genetic cascade at the level of complex structures such as limbs or (it appears) brain areas'.

The first simulative model on the evolution of modularity using an evolutionary connectionism approach, developed at Yale University (Calabretta & Wagner, 1997; Calabretta *et al.*, 1997), had aimed at understanding the role of genetic duplication for the evolution of brain modules, revealed unexpected evolutionary dynamics (Calabretta *et al.*, 1998a, 1998b), and led to the discovery of a new mechanism for the origin of functional modularity (Calabretta *et al.*, 2000): functional modularity not as an adaptation, but as a dynamic side effect of the genetic neural module duplication. To our knowledge, 'it was the first example where the origin of modularity was simulated in a computational model' (Gunter Wagner, personal communication, 19 March 2001).

This important and surprising result led us to go ahead with this research project and in 1999 to focus the study on the origin of structural modularity (Di Ferdinando & Calabretta, 2000). To do that, as a model of modules present in the brain of real organisms, we chose the two cortical visual streams (dorsal and ventral) for the 'What' and 'Where' task, identified in primates by Ungerleider & Mishkin (1982).

The choice to use a simulative model that takes inspiration from the real structure of the primate brain turned out to be a good one. Simulative studies produced several intriguing results and, on this basis, we proposed a new hypothesis for the evolutionary origin of brain architecture: evolution that takes care of the architecture, and learning that takes care of neural network connection weights (Di Ferdinando *et al.*, 2000, 2001; Calabretta *et al.*, 2003). To our knowledge, it was the first simulative model in which the evolution of structural modularity was achieved. The evolution of structural modularity was independently achieved in the model of RNA secondary structure published by Ancel & Fontana (2000).

The goal of this chapter is to review ten years of simulative research on the evolution of brain modularity, stressing the more recent research, on sequential learning and generalisation in modular and nonmodular neural networks. In conclusion, some possible improvements of the dorsal and ventral streams study model will be sketched, which take into consideration up-to-date theoretical positions in neuroscience (visual streams of the 'What' and 'How' task; Milner & Goodale, 1995, 2008).

6.2 The study model: the two cortical visual systems for the 'What' and 'Where' task

In their seminal paper 'Two cortical visual systems', Ungerleider & Mishkin (1982) distinguished two primate cortical visual systems: the dorsal stream, which progresses dorsally from the visual cortex to the posterior parietal cortex and is functionally involved in object spatial localisation (specialised for spatial vision: 'Where' task) and the ventral stream, which progresses ventrally from visual cortex to the inferotemporal cortex and is functionally involved in object identification (specialised for object vision: 'What' task). (Today a different account for the 'Where' dorsal stream is widely accepted, which is considered as involved in a vision for action, i.e. how the organism has to act on an object: a 'How' task; see Section 6.4.)

Rueckl *et al.* (1989) explored the computational properties of these two systems by constructing a simulation connectionist model. They compared modular and nonmodular neural networks that have to recognise the identity (what type of object it is: 'What' task) and the spatial location (where the object is located: 'Where' task) of perceived objects by means of a learning algorithm (back-propagation procedure; Werbos, 1974; Rumelhart & McClelland, 1986).

These two tasks require neural networks to extract two different types of information from the same input. In their simulations, 9 different objects are used which are represented as patterns of 3×3 black and white cells in a 5×5 retina (see Figure 6.1).

The objects are presented in 9 different locations by moving the same 3×3 pattern (same object) to 9 different positions in the 5×5 retina. Both modular and nonmodular networks have 25 input units for encoding the 5×5 cells of the retina and 18 output units, 9 localistically encoding the 9 different responses to the 'Where' question 'Where is the object?' and the other 9 units localistically encoding the 9 different responses to the 'What' question 'What is the object?'. Both modular and nonmodular networks have a single layer of 18 internal units and each of the 25 input units projects to all 18 internal units. Where the modular and the nonmodular architectures are different is in the connections from the internal units to the output units. In the modular networks only 4 internal units project to the 9 'Where' output units, while the remaining 14 internal units project to the 9 'What' output units. In the nonmodular networks all 18 internal units project to both the 9 'Where' output units and the 9 'What' output units (see Figure 6.2).

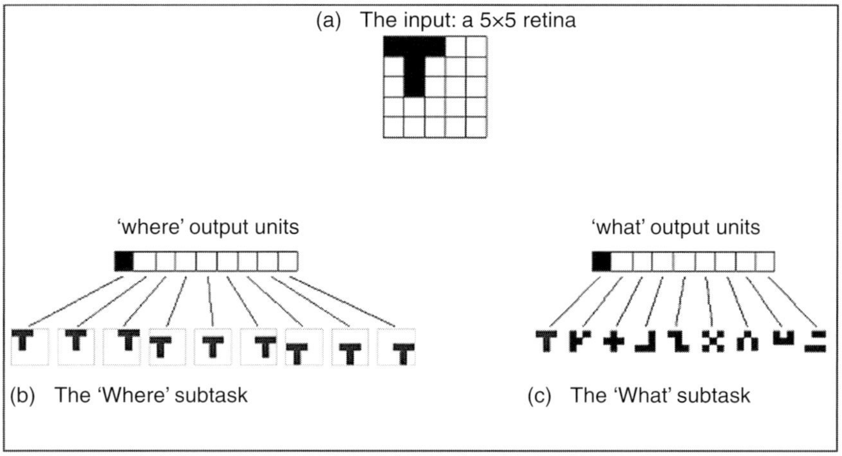

Figure 6.1. The 'What' and 'Where' task: (a) the input retina; (b) the 'Where' subtask; (c) the 'What' subtask. Modified from Calabretta *et al.* (2004).

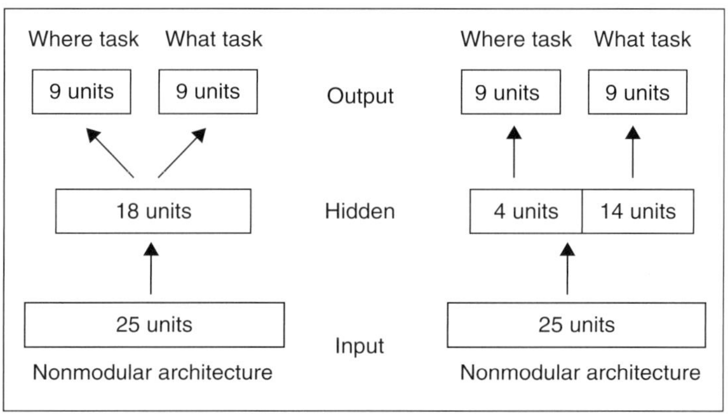

Figure 6.2. Nonmodular (left) and modular (right) network architectures for learning the 'What' and 'Where' task. Modified from Calabretta *et al.* (2008).

The fact that in nonmodular architecture some connections are involved in both tasks is important in that it raises the problem of neural interference (see Calabretta & Parisi, 2005, figure 14.4). The interference derives from the fact that the correct accomplishment of the first task may require the initial weight value of the connection to be increased during learning, while the correct accomplishment of the second task may require the initial value to be decreased (Plaut & Hinton, 1987; see Figure 6.2). In the modular architecture, there is no neural interference between the two tasks: it is in fact two separate architectures, with two nonoverlapping subsets of connection weights each dedicated to only one task. (For an operative definition of modularity and nonmodularity, see Calabretta, 2007, p. 404.)

Rueckl *et al.* (1989) compared performances of modular and nonmodular neural networks, showing that the architecture with the best performance in the two tasks is a modular one in which there are more resources (i.e. hidden units) for the more complex task, that is, the 'What' task, than for the simpler task, that is, the 'Where' task. More precisely, the best architecture is that in which 14 hidden units are dedicated to the 'What' task, and 4 hidden units are dedicated to the 'Where' task.

They have put forward a few hypotheses to account for these results and have argued that the difference in performance among modular and nonmodular architectures might explain the evolutionary reason for which these two separate neural pathways evolved for the two tasks.

A very simple test for this hypothesis would be to modify Rueckl *et al.*'s model set-up by adding the genetic algorithm as a model of evolution (Holland, 1992). In this way, it becomes possible to simulate the evolution of modular architectures starting from nonmodular architectures.

This was done by Di Ferdinando *et al.* (2000, 2001) with quite surprising results, as we will see in the next section.

6.3 Simulation results

6.3.1 Genetic interference and a new hypothesis on the origin of brain architecture

In Rueckl *et al.*'s (1989) simulations the two network architectures are hardwired by the researcher. It is the researcher who designs both modular and nonmodular neural architectures and then shows that modular architectures produce better results than nonmodular ones. However, if networks (organisms) that must perform a plurality of tasks exhibit better performances with modular rather than nonmodular architectures, nature is likely to play the role of the researcher. In other words, there must be ways for modular architectures to emerge spontaneously in opposition to nonmodular ones if there is a chance for nature to choose between the two architectures.

Another possibility is to imagine that biological evolution takes care of the problem of finding the appropriate modular architecture for networks that must learn two distinct tasks, and to use the genetic algorithm as a model of evolution (Holland, 1992; Mitchell, 1996). In this case network architecture emerges not during an individual's life, but during a succession of generations in a population of individuals. We have conducted a series of simulations using the genetic algorithm to develop neural networks that are able to accomplish the 'What' and 'Where' task. In spite of its simplicity, this model has shown itself to be very sensitive in that it reveals some very interesting phenomena related to the evolvability of artificial systems.

In one first set of simulations, we repeated the experiments of Rueckl *et al.*, but, instead of comparing the performance of neural networks with different architectures, we used a genetic algorithm to evolve the architecture and weights of neural networks (Di Ferdinando *et al.*, 2001).

Imagine a population of organisms living in an environment in which the reproductive chances of each individual depend on the individual's performance in the 'What' and 'Where' task. In other words, the individuals that have a smaller error on the 'What' and 'Where' task are more likely to reproduce than the individuals with a larger error.

An individual is born with an inherited genotype that specifies both the architecture of the individual's neural network and the network's connection weights. Some general features of the architecture are fixed and identical in all individuals (and therefore are not encoded in the genotype and do not evolve). All architectures have three layers of units with 25 input units, 18 hidden units and 18 output units, and in all architectures each input unit is linked to all the hidden units (lower connections). What can vary from an architecture to another are the higher connections between the hidden units and the output units. The genotype is divided into two parts. The first part encodes the network architecture and it includes 18 genes, one for each hidden unit. Each of these architectural genes has three possible values that specify if the corresponding hidden unit is connected (a) to (all) the What output units, (b) to (all) the Where output units, or (c) to both the What and the Where output units. The third possibility, (c), is included to allow for the evolution of nonmodular architectures. The second part of the genotype encodes the connection weights and it includes one gene for each possible connection weight (weight genes). The weight genes are encoded as real numbers.

At the beginning of the simulation a population of 100 individuals is created and each individual inherits a genotype with random values for both the architectural and the weight genes. The values of the weight genes are randomly chosen in the interval between –0.3 and +0.3. Each individual is presented with the 81 input patterns of the 'What' and 'Where' task and the individual's fitness is measured as the opposite of the summed squared error on these patterns. The 20 best individuals are selected for reproduction. Each of these individuals generates five offspring which inherit the genotype of their single parent with the addition of some random mutations. The architectural genes are mutated by replacing the value of a gene with a new randomly chosen value with a probability of 5%. The weight genes are mutated by adding a quantity randomly chosen in the interval between –1 and +1 to 10% of the genes. The simulation is terminated after 10 000 generations. Ten replications of the simulation were run with randomly chosen initial conditions.

In summary, the simulation results show that the genetic algorithm is unable to find the appropriate neural network architecture and weights for solving both the 'What' and the 'Where' task. More specifically, the neural networks are able to solve the easier 'Where' task and not the more difficult 'What' task.

There are two different processes that are going on in our simulations: (1) the process of the change of the neural network architecture and (2) the process of the change of the neural network weights. We decoupled these two processes and conducted another set of simulations in which the neural network is fixed and is the optimal one for the 'What' and 'Where' task (14 hidden units for the 'What' subtask and 4 hidden units for the 'Where' subtask) and what is required from the genetic algorithm is simply to evolve the appropriate connection weights for this architecture. We compared the results obtained with this architecture with those obtained with a fixed nonmodular architecture (Calabretta *et al.*, 2003). In these simulations (terminated after 50 000 generations instead of the 10 000 of Di Ferdinando *et al.*, 2001), we varied the mutation rate (from a value of 0.0016% to a value of 10%), the fitness formula and the kind of reproduction.

Simulation results not only confirmed the existence of the phenomenon of neural interference in nonmodular network architectures, but also for the first time revealed the existence of another kind of interference at the genetic level, i.e. genetic interference (Calabretta *et al.*, 2003; Calabretta, 2007). Genetic interference is an interesting general phenomenon in that it is unexpected according to models of population genetics and is independent from the network architecture (for a detailed discussion see Calabretta *et al.*, 2003, p. 258).

In fact, genetic interference is also present in networks with modular architecture, in which the genotype encodes multiple separate modules underlying the two tasks. It is a consequence of mutations affecting the two different tasks, and tends to appear when the 'Where' task is already optimised. In this case, the probability of deleterious mutations to the 'Where' task increases and therefore the probability that a positive mutation falling on the genetic module underlying the 'What' task is accompanied by an additional negative mutation on the module underlying the 'Where' task, increases as well. In the case of nonsexual reproduction, in which the possibility of genetic recombination among modules is not present, either the evolution retains both the mutations or discharges them. In both cases, the selection process will not be very effective (see Figure 6.3). This can completely prevent the optimisation of the 'What' task.

In order to understand better the nature of genetic interference, we carried out an extensive series of analyses and showed that it is a new class of genetic constraints requiring both high mutation rates (see Calabretta *et al.*, 2003, figure 6.6) and linkage between favourable and unfavourable mutations affecting distinct tasks to occur. These tasks can be either identical or different, with either the same or different difficulty (see Calabretta *et al.*, 2003, p. 253).

In another set of simulations, we explored the interaction between evolution and learning in facing the problem of genetic interference. The genetic algorithm was used to evolve the network architecture, but the connection weights of each neural network were randomly chosen in each generation and learned by the individual networks during their 'life' using the back-propagation procedure.

The simulation results were very interesting: evolution was able to select the optimal modular architecture and learning was able to identify the appropriate weights for the inherited architecture. At the end of the evolutionary process, as a result of complex interaction and cooperation between genetic adaptation and individual learning, neural networks were able to solve both the 'What' and the 'Where' tasks.

This important result shows a way to resolve the traditional dichotomy in cognitive science between nativism (the view according to which mind is modular and this modularity is basically innate; see Keil, 1999, p. 583) and empiricism (the view according

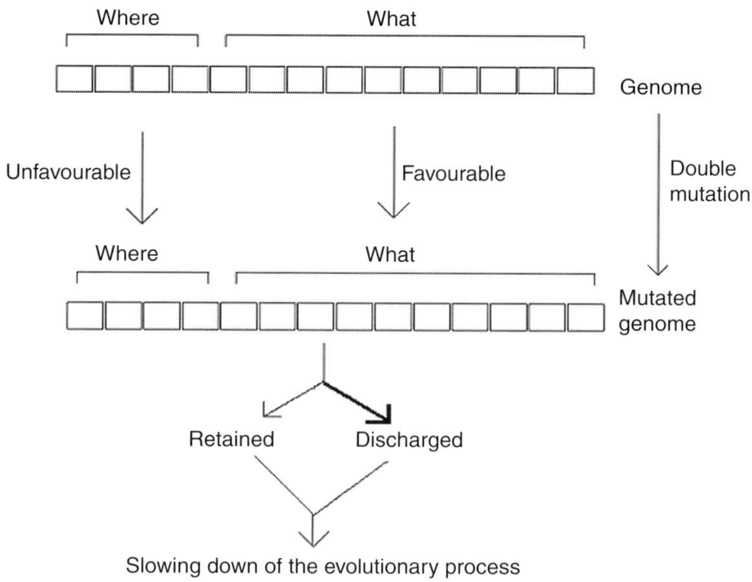

Figure 6.3. Genetic interference in asexual populations. A favourable mutation can fall on the separate portion of the genotype encoding the connection weights for the 'What' neural module, and can be accompanied by an unfavourable mutation that falls on the separate portion of the genotype encoding the connection weights for the 'Where' neural module, thereby slowing down the evolutionary process. Modified from Calabretta *et al.* (2003).

to which mind is the result of learning and experience during life; see Schwartz, 1999, p. 703), and can represent a new hypothesis for the origin of brain structure: for an organism to be able to acquire complex abilities, evolution and learning cooperate by respectively selecting the appropriate architecture and finding the appropriate connection weights (Calabretta *et al.*, 2003; see also Wagner *et al.*, 2005).

6.3.2 Sequential learning of the 'What' and 'Where' task in nonmodular neural networks

The results of Rueckl *et al.*'s (1989) simulations show that fixed modular network architectures perform better than fixed nonmodular ones in learning the 'What' and 'Where' tasks. In a modular architecture each module is dedicated to one particular task so that the synaptic weights of the module can be adjusted without interfering with the other task. Therefore, the problem of neural interference between the two tasks, which is present in nonmodular architectures (Figure 6.4a), can be avoided. (In modular architectures a form of intra-task neural interference that also exists in learning a single task is still active (Plaut & Hinton, 1987), but it is much weaker than inter-task neural interference.)

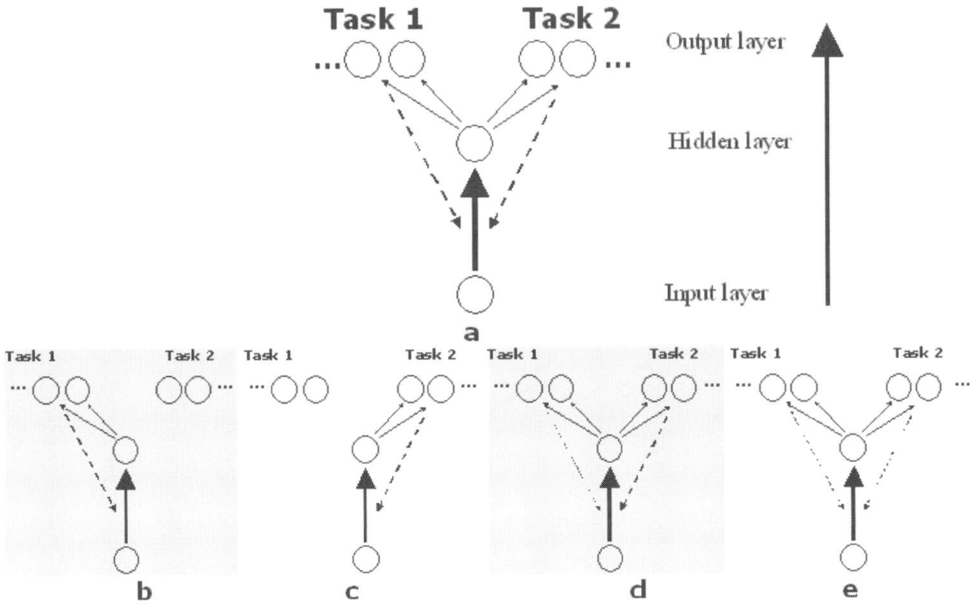

Figure 6.4. Neural interference between the 'What' task and the 'Where' task in nonmodular network architecture. The solid lines represent connection weights, while the dotted lines represent error messages (the thicker the dotted line, the bigger the error message). (a) shows an early stage of simultaneous learning of the two tasks, while (b, c, d and e) show four succeeding stages of sequential learning (for more details, see text). Modified from Calabretta *et al.* (2008).

An important question is the following: is there a way for nonmodular networks to acquire both tasks equally well, thereby avoiding the problem of neural interference?

One solution is to opportunely change the learning algorithm; for example, by modifying the equations of the back-propagation algorithm, as we did in Calabretta *et al.* (2008). This solution is efficient but also tied to the specific algorithm used; moreover, it does not eliminate neural interference in nonmodular networks, but eliminates its negative effects.

Another solution is to use sequential learning: the two tasks are learned sequentially, but, in order to avoid catastrophic forgetting (French, 1999), the teaching input for first task is continued to be provided also when the network starts learning the second task. By collaborating with Frank Keil, a psychologist at Yale University, we found that, surprisingly, it is better to learn the more difficult 'What' task first and then the easier 'Where' task rather than the other way round (Calabretta *et al.*, 2002). We performed different analyses in order to explain this counterintuitive pattern of results (Calabretta *et al.*, 2008).

In a first phase, the network learns the first task alone and, therefore, in absence of the neural interference between the two tasks (Figure 6.4b). In a second phase, when the first task is learned, the second task is added, but neural interference between the two tasks is still not present because the first task has already been learned and the error messages coming from it (and that could cause interference with the learning of the second task) are therefore very small (if not nonexistent; Figure 6.4c). When the learning algorithm starts to modify the connection weights for solving the second task, the first task is damaged and therefore begins again to send error messages, but in any case the entity of the error for the first task is not as big as that for the second task (Figure 6.4d). Moreover, as the first task is damaged, the second task is improved; as a consequence there is never the simultaneous presence of two big errors (Figure 6.4e).

If, for example, we let the neural network first learn the 'Where' task and then the 'What' task is added, one can see that the error is never contemporarily big in the two tasks. The result is more clear if the network learns first the more difficult 'What' task, and then the 'Where' task is added. In this kind of sequential learning a nonmodular network reaches a performance comparable to that of a modular network learning both tasks at the same time. The reason for these counterintuitive results is that learning the 'What' task presents points of particular difficulty, which can become impossible to overcome if the network has to continue to learn the 'Where' task, as happens when the 'What' task is learned as the second task; instead, learning the 'What' task as the first task allows the network to cross the points of particular difficulty without any interference, and then to learn the 'Where' task as the second task.

6.3.3 Generalisation in modular and nonmodular neural networks

From the collaboration with Frank Keil on sequential learning in 2001 sprang the idea of testing modular and nonmodular architectures for generalisation of the 'What' and 'Where' tasks. We did two kinds of simulation (Calabretta *et al.*, 2004): in the first one, we tested neural networks in a test of generalisation both for the 'What' and the 'Where' task by using the usual model of Rueckl *et al.* (1989); in the second one, we used ecological neural networks, that is, neural networks that have a body and that live and learn in an physical environment (Parisi *et al.*, 1990).

6.3.3.1 Generalisation in the 'What' and 'Where' tasks

In the Rueckl *et al.*(1989) simulations modular networks trained using the back-propagation procedure learn both tasks, whereas nonmodular networks learn the easier 'Where' subtask, but are unable to learn the more difficult 'What' subtask. When asked where a presented object is, nonmodular networks give a correct answer, but they make errors when asked what the object is.

We have carried out some new simulations in which by using the back-propagation algorithm we tested modular and nonmodular neural networks in a test of generalisation both for the 'What' and the 'Where' tasks. In order to do that, for the training test, we used 63 out of the 81 patterns used by Rueckl *et al.* (these patterns were chosen in such a way that each object and each position was presented seven times to the networks); for the generalisation task, we used the remaining 18 patterns.

Our initial hypothesis was that modular networks are also better than nonmodular ones with respect to generalisation, which is a critical capacity for organisms (see Ghirlanda & Enquist, 2003).

Surprisingly, both the modular and the nonmodular networks were unable to perform the 'What' generalisation task. We instead discovered a good ability on the part of modular and nonmodular networks to generalise on the 'Where' task.

We confirmed and strengthened this result by using a 6×6 retina (instead of the 5×5 retina), in which 16 different objects (the 9 objects used by Rueckl *et al.* (1989), plus 7 new objects) can appear in 16 different positions. For the training test we used 192 patterns (these patterns were chosen in such a way that each object and each position were presented 12 times to the networks); for the generalisation task we used the remaining 64 patterns.

It was confirmed (results not published) that both modular and nonmodular networks were able to perform the 'Where' generalisation task, and were unable to perform the 'What' generalisation task.

The failure to generalise in the 'What' task surprised us, in that the ability of the neural networks to generalise is well known (see Hummel, 1995).

6.3.3.2 Generalisation in the ecological task

We wondered whether the difficulty on the part of the neural networks to generalise the 'What' task depends on some parameters of the simulation or on the specific nature of the task. In order to answer this question we carried out the following ecological simulation.

An artificial organism has and can move a single eye and a single 2-segment arm in a bidimensional world. In front of the organism there is a screen, which is composed by 6×2 = 12 cells, and 5 buttons. At any given time, one out of four possible objects different in shape (Figure 6.5) appears in the screen. The organism with its arm and its visual field (the grey rectangle), and the screen (the black rectangle) are illustrated in Figure 6.6.

When an object appears in the screen, it can be located either in the left or in the central or in the right portion of the screen. Each portion is constituted by a small grid of 2×2 = 4 cells. At this moment, the organism's eye is in the central position and, as a consequence, the organism's visual field (i.e. the 'retina') corresponds exactly to the screen. Then, the organism can move the eye horizontally to the left or to the right, and as a result, there is a displacement of the visual field corresponding to two cells. In this situation, the organism sees only 2/3 of the screen (see Figure 6.6).

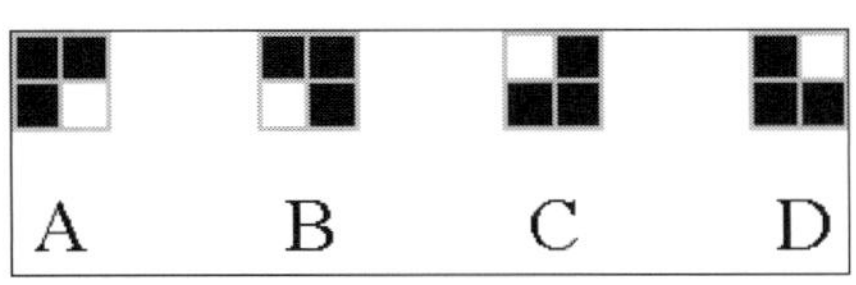

Figure 6.5. The four objects. Modified from Calabretta *et al.* (2004).

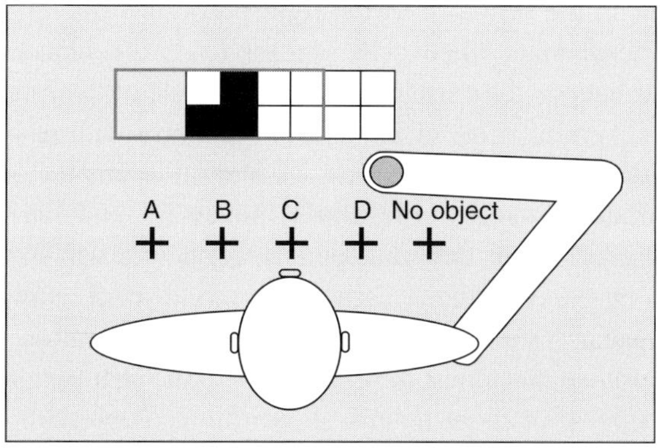

Figure 6.6. The organism with its total visual field (grey rectangle) and its 2-segment arm in front of a screen (black rectangle). The organism has moved its eye to the left so that the object is seen at the centre ('fovea') of its visual field. The five buttons are represented as crosses localised in different positions in space. Modified from Calabretta *et al.* (2004).

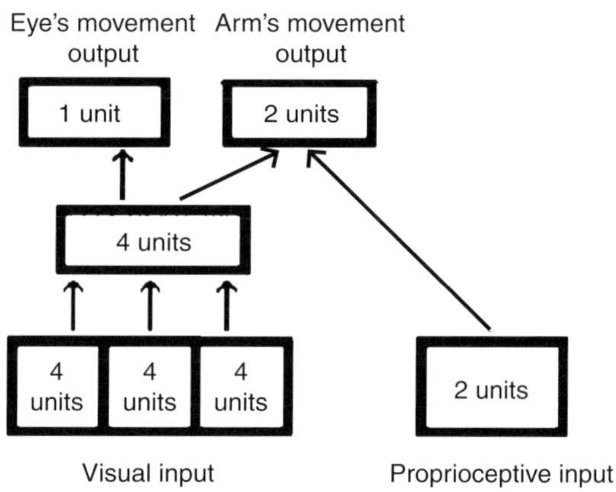

Figure 6.7. The network architecture. Modified from Calabretta *et al.* (2004).

The task for the organism is to recognise the identity of the object that it sees in its visual field by pressing with its arm the appropriate button out of the four buttons located below the screen. When no object appears in the screen, the organism has to press a fifth button.

The entire neural architecture controlling the organism's behaviour is illustrated in Figure 6.7. The input layer is composed of two distinct sets of units, one for the visual input and one for the proprioceptive input. Each cell of the retina corresponds to one input unit and, therefore, the visual input units are 12 (4 units corresponding to the 4 cells of the left portion of the retina, 4 units corresponding to the 4 cells of the central portion, and 4 units corresponding to the right portion). Objects are encoded as patterns of filled cells appearing in the left, in the centre or in the right portion of the retina (see Figure 6.6): input units that correspond to filled cells are set to 1, the remainder to 0. The two additional input units encode the current value of the angle of the forearm with respect to the shoulder, and the current value of the angle of the arm with respect to the forearm.

The network's internal (hidden) layer includes four units, and the output layer includes two distinct sets of units, one for the arm's movement and the other of the eye's movement. The first set contains two units which encode the arm's movements, one unit for the arm and the other unit for the forearm. The second set of output units contains one unit which encodes the eye's movements. The continuous activation value of this unit is mapped into three possible outcomes: the eye moves to the left; the eye doesn't move; the eye moves to the right. (For more details about this ecological model, see Calabretta *et al.*, 2004, p. 43.)

We have studied four experimental conditions:

(1) Condition 1: the network can move the eye and is trained on all the 12 cases.
(2) Condition 2: the network can move the eye and is trained on only 10 cases.
(3) Condition 3: the network cannot move the eye and is trained on all the 12 cases.
(4) Condition 4: the network cannot move the eye and is trained on only 10 cases.

As we already said, at the beginning the eye is in the central position and is oriented in such a manner that the visual field matches the screen. Then, while in conditions 1 and 2, the eye can be moved, in conditions 3 and 4 it remains fixed. In this position, however, the organism sees all the content of the screen.

In the conditions 1 and 3, organisms experience all the 12 cases (i.e. 4 objects multiplied for 3 positions), while in conditions 2 and 4 they experience only 10 cases and the remaining 2 cases are used for a generalisation test.

This generalisation task is simpler than the preceding one: there are only 10 possible inputs for training and 2 inputs for generalisation, but the proportion between training and generalisation inputs is similar to that of the task used in the other generalisation simulation (about 20%). We also tested results for statistical significance. We used a genetic algorithm for selecting the appropriate connection weights of the fixed neural network architecture.

We chose to adopt a simpler task because in ecological models a genetic algorithm works better than back-propagation used in the preceding simulation for finding the network weights, and a genetic algorithm functions better with smaller networks. In any case, we repeated the simulation using the back-propagation with this easier task and obtained the same result: to generalise the 'What' task is very difficult for both modular and nonmodular networks.

We obtained the following results:

(1) From a performance point of view, there is no difference in fitness between an evolved organism with a mobile eye and an organism with a fixed eye. In other words, the capacity to move the eye does not provide advantages in fitness. This applies both in the comparison between conditions 1 and 3 and in the comparison between conditions 2 and 4.

(2) If one pays attention to the evolved organisms' behaviour in conditions 1 and 2, one sees that when an object is presented to the organism most organisms first move their eyes in order to have the object in the central portion of the perceptual field and then they press one of the buttons.
(3) Whereas there is no difference in fitness between organisms evolved in condition 2 and organisms evolved in condition 4, there is a big difference in terms of generalisation.

The organisms that are not allowed to move their eyes are able to recognise visual patterns that they saw during the evolutionary process, but are not able to recognise visual patterns that they did not see during the evolutionary process. In other words, they are not able to generalise (Figure 6.8, right).

In contrast, organisms that can move their eye use this ability and, as a consequence, are able to recognise both the 10 visual patterns that they saw during the evolutionary process and the 2 visual patterns that they did not see during the evolutionary process. In other words, they are also able to generalise (Figure 6.8, left), by adopting the attentional strategy of moving their eye. More specifically, successful organisms adopt a 2-step spontaneous winning strategy that allows them to recognise the identity of objects by means of the specialisation of different portions of the retina: the simple information concerning the object's presence in the periphery causes the eye to move and to bring the object in front of the central portion of the retina, the fovea, where the object can be recognised.

The central portion of the retina evolved the ability to identify objects, like a fovea, while the two peripheral portions of the retina evolved the ability to detect the presence of the object and to trigger the eye movements. These simulation results can contribute to explaining why the central portion of the retina (fovea) in real organisms evolved to have densely packed cones and therefore more computational resources than the periphery of the retina.

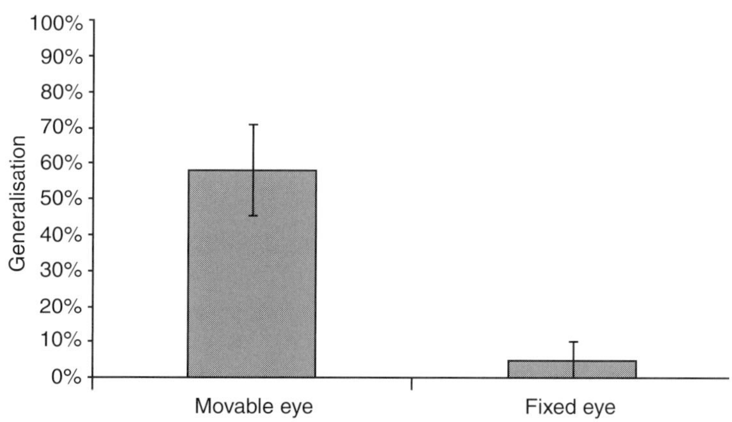

Figure 6.8. Percentage of correct responses in the generalisation test in the condition with the eye fixed and in the condition with the eye movable. Modified from Calabretta *et al.* (2004).

6.4 Conclusion: possible improvements of our study model and new simulations

In 1982, Ungerleider & Mishkin hypothesised the existence in the primate brain of two modular cortical visual systems, involved in the analysis of the visual array and specifically in the individuation of spatial localisation and object identification: respectively, the dorsal stream ('Where' task) and the ventral streams ('What' task).

Subsequent research confirmed that the two visual systems in the cortex are anatomically separate, but Milner & Goodale (1995, p. 24) considered it 'unlikely that they evolved simply to handle different aspects of stimulus array' and proposed that the anatomical distinction corresponds to the distinction between perceptual representation and visuomotor control. According to them, 'the reason there are two cortical pathways is that each must transform incoming visual information for different purposes': the ventral stream transforms visual inputs in perceptual representations that embody the enduring features of objects (representing what an object is: 'What' task, as in the Urgeleider & Mishkin position), and the dorsal stream transforms visual inputs into the appropriate moment-to-moment egocentric coordinates for the programming and on-line control of actions such as object reaching and grasping (representing how the organism has to act on an object: 'How' task).

Milner & Goodale (1995, p. 41) claim that 'separate processing "modules" have evolved in non-human vertebrates to mediate the different visually guided behaviours that such organisms exhibit' (in the frog, for example, there are independent modules controlling visually elicited prey catching and visually guided behaviours during locomotion); and that 'it is probably not until the evolution of primates, at a late stage of phylogenetic history, that we see the arrival on the scene of fully developed mechanisms for perceptual representation' (Milner & Goodale, 1995, p. 65).

Recently, Milner & Goodale (2006) published the second edition of their influential book with the addition of an epilogue chapter, in which they summarise developments in neuroscience and psychology made thanks to functional neuroimaging and confirm their model of cortical visual processing based on the distinction between vision for perception (ventral stream) and vision for action (dorsal stream). In Milner & Goodale (2008), the same authors clarify and refine the formulation of their model, answering the challenges posed by a number of authors (e.g. Glover, 2004). (Jacob & Jeannerod, 2007, call the two visual systems 'semantic' and 'pragmatic', instead of 'perception' and 'action'.)

In order to study the evolution of brain structural modularity, we took inspiration from the computational work on the 'What' and 'Where' task by Rueckl *et al.* (1989). If we wanted to reproduce in simulation the evolution of the dorsal and ventral streams according to the more recent and prevalent position that considers these streams to be involved in the 'What' and 'How' tasks instead of the 'What' and 'Where' tasks (Milner & Goodale, 1995, 2008), we should change our simulative model and build an ecological task (i.e. a task in which the neural network controls movements of a body situated in a physical environment, and in which body movements determine partly the inputs that reach the neural network from the environment).

We should imagine an artificial organism provided with an arm that can perform not only two, but several different complex tasks (e.g. object reaching and grasping) regarding 9 objects in 9 different spatial locations. The arm position coordinates would vary as they change during the accomplishment of the tasks. The action the organism is supposed to perform will be different for each object, otherwise it would be an ecological version of the 'Where' task. In this new scenario, we should repeat the many different simulations we performed in the case of the 'What' and 'Where' tasks, starting from that in which the genetic algorithm evolves both the network architecture and connection weights.

Coming back to the simulations realised in the past ten years, we can first of all say that they concern in general the evolution of brain modularity, and that the 'What' and 'Where' tasks are an example of tasks interfering each other, among the many such tasks. On the other hand, it is important to notice that the existence of dorsal and ventral streams for the 'What' and 'Where' tasks has been hypothesised recently for the processing of auditory patterns and auditory spatial information respectively (Rauschecker, 1998).

In general, it can be said that almost all the results of our ten years of research still remain valid: the identification of the genetic population mechanism of genetic interference; the fact that, for several reasons, evolution alone is not able to evolve both the network architecture and the connection weights; the difficulty for both modular and nonmodular networks of generalising the 'What' task, and the hypothesis of an ecological solution to this problem based on the movement of the eye and on the evolution of a fovea; the importance of some types of sequential learning (first, the more difficult 'What' task, then, the easier 'Where' task without discontinuing the learning of the 'What' task), which allows nonmodular architecture to achieve similar performance to modular architecture.

Another important result of our studies is that if we use the genetic algorithm in order to evolve the connection weights of a fixed modular architecture, sexual reproduction mitigates (but does not eliminate) the problem of genetic interference by recombining portions of genotypes and finding new genotypes that incorporate only favourable mutations or only unfavourable mutations. This can explain the evolutionary prevalence of sexual reproduction in complex organisms, and the observed fact that sexual populations tend to have higher mutation rates than asexual populations (Maynard Smith, 1978).

What do all these results mean with regard to the issue of the evolution of brain modularity?

As we have already said, a very important and surprising discovery of our research (Di Ferdinando *et al.*, 2000, 2001; Calabretta *et al.*, 2003) is that brain structure can evolve as the result of the adaptive cooperation between evolution, which takes care of the architecture, and learning, which takes care of the connection weights (cited among the models about the origin of modularity in the journal *Nature Reviews Genetics* by Wagner *et al.*, 2007, Box 2, p. 927). This discovery can contribute to overcoming the old dichotomy between nativism and empiricism (see Calabretta & Parisi, 2005).

If one would like to take into consideration the hypothesis that structural modularity evolved not for its evolutionary advantage, then a possible hypothesis is the same proposed regarding the evolution of functional modularity (Calabretta *et al.*, 1997, 1998a, 1998b, 2000): brain modularity as a dynamic side effect of genetic duplication. This hypothesis regarding the evolution of module functional specialisation has recently found strong confirmation in analyses of module duplication in *Saccharomyces cerevisiae* protein complexes, made at the Laboratory of Molecular Biology, Structural Studies Division at Cambridge University by José Pereira-Leal and Sarah Teichmann, who had been inspired by our simulative results (Pereira-Leal & Teichmann, 2005, pp. 552, 557; Pereira-Leal *et al.*, 2006, p. 512).

Is it plausible to hypothesise that the evolutionarily more recent ventral stream for the 'What' task originated from duplication of the more ancient dorsal stream for the 'How' task? This was a personal conjecture, proposed in 2000 and shared with colleagues (personal communication between Raffaele Calabretta and Dan McShea, 10 November 2000); only subsequently did we discover that it can be linked to the hypothesis of duplication of cortical areas proposed by Allman & Kaas (1971; see also Kaas, 1984). According to Striedter (2005), Allman & Kaas's hypothesis is plausible but with limitation, and its most appealing aspect 'is that the functional redundancy created by a duplication would allow one of the duplicates to diverge in structure and function (during subsequent generations) while the original function is maintained'.

This in fact is the stepwise mechanism we found in our simulation on genetic module duplication (Calabretta *et al.*, 2000). To verify this hypothesis applied to the evolution of dorsal and ventral streams, one should build the ecological simulations of the 'What' and 'How' task described above, with the addition of the genetic operator of duplication.

Acknowledgements

The author wishes to thank Andrea Di Ferdinando and Domenico Parisi for helpful discussions.

References

Allman, J. M. & Kaas, J. H. 1971. A representation of the visual field in the caudal third of the middle temporal gyrus of the owl monkey (*Aotus trivirgatus*). *Brain Res* **31**, 85–105.

Ancel, L. W. & Fontana, W. 2000. Plasticity, evolvability, and modularity in RNA. *J Exp Zool B: Molec Dev Evol* **288**(3), 242–283.

Calabretta, R. 2002. Connessionismo evolutivo e origine della modularità. In *Scienze della Mente* (ed. A. Borghi & T. Iachini), pp. 47–63. Il Mulino.

Calabretta, R. 2007. Genetic interference reduces the evolvability of modular and nonmodular visual neural networks. *Phil Trans R Soc B* **362**(1479), 403–410.

Calabretta, R. & Parisi, D. 2005. Evolutionary connectionism and mind/brain modularity. In *Modularity. Understanding the Development and Evolution of Complex Natural Systems* (ed. W. Callebaut & D. Rasskin-Gutman), pp. 309–330. MIT Press.

Calabretta, R. & Wagner G. P. 1997. Evoluzione della modularità in reti neurali. In *Atti del Congresso Nazionale della Sezione di Psicologia Sperimentale* (ed. G. Nigro), pp. 35–36. Associazione Italiana di Psicologia (AIP).

Calabretta, R., Wagner, G. P., Nolfi, S. & Parisi, D. 1997. *Evolutionary Mechanisms for the Origin of Modular Design in Artificial Neural Networks*. Technical Report # 51, Yale Center for Computational Ecology, Yale University. Also in *Abstract Book of the First International Conference on Complex Systems* (ed. Y. Bar-Yam). Nashua (NH), 23 September, New England Complex Systems Institute.

Calabretta, R., Nolfi, S., Parisi, D. & Wagner, G. P. 1998a. A case study of the evolution of modularity: towards a bridge between evolutionary biology, artificial life, neuro- and cognitive science. In *Artificial Life VI* (ed. C. Adami, R. Belew, H. Kitano & C. Taylor), pp. 275–284. MIT Press.

Calabretta, R., Nolfi, S., Parisi, D. & Wagner, G. P. 1998b. Emergence of functional modularity in robots. In *From Animals to Animats 5* (ed. R. Pfeifer, B. Blumberg, J.-A. Meyer & S.W. Wilson), pp. 497–504. MIT Press.

Calabretta, R., Nolfi, S., Parisi, D. & Wagner, G. P. 2000. Duplication of modules facilitates the evolution of functional specialization. *Artificial Life* **6**, 69–84. Also Technical Report # 59, Yale Center for Computational Ecology, Yale University.

Calabretta, R., Di Ferdinando, A., Keil, F. C. & Parisi, D. 2002. Sequential learning in nonmodular neural networks. Abstract in *Proceedings of the Special Workshop on Multidisciplinary Aspects of Learning*, European Society for the study of Cognitive Systems, Clichy, 17–19 January.

Calabretta, R., Di Ferdinando, A., Wagner, G. P. & Parisi, D. 2003. What does it take to evolve behaviorally complex organisms? *BioSystems* **69**, 245–262.

Calabretta, R., Di Ferdinando, A. & Parisi, D. 2004. Ecological neural networks for object recognition and generalization. *Neural Processing Letters* **19**, 37–48.

Calabretta, R., Di Ferdinando, A., Parisi, D. & Keil, F. C. 2008. How to learn multiple tasks. *Biol Theory* **3**, 1 (MIT Press). Also Technical Report 29 July 2003. Institute of Cognitive Science and Technologies, Italian National Research Council (CNR).

Cosmides, L. & Tooby, J. 1994. The evolution of domain specificity: the evolution of functional organization. In *Mapping the Mind: Domain Specificity in Cognition and Culture* (ed. L. A. Hirschfeld & S. A. Gelman). MIT Press.

Di Ferdinando, A. & Calabretta, R. 2000. Evoluzione di reti neurali modulari per compiti di "what" e "where". In *Atti del Congresso Nazionale della Sezione di Psicologia Sperimentale, Associazione Italiana di Psicologia (AIP)* (ed. B. Pinna), pp. 90–92. Carlo Delfino Editore.

Di Ferdinando, A., Calabretta, R. & Parisi, D. 2000. Evolution of modularity in a vision task. Technical Report # IP/CNR SNVA 00–2. Institute of Psychology, Italian National Research Council (CNR), Rome.

Di Ferdinando, A., Calabretta, R. & Parisi, D. 2001. Evolving modular architectures for neural networks. In *Connectionist Models of Learning, Development and Evolution* (ed. R. M. French & J. P. Sougné), pp. 253–262. Springer-Verlag.

Elman, J. L., Bates, E. A., Johnson, M. H. *et al.* 1996. *Rethinking Innateness. A Connectionist Perspective on Development*. MIT Press.

Fodor, J. A. 1983. *The Modularity of Mind*. MIT Press.

French, R. M. 1999. Catastrophic forgetting in connectionist networks. *Trends Cogn Sci* **3**(4), 128–135.

Ghirlanda, S. & Enquist, M. 2003. One century of generalization. *Anim Behav* **66**, 15–36.

Glover, S. 2004. Separate visual representations in the planning and control of action. *Behav Brain Sci* **27**, 3–78.

Gould, S. J. 1997. Evolution: the pleasures of pluralism. *N Y Rev Books*, 26 June.

Gould, S. J. & Vrba, E. S. 1982. Exaptation – a missing term in the science of form. *Paleobiology* **8**(1), 4–15.

Holland, J. H. 1992. *Adaptation in Natural and Artificial Systems: an Introductory Analysis with Applications to Biology, Control, and Artificial Intelligence*. MIT Press.

Hummel, J. E. 1995. Object recognition. In *Handbook of Brain Theory and Neural Networks* (ed. M. A. Arbib), pp. 658–660. MIT Press.

Kaas J. H. 1984. Duplication of brain maps in evolution. *Behav Brain Sci* **7**(3), 342–343.

Karmiloff-Smith, A. 2000. Why babies' brains are not Swiss army knives. In *Alas, poor Darwin* (ed. H. Rose & S. Rose), pp. 144–156. Jonathan Cape.

Keil, F. C. 1999. Nativism. In *The MIT Encyclopedia of the Cognitive Sciences* (ed. R. A. Wilson & F. C. Keil), pp. 583–586. MIT Press.

Jacob, P. & Jeannerod, M. 2007. Précis of ways of seeing, the scope and limits of visual cognition. *PSYCHE* **13**(2), April. http://psyche.cs.monash.edu.au

Marcus, G. 2006. Cognitive architecture and descent with modification. *Cognition* **101**(2), 443–465.

Maynard Smith, J. 1978. *The Evolution of Sex*. Cambridge University Press.

Milner, A. D. & Goodale, M. A. 1995. *The Visual Brain in Action.* Oxford University Press.

Milner, A. D. & Goodale, M. A. 2006. *The Visual Brain in Action.* 2nd edn. Oxford University Press.

Milner, A. D. & Goodale, M. A. 2008. Two visual systems re-viewed. *Neuropsychologia* **46**(3), 774–785.

Mitchell, M. 1996. *An Introduction to Genetic Algorithms*. MIT Press.

Ohno, S. 1970. *Evolution by Gene Duplication*. Springer Verlag.

Parisi, D., Cecconi, F. & Nolfi, S. 1990. Econets: neural networks that learn in an environment. *Network* **1**, 149–168.

Pereira-Leal, J. B. & Teichmann, S. A. 2005. Novel specificities emerge by stepwise duplication of functional modules. *Genome Research* **15**, 552–559.

Pereira-Leal, J. B., Levy, E. D. & Teichmann, S. A. 2006. The evolution and origins of modularity: lessons from protein complexes. *Phil Trans R Soc B* **361**(1467), 507–517.

Pinker, S. & Prince, A. 1988. On language and connectionism: Analysis of a parallel distributed processing model of language acquisition. *Cognition* **28**, 73–193.

Plaut D. C. & Hinton, G. E. 1987. Learning sets of filters using backpropagation. *Comp Speech Lang* **2**, 35–61.

Rauschecker J. P. 1998. Cortical processing of complex sounds. *Curr Opin Neurobiol* **8**(4), 516–521.

Rueckl, J. G., Cave, K. R. & Kosslyn, S. M. 1989. Why are "what" and "where" processed by separate cortical visual systems? A computational investigation. *J Cogn Neurosci* **1**, 171–186.

Rumelhart, D. & McClelland, J. 1986. *Parallel Distributed Processing: Explorations in the Microstructure of Cognition*. MIT Press.

Schwartz, R. 1999. Rationalism and empiricism. In *The MIT Encyclopedia of the Cognitive Sciences* (ed. R. A. Wilson & F. C. Keil), pp. 703–705. MIT Press.

Striedter, G. 2005. *Principles of Brain Evolution*. Sinauer.

Ungerleider, L. G. & Mishkin, M. 1982. Two cortical visual systems. In *The Analysis of Visual Behavior* (ed. D. J. Ingle, M. A. Goodale & R. J. W. Mansfield), pp. 549–586. MIT Press.

Wagner, G. P., Pavlicev, M. & Cheverud, J. M. 2007. The road to modularity. *Nat Rev Gen* **8**, 921–931.

Wagner, G. P., Mezey, J. & Calabretta, R. 2005. Natural selection and the origin of modules. In *Modularity. Understanding the Development and Evolution of Complex Natural Systems* (ed. W. Callebaut & D. Rasskin-Gutman), pp. 33–49. MIT Press.

Werbos, P. J. 1974. *Beyond Regression: New Tools for Prediction and Analysis in the Behavioral Sciences*. PhD thesis. Harvard University.

7

Effects of network structure on associative memory

Hiraku Oshima and Takashi Odagaki

7.1 Introduction

The brain has various functions such as memory, learning, awareness, thinking and so on. These functions are produced by the activity of neurons that are connected to each other in the brain. There are many models to reproduce the memory of the brain, and the Hopfield model is one of the most studied (Hopfield, 1982). The Hopfield model was proposed to reproduce associative memory, and it has been studied extensively by physicists because this model is similar to the Ising model of spin glasses. This model was studied circumstantially, for example, the storage capacity was analysed by the replica method (Amit, 1989; Hertz *et al.*, 1991). However, in these studies, the neural networks are completely connected, i.e. each neuron is connected to all other neurons. It was not clear how the properties of the model depend on the connections of neurons until recently (Tosh & Ruxton, 2006a, 2006b).

In recent years the study of complex networks has been paid much attention. A network consists of nodes and links. A node is a site or point on the network such as a neuron; the nodes are connected by links such as an axon or synapse of a neuron. Several characteristic network structures have been proposed, and the small-world and the scale-free networks have been studied heavily in recent years. Small-world networks have the properties that the characteristic path length is very short, and simultaneously the clustering coefficient is large (Watts & Strogatz, 1998). Here, the characteristic path length is the average path length between any two nodes on the network, and the clustering coefficient represents the degree of its connectivity among the neighbours of a given node. Scale-free networks have the distribution of degrees following a power law, where the degree is the number of neighbours of each node (Barabási & Albert, 1999). These characteristic networks have been identified in various real networks such as acquaintance networks, the World Wide Web, power grids and neural networks, etc. (Albert & Barabási, 2002; Newman, 2003). A model reproducing the small-world networks was proposed by Watts & Strogatz (WS) (1998). A model reproducing scale-free networks was proposed by Barabási & Albert (BA) (Barabási *et al.*, 1999). These models are often used to study complex networks.

Modelling Perception with Artificial Neural Networks, ed. C. R. Tosh and G. D. Ruxton. Published by Cambridge University Press.

In the brain, it is believed that the neural network also forms a small-world network (McGraw & Menzinger, 2003; Strogatz, 2003). For example, the neural network of the nematode *Caenorhabditis elegans* (*C. elegans*) is small-world. It is an important question why the neural network is a small-world network. It is possible that neural networks are small world in order to optimise some of their functions. To answer this question, it is necessary to understand first how the network structure affects the functions. Some researchers focus on the associative memory of neural networks to understand the effects of the network structure, and they have investigated either small-world or scale-free networks (Bohland & Minai, 2001; McGraw & Menzinger, 2003; Stauffer *et al.*, 2003; Kim, 2004; Morelli *et al.*, 2004; Lu *et al.*, 2006). According to these studies, the stability of stored patterns depends on the controlling parameter of the network structure in the WS model (McGraw & Menzinger, 2003) and on the clustering coefficient of the network (Kim, 2004). In these investigations, the effects of network structure on memory have been studied through the stability of the stored patterns on several networks. We review this research in Section 7.2 briefly. However, it is not clear how many patterns the system can store in various networks, and how the number of patterns that can be stored depends on the characteristic path length and the clustering coefficient. This upper limit of the number of patterns that can be stored is called the storage capacity, and this quantity is important because the system changes from the storable phase to a spin-glassy phase at this point. In addition, the effects of the network structure on the time required for the system to retrieve patterns have not been studied yet. We define this time as the retrieval time, and it is worth studying this quantity because it is not useful for the neural network to store many patterns if retrieving them takes a long time.

In this chapter, we investigate how the storage capacity and the retrieval time of the associative memory of a Hopfield-type model depend on the characteristics of the network. The connection of neurons is determined by the structure of the network, in contrast to the Hopfield model where all neurons are connected mutually. In Oshima & Odagaki (2007), we focused on the regular, small-world and random networks generated by the WS model and the neural network of the nematode *C. elegans*. *Caenorhabditis elegans* is a living organism, the neural network of which has been investigated experimentally. In addition, we also study the memory on the regular-random network and the scale-free network generated by the Dorogovtsev–Mendes–Samukhin (DMS) model (Dorogovtsev *et al.*, 2000; Boccaletti *et al.*, 2006), and compare these six types of network structures in this chapter. The regular-random network is a network in which all nodes have the same degree and all links are connected randomly. This network is useful to understand the effect of degree dispersion on the memory. In addition, the scale-free networks generated by the DMS model are useful to clarify the relation between the memory and the exponent of the degree distribution because the exponent of this model can be controlled. We show that (1) as the randomness of network is increased in the WS network, its storage capacity is enhanced; (2) although the retrieval time of WS networks does not depend on the controlling parameter r, the retrieval time of *C. elegans*'s neural network is longer than that of WS networks; (3) the storage capacity of the *C. elegans* network is smaller than that of networks generated by the WS model with the same number of neurons and average degree; (4) the storage capacity does not depend much on the degree distribution when the degree follows the Poisson distribution; (5) the storage capacity of the scale-free network is an increasing function of the exponent of the degree distribution.

The paper is organised as follows. In Section 7.2, we introduce some research related to the Hopfield model and complex networks. In Section 7.3, we explain the model that we use to understand the relation between memory and the network structure, and we show the results of a computer simulation in Section 7.4. In Section 7.5, the results are analysed. We discuss these results in Section 7.6.

7.2 Neural networks and complex networks

In order to understand the associative memory of the brain it is reasonable to study conceptually simple models. The Hopfield model is one of simplest models that reproduces the function for memory. Recently, the effects of the network structure of neurons on this model have been focused on, because the neural network in the brain is considered the complex network such as a small-world or scale-free network (McGraw & Menzinger, 2003; Strogatz, 2003; Eguíluz *et al.*, 2005). However, the original model has been studied with a fully connected or randomly connected network.

McGraw & Menzinger (2003) studied the stability of the memorised patterns on different networks. The WS model and the BA model are used as the network structure. They showed that the memorised patterns are more stable on the more randomised network or scale-free network. In the scale-free network, the higher degree nodes are able to remember the memory correctly while the lower degree nodes are prone to errors. Therefore the scale-free network achieves a very strong partial recognition.

Kim (2004) focused on the role of the clustering in order to understand the effect of network structure on the stability of stored patterns. He used the link exchange method to control the clustering coefficient. The important property of this method is that the degree of each node is not changed, and so the effect of the degree distribution is eliminated. The WS model, the BA model and the *C. elegans* network are generated initially, and control the clustering coefficient of these networks by using the link exchange method. He showed that the performance of each network monotonically decreases as the clustering coefficient becomes stronger. This result suggests the clustering coefficient of the network structure has a significant impact on the memory.

Lu *et al.* (2006) studied the performance of the Hopfield model in the WS model and the Klemm–Eguíluz (KE) model which is one of the scale-free models. In the KE model, for each of m links of the new node it is decided randomly, with probability u, whether the link connects to a random node or not. For $u=0$ it is a high clustering model, while for $u=1$ it corresponds to the BA model. They also defined the performance of the neural network by the stability of the stored patterns and found that the performance of the associative memory becomes higher as the network is more disordered, and that the larger the clustering coefficient is, the worse the performance is. In addition, they showed that there exists a constant c; the performance of random structure is better than that of scale-free topology when $p/\langle k \rangle c$, it is opposite otherwise. Here p is the number of the stored patterns and $\langle k \rangle$ is the average degree.

The associative memory at the critical point in the WS model was studied by Morelli *et al.* (2004). They defined the efficacy as the fraction of realisations in which one of the stored patterns is perfectly retrieved. The efficacy has a transition as the controlling parameter of WS model increases. At the critical point, they derived a scaling function of the efficacy with respect to the controlling parameter and the number of neurons N. In addition, they showed that the system of a regular network is always attracted to non-symmetric mixture states.

In the BA model, each newly added neuron is connected to only m other neurons. Stauffer *et al.* (2003) found that although the quality of retrieval decreases for small m, associative memory is effective for $1<<m<<N$. For $1<<m<<N$ the number of connections is mN while N^2 in the fully connected case. Thus not too many connections are neededfor relatively good retrieval in the BA network.

According to the studies explained above, the network structure has a significant effect on the associative memory. However, it is not clear yet what properties of the network structure are related with memory and what influence they have. In order to understand these questions, we studied the storage capacity and the retrieval time of the Hopfield model.

7.3 Model

7.3.1 Hopfield model on a network

We use McCulloch–Pitts neurons (McCulloch & Pitts, 1990), each element of which can have only two states, $S_i = 1$ (firing) and $S_i = -1$ (not firing). Each neuron is connected to some other neurons via synaptic connections, which are characterised by their synaptic weights J_{ij} determined by stored patterns. To represent connectivity among neurons, we use the adjacency matrix $\{a_{ij}\}$, where $a_{ij} = 1$ when there is a connection between i and j, $a_{ij} = 0$ otherwise. The adjacency matrix $\{a_{ij}\}$ uniquely determines the structure of the network.

The stored patterns in the neural network are represented by network states ξ^{μ}, where $\mu = 1, \ldots, p$, p being the number of stored patterns. These patterns are randomly generated with probabilities $\text{Prob}(\xi^{\mu}{}_{i} = \pm 1) = 1/2$. Using these patterns, the synaptic weights J_{ij} are given by the Hebbian rule

$$J_{ij} = \frac{1}{\langle k \rangle} a_{ij} \sum_{\mu=1}^{p} \xi_i^{\mu} \xi_j^{\mu} \tag{1}$$

for $i,j = 1, \ldots, N$. Here, N is the number of neurons in the network, and $\langle k \rangle$ is the average degree of the network. The degree is defined by the number of neighbours of a node. We set $J_{ij} = 0$, so that no connection within a neuron is allowed.

The state of neurons is updated asynchronously by the Monte Carlo method. A neuron is chosen at random with the probability $1/N$, and the state of the neuron is updated according to

$$S_i = \mathrm{sign}\left(\sum_{j=1}^{N} J_{ij} S_j\right). \tag{2}$$

The Monte Carlo time is advanced by 1 after N neurons have been accessed by the updating process. Although the same neuron can be selected twice in this algorithm, its probability is very small when N is large.

7.3.2 Storage capacity

To quantify pattern retrieval, we use the overlap order parameters (Amit, 1989)

$$m^{\mu} = \frac{1}{N}\sum_{i=1}^{N} S_i \xi_i^{\mu}, \tag{3}$$

where μ is the pattern number. The overlap is a parameter which represents the degree of similarity between the state of the system S and the stored pattern ξ^{μ}. For example, $m^1 = 1$ if the state is the same as the pattern ξ^1, while $m^1 = 0$ if the state is far from the pattern ξ^1.

We measure the stability of the pattern ξ^1 to check whether it is stored or not. First, we construct the structure of the network $\{a_{ij}\}$. Second, we store p random patterns in the network by setting J_{ij} according to the Hebbian rule (Eq. 1). We change p from 1 to 50 to judge the storage capacity. Third, we initialise the system to the pattern ξ^1 at $t = 0$, and evolve the state in time until the dynamics converges. We think of the convergence as the state that does not change with respect to the time. Finally, to check whether the pattern ξ^1 is stored or not, we measure the overlap m^1 between the final state and ξ^1. The pattern ξ^1 is considered as stored if the overlap m^1 is no less than 0.9. When $m^1 \geq 0.9$, the number of error bits defining the difference between the final state of network and the pattern ξ^1 is no more than 5% of all bits. It is practically reasonable that a pattern within 5% error is considered as being stored.

Repeating the above procedure several times by using different samples of networks and stored patterns, we count up the number of times that the pattern is successfully stored, and calculate the fraction of the number of trials that is successfully stored. These calculations are the same as those used for all-to-all connection networks in Forrest & Wallace (1991). We define this fraction as the retention rate. The retention rate is unity when p is small, while it is 0 when p is large. There exists a transition point where the retention rate changes from 1 to 0 sharply when p is increased. Therefore, we define the storage capacity $a_c \equiv p_c/\langle k\rangle$ as the number of patterns when the retention rate becomes 0.5.

7.3.3 *Retrieval time*

The retrieval time is defined as the time required for the system to retrieve a stored pattern. The time t is measured in Monte Carlo time steps. First, we initialise the system to the state of the pattern ξ^1 with noise. To add noise to ξ^1, we select some neurons of ξ^1 randomly and flip them, and $m^1 = (t = 0)$ is set at 0.9, 0.8, 0.7, 0.6 in this way. We do not use smaller $m^1(0)$ than these values, because the system tends to go to a stable point different from the pattern ξ^1, when $m^1(0)$ is small. Next, we measure $m^1(t)$ at each time step until the dynamics converges. This convergence time is denoted as t_f. After repeating this procedure several times for other samples, we scale the overlap $m^1(t)$ and calculate its average according to

$$f(t) = \left\langle \frac{m^1(t) - m^1(t_f)}{m^1(0) - m^1(t_f)} \right\rangle, \tag{4}$$

where $f(0) = 1$ and $f(t > t_f) = 0$. We define the retrieval time as the reciprocal of the slope of $-\ln f(t)$, i.e.

$$\frac{1}{\tau} = -\frac{d \ln f(t)}{dt} \tag{5}$$

because $f(t)$ decays exponentially.

7.4 Results

First, we compared networks which are generated by the three network models in order to understand the effects of network structure. Next, we compared the neural network of the nematode *C. elegans* (Oshio *et al.*, 2003) and WS networks in order to clarify differences between the real and model networks.

7.4.1 *Results for the network models*

We carried out a Monte Carlo simulation of the Hopfield model for several networks with $N = 10\,000$ and average degree $\langle k \rangle = 100$. First, we constructed a network by using the network models explained below. Second, we stored p random patterns, from 1 up to 50, by setting J_{ij} according to Eq. 1. Third, we evolved the system in time and measured the overlap m^1 after the dynamics converged. We repeated this procedure 1000 times using different samples of networks and patterns with the same p.

Three network models were investigated. The first one is the WS model that can generate the regular, small-world and random network (Watts & Strogatz, 1998). A network is the rewiring with probability r, i.e. the links on the network are selected by probability r, and they are rewired randomly. The network is regular, random, and small world when $r = 0$, $r = 1$, $0 < r << 1$, respectively. The second model is the regular-random network. The regular-random network is a random network in which each link is randomly connected and each neuron has the same degree. We generated this network by connecting two neurons selected randomly under the condition that each neuron has the same number of neighbours. The third model is the scale-free networks generated by the Dorogovtsev–Mendes–Samukhin (DMS) model (Dorogovtsev *et al.*, 2000; Boccaletti *et al.*, 2006). In this model a new node j with m links is added to the network in the same way of the BA model, while the probability that a link will connect j to an existing node i is set as

$$\Pi_{j\to i} = \frac{k_i + k_0}{\sum_l (k_l + k_0)}. \tag{6}$$

For this attachment probability one gets a power law degree distribution with an exponent $\gamma = 3 + k_0/m$. Especially, for $k_0 = 0$ one gets the BA model. We set $m = 50$ and investigated them for $k_0 = 0,100$ and 200, i.e. $\gamma = 3$, 5 and 7 respectively.

Figure 7.1 shows the retention rate as a function of $a = p/\langle k\rangle$. The storage capacity a_c is defined by a when the retention rate becomes 0.5. From Figure 7.1, we find

$$a_c = \begin{cases} 0.207 & (\text{for regular of } WS,\ r = 0) \\ 0.234 & (\text{for small-world of WS}, r = 0.1) \\ 0.342 & (\text{for regular of WS}, r = 1) \\ 0.269 & (\text{for DMS}, k_0 = 0) \\ 0.284 & (\text{for DMS}, k_0 = 100) \\ 0.288 & (\text{for DMS}, k_0 = 200) \\ 0.343 & (\text{for regular-random}). \end{cases} \tag{7}$$

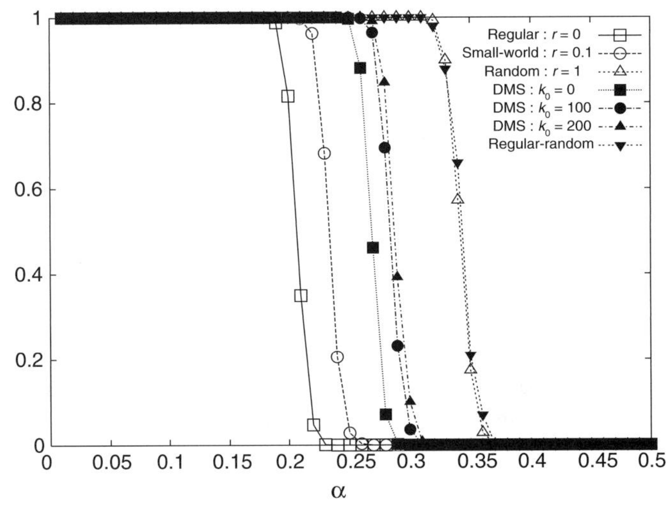

Figure 7.1. Retention rate as a function of $a = p/\langle k\rangle$. The squares, circles and triangles represent regular ($r = 0$), small-world ($r = 0.1$), and random networks ($r = 1$) in the WS model, respectively. The filled squares, filled circles and filled triangles represent $k_0 = 0$, 100 and 200 in the DMS model, respectively. The inverted and filled triangles are the results of the regular-random network. All networks have $N = 10\,000$ and $\langle k\rangle = 100$.

We estimated these values by fitting the lines with $\{1+\tanh[a(\alpha_c-\alpha)]\}/2$, where a and α_c are fitting parameters. We find that the storage capacity is an increasing function of r in the WS model or k_0 in the DMS model. In addition, the regular-random network has approximately the same storage capacity as the random network ($r=1$) in the WS model.

Next, we measured the retrieval time of the WS model, which is shown in Figure 7.2 for four different initial states, i.e., $m^1(t=0)=0.9, 0.8, 0.7, 0.6$, where three patterns were stored. These data were calculated using 10000 different samples of networks and patterns. We find that the retrieval time of all networks is approximately one Monte Carlo step because the slope of $\ln f(t)$ is unity. Therefore, it does not strongly depend on network structure. We confirmed that this result is independent of N and $\langle k \rangle$.

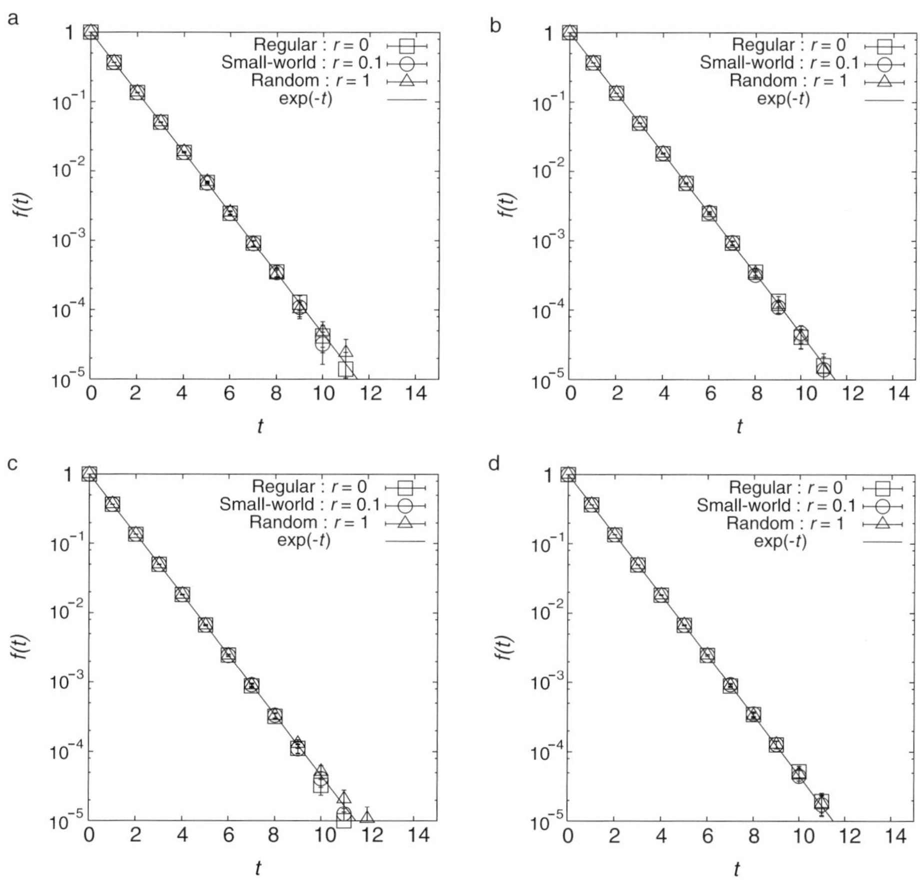

Figure 7.2. Process of retrieval in different networks, regular (squares), small-world (circles), and random networks (triangles), when $p=3$. All networks have $N=1000$ and $\langle k \rangle = 100$. The initial state of the system is $m^1(0)=$ (a) 0.9, (b) 0.8, (c) 0.7 and (d) 0.6.

Table 7.1. The characteristic path length L and clustering coefficient C of the *C. elegans*, the random ($r=1$) and the small-world network ($r=0.3$) generated by the WS model. *Caenorhabditis elegans*'s network has about the same L as the random network, while it has larger C. The small-world network with $r=0.3$ has roughly the same L and C as *C. elegans*'s network.

Network	L	C	$<k>$
C. elegans	2.65	0.245	≅14
Random ($r=1$)	2.38	0.0525	14
Small-world ($r=0.3$)	2.54	0.261	14

7.4.2 Results for C. elegans

Caenorhabditis elegans is a living organism used frequently in biological experiments. Although its neural network consists of only 302 neurons, it has abilities of learning and memory. The connections in this network have been investigated experimentally, and these data are available from the database on the web (Oshio *et al.*, 2003).

We reconstructed *C. elegans*'s network by using the connection data from the database (Oshio *et al.*, 2003). A connection between two neurons by either a synapse or a gap junction is assumed to be an undirected link. If there are two or more connections between two neurons, we regarded them as one connection. *C. elegans*'s network is known to be a small-world network with $N=251$ and $\langle k\rangle \cong 14$, because it has about the same L as but larger C than the random network (Table 7.1).

We carried out the same procedure as in the previous section for the neural network of *C. elegans*. We compared its storage capacity with that of three WS networks $r=0,0.3,1$, which have the same number of neurons ($N=251$) and approximately the same average degree ($\langle k\rangle = 14$). The WS network with $r=0.3$ has roughly the same characteristic path length L and clustering coefficient C as *C. elegans*'s network. Figure 7.3 shows the retention rate as a function of a. The storage capacity of *C. elegans*'s network is the smallest among the networks examined,

$$a_c = \begin{cases} 0.395 & (\text{for regular}, r=0) \\ 0.431 & (\text{for small-world}, \ r=0.3) \\ 0.460 & (\text{for random}, \ r=1) \\ 0.329 & (\text{for } C.elegans). \end{cases} \tag{8}$$

Here, the retention rate of *C. elegans* for $a=0.43$, i.e. $p=6$, is larger than that for $p=5$. This comes from the fact that there are neurons having only one neighbour in the neural network of *C. elegans*. For $p=6$, there are some links with $J_{ij}=0$ because the number of stored patterns is even. If a link connecting to the neuron i with $k_i=1$ has J_{ij}, it is the same that the neuron i has no neighbour. The state of the neuron having no neighbour is stable

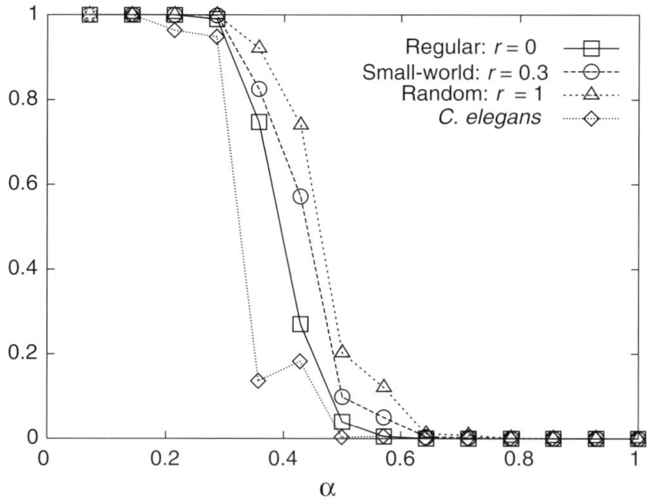

Figure 7.3. Retention rate in *C. elegans* neural network (diamonds). Data for three WS networks are also shown for comparison. These networks are regular (squares), small world (circles) and random (triangles), which have the rewiring probability $r = 0$, 0.3 and 1, respectively. All networks have $N = 251$ and $\langle k \rangle \cong 14$.

because its neuron interacts with no other neurons. This case occurs frequently for $p = 6$ because the probability that the synaptic weight of a link connecting to a neuron with $k = 1$ is equal to 0, is ${}_6C_3/2^6 \cong 0.312$ for $p = 6$. Therefore the stored patterns for $p = 6$ are more stable than that for $p = 5$ relatively.

The retrieval time is shown in Figure 7.4 for different initial states $m^1(t = 0) = 0.9$, $0.8, 0.7, 0.6$, when one pattern was stored. The retrieval time of WS networks is one Monte Carlo step, as in the previous section. The neural network of *C. elegans* retrieved a stored pattern more slowly than other networks; its retrieval time is approximately 1.1 Monte Carlo steps. We conclude that the retrieval time of WS networks does not depend on the structure of the network, but the retrieval time of *C. elegans* is longer than that of WS networks.

7.5 Analysis

We analysed the relations between the storage capacity and network properties in order to understand what characteristics of the structure of the network determine the storage capacity. We calculated the storage capacities of the neural networks generated by the WS model with several probabilities, and plotted them against the rewiring probability r (Figure 7.5), the characteristic path length L (Figure 7.6) and the clustering coefficient C of each network (Figure 7.7). Each set of data is fitted by a curve as a guide for the eyes. According to Figure 7.6, the characteristic path length may have a large effect on

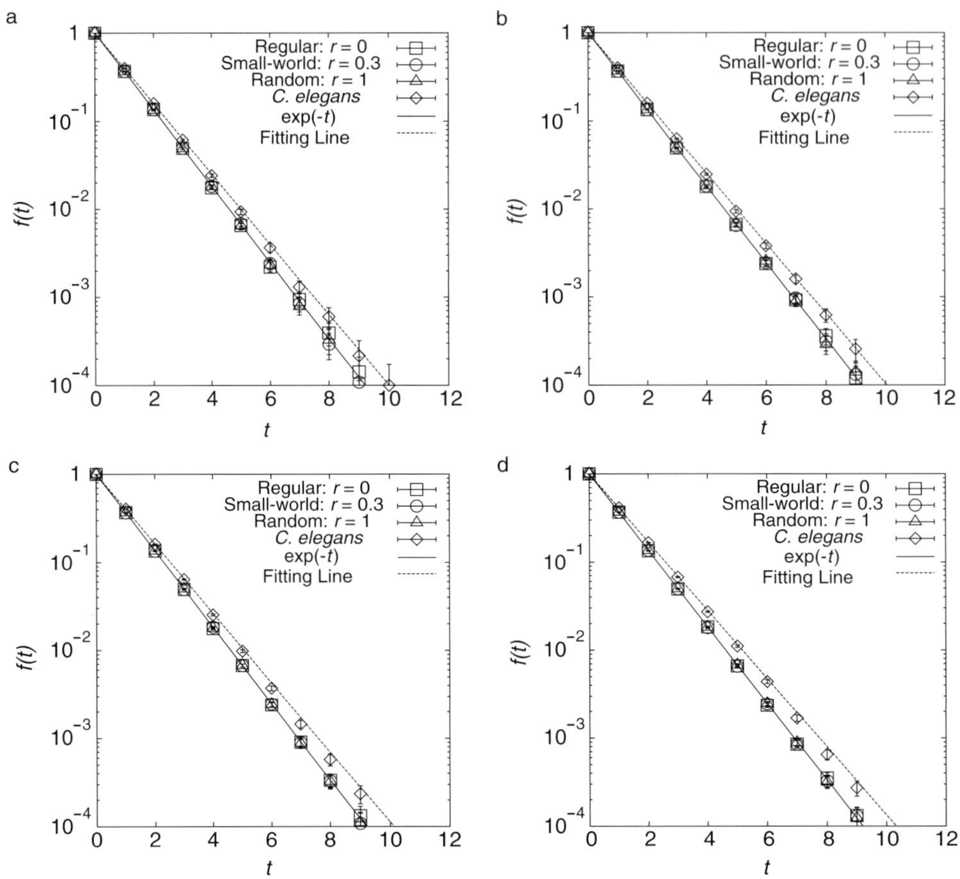

Figure 7.4. Process of retrieval in different networks, i.e. regular (squares), small world (circles), random (triangles) and *C. elegans* (diamonds), when $p = 1$. All networks have $N = 251$ and $\langle k \rangle \cong 14$. The data of *C. elegans* are fitted with a line of slope approximately 1/1.1. The initial state of the system is $m^1(0) =$ (a) 0.9, (b) 0.8, (c) 0.7 and (d) 0.6.

the storage capacity because the storage capacity changes significantly in the area of $L \approx 2$. The data are fitted well by $\alpha_c(L) = a_L(L - b_L)^{-c_L} + d_L$. Here the fitting parameters are $a_L = 0.01$, b_L $a_L = 0.01$, $b_L = 2$, $c_L = 2.5$, $d_L = 0.205$. On the other hand, according to Figure 7.7, the storage capacity depends on the clustering coefficient linearly. The data are fitted well by $\alpha_c(C) = a_C C + b_C$, and the fitting parameters are $a_C = -0.186$, $b_C = 0.341$. Therefore the storage capacity may depend on both L and C. In Figure 7.5, α_c is saturated at large r because C is saturated at large r. The data are fitted well by $\alpha_c(r) = a_r \tanh(b_r r) + c_r$, and the fitting parameters are $a_r = 0.143$, $b_r = 2$, $c_r = 0.208$.

Next, in order to investigate the dependence of the degree distribution of neural networks, we compared the storage capacities of the random network in the WS model and the regular-random network. In the regular-random network each neuron has the same

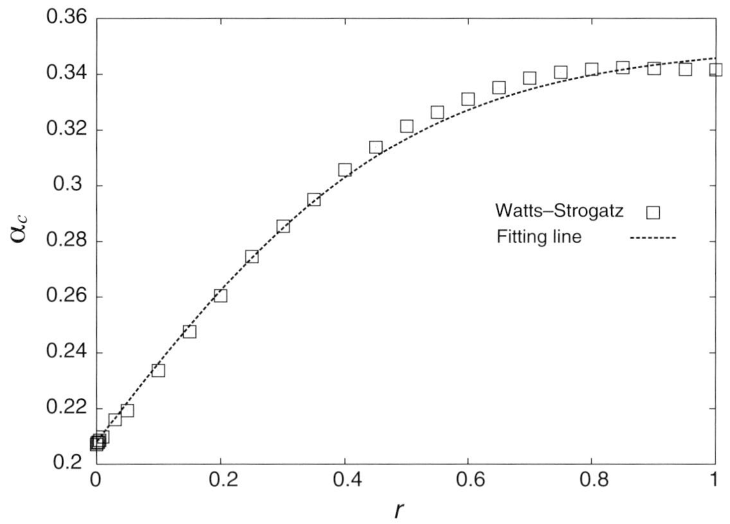

Figure 7.5. Storage capacity as a function of the rewiring probability r. The parameters are N=10000 and $\langle k \rangle$=100. The curve is a guide for the eyes.

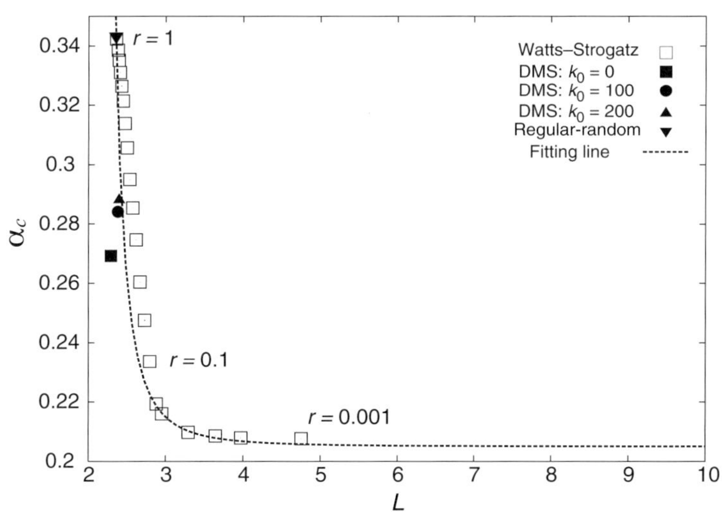

Figure 7.6. Storage capacity as a function of the characteristic path length of different networks: that generated by the Watts–Strogatz model (squares), the regular-random network (inverted and filled triangle) and generated by the DMS model (filled square: k_0=0, filled circle: k_0=100, filled triangle: k_0=200). The parameters are N=10000 and $\langle k \rangle$=100. The curve is a guide for the eyes.

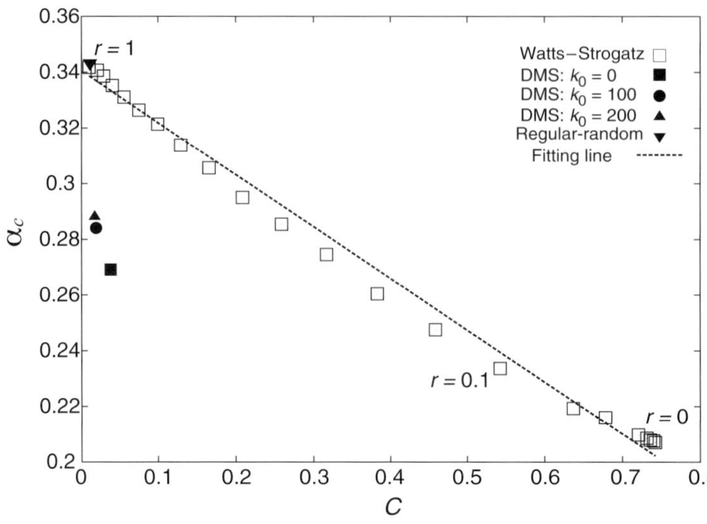

Figure 7.7. Storage capacity as a function of the clustering coefficient of different networks: that generated by the Watts–Strogatz model (squares), the regular-random network (inverted and filled triangle) and generated by the DMS model (filled square: $k_0 = 0$, filled circle: $k_0 = 100$, filled triangle: $k_0 = 200$). The parameters are $N = 10\,000$ and $\langle k \rangle = 100$. The curve is a guide for the eyes.

degree. On the other hand, the degree distribution of the random network generated by the WS model with $r = 1$ obeys the Poisson distribution approximately. By comparing the results for these networks, we can understand the effects of a neuron that has a lower or higher degree than average. The relation between the storage capacity and the network properties of this network are plotted in Figures 7.6 and 7.7. There is little difference between the storage capacities of the regular-random network and random network. We conclude that the storage capacity does not depend much on the degree distribution when the degree follows the Poisson distribution.

In addition, we focused on the dependence of a_c on L and C in the DMS model. a_c increases as L increases in the DMS model, while a_c increases as L decreases in the WS model (Figure 7.6). On the other hand, a_c increases as C decreases in both the DMS and WS model (Figure 7.7). These results suggest that the effect of the characteristic path length on the storage capacity is small, while that of the clustering coefficient is relatively large. This supports Kim's (2004) results that the clustering coefficient has a significant effect on the memory.

7.6 Discussion

We investigated the storage capacity and the retrieval time of the Hopfield model on several different networks. We showed that the storage capacity depends on the randomness of networks in the WS network. We found that the retrieval time of WS networks does not depend on the network structure, but the retrieval time of *C. elegans*'s neural network is longer than that of WS networks. According to comparisons between the random network generated by the WS model and the regular-random network, the storage capacity does not depend much on the degree distribution when the degree follows the Poisson distribution. The results of the DMS network showed that the storage capacity is an increasing function of the exponent of the degree distribution. On the other hand, they also suggest that the effect of the characteristic path length on the storage capacity is small, while that of the clustering coefficient is relatively large.

The storage capacity may depend on the characteristic path length, the clustering coefficient and the exponent of the degree distribution. However, it is not clear which effect is the largest because these properties cannot be changed independently in the network models. The effect of C was studied in Kim (2004), while the effect of L still has not been studied independently. Although we studied the effect of C and L in Section 7.5, the effect of the exponent of the degree distribution is not clear. The heterogeneity of the network structure has not been studied independently by our study because in the DMS model the exponent is changed if k_0 is changed. It is a topic for future research to clarify the effects of network properties independently. Furthermore, the storage capacity of the *C. elegans* network is smaller than that of the network generated by the WS model. We consider that the *C. elegans* network has a small storage capacity because it has many neurons whose degree is 1. The error often occurs on a neuron with $k = 1$, because its state becomes different from a stored pattern whenever its neighbourhood goes to the different state from stored.

The neural network in the brain is thought to be small-world. However, the results in this paper show that more randomised networks have more storage capacity, and the real neural network of *C. elegans* has a longer retrieval time than WS networks. Our work implies that the neural network of the brain is not optimised only for maximising the storage capacity and minimising the retrieval time. The reason why the brain's network is not random but small world may be that making long-distance connections, as in a random network, costs more energy or proteins than making short-distance connections (McGraw & Menzinger, 2003). The question why the neural network is small world is still an open one. The other possibility is that the Hopfield model does not explain the memory of the brain well. Then we will need a better model than the Hopfield model in order to reproduce the memory of the brain.

References

Albert, R. & Barabási, A. L. 2002. Statistical mechanics of complex networks. *Rev Mod Phys* **74**, 47–97.

Amit, D. J. 1989. *Modeling Brain Function: The World of Attractor Neural Networks.* Cambridge University Press.

Barabási, A. L. & Albert, R. 1999. Emergence of scaling in random networks. *Science* **286**, 509–512.

Barabási, A. L., Albert, R. & Jeong, H. 1999. Mean-field theory for scale-free random networks. *Physica A* **272**, 173–187.

Boccaletti, S., Latora, V., Moreno, Y., Chavez, M. & Hwang, D. 2006. Complex networks: structure and dynamics. *Phys Rep* **424**, 175–308.

Bohland, J. W. & Minai, A. A. 2001. Efficient associative memory using small-world architecture. *Neurocomputing* **38**–40, 489–496.

Dorogovtsev, S. N., Mendes, J. F. F. & Samukhin, A. N. 2000. Structure of growing networks with preferential linking. *Phys Rev Lett* **85**, 4633–4636.

Eguíluz, V. M., Chialvo, D. R., Cecchi, G. A., Baliki, M. & Apkarian, A. V. 2005. Scale-free brain functional networks. *Phys Rev Lett* **94**, 018102.

Forrest, B. M. & Wallace, D. J. 1991. Storage capacity and learning in Ising–Spin neural networks. In *Models of Neural Networks* (ed. E. Domany, J. L. van Hemmen & K. Schulten), pp. 121–148. Springer-Verlag.

Hertz, J., Krogh, A. & Palmer, R. G. 1991. *Introduction to the Theory of Neural Computation*. Addison-Wesley Publishing Co.

Hopfield, J. J. 1982. Neural networks and physical systems with emergent collective computational abilities. *Proc Natl Acad Sci USA* **79**, 2554–2558.

Kim, B. J. 2004. Performance of networks of artificial neurons: the role of clustering. *Phys Rev E* **69**, 045101.

Lu, J., He, J., Cao, J. & Gao, Z. 2006. Topology influences performance in the associative memory neural networks. *Phys Lett A* **354**, 335–343.

McCulloch, W. S. & Pitts, W. 1990. A logical calculus of the ideas imminent in nervous activity. *Bull Math Biol* **52**, 99–115.

McGraw, P. N. & Menzinger, M. 2003. Topology and computational performance of attractor neural networks. *Phys Rev E* **68**, 047102.

Morelli, L. G., Abramson, G. & Kuperman, M. N. 2004. Associative memory on a small-world neural network. *Eur Phys J B* **38**, 495–500.

Newman, M. E. J. 2003. The structure and function of complex networks. *SIAM Rev* **45**, 167–256.

Oshima, H. & Odagaki, T. 2007. Storage capacity and retrieval time of small-world neural networks. *Phys Rev E* **76**, 036114.

Oshio, K., Iwasaki, Y., Morita, S. *et al.* 2003. *Database of Synaptic Connectivity of* C. elegans *for Computation*. Technical Report of CCeP, Keio Future No. 3. Keio University. http://www.bio.keio.ac.jp/ccep/

Stauffer, D., Aharony, A., Costa, L. D. & Adler, J. 2003. Efficient Hopfield pattern recognition on a scale-free neural network. *Eur Phys J B* **32**, 395–399.

Strogatz, S. H. 2003. *Sync: The Emerging Science of Spontaneous Order*. Hyperion.

Tosh, C. R. & Ruxton, G. D. 2006a. Artificial neural network properties associated with wiring patterns in the visual projections of vertebrates and arthropods. *Am Natl* **168**, E38–E52.

Tosh, C. R. & Ruxton, G. D. 2006b. Introduction. The use of artificial neural networks to study perception in animals. *Phil Trans R Soc B* **362**, 337–338.

Watts, D. J. & Strogatz, S. H. 1998. Collective dynamics of ‘small-world’ networks. *Nature* **393**, 440–442.

8

Neural networks and neuro-oncology: the complex interplay between brain tumour, epilepsy and cognition

L. Douw, C. J. Stam, M. Klein, J. J. Heimans and J. C. Reijneveld

8.1 Introduction

The human brain is by far the most complex network known to man. Neuroscience has for a long time focused on a reductionistic approach when studying the brain, in part precisely because of its daunting complexity. Although highly important insights have been obtained by using a localisational method, this type of research has failed to elucidate the elaborate mechanisms involved in higher brain functioning and perception. As a consequence, an increasing body of research regarding the brain's functional status has become founded on modern network theory. In this subdivision of mathematics and physics, emphasis is placed on the manner in which several parts of the brain interact, instead of on which specific part of the cortex is responsible for a certain task. The first studies using networks to investigate the brain have made use of computational models and animal studies. Due to the great research advances in recent years, network theory is now being readily applied to the human brain. Studies are being performed in both the healthy population and several patient groups, in order to find out what constitutes a healthy versus a diseased brain (for an introduction into brain networks, see Watts & Strogatz, 1998; Bassett & Bullmore, 2006; Reijneveld *et al.*, 2007).

Brain tumours almost invariably cause highly burdensome symptoms, such as cognitive deficits and epileptic seizures. The tumour has significant impact on the brain, since it forces the non-tumoural tissue to adapt to the presence and constant expansion of a foreign entity. How this presumably whole-brain adaptation process takes place has yet to be elucidated. Hence, brain tumour patients present us with a great challenge: to clarify the underlying mechanisms accounting for the changes in patients' functional status throughout the brain, for cognitive impairments, and epileptic seizures. When reviewing the current state of the art in neuro-oncology, there is a growing need for theory regarding the complex relations between the tumour, epilepsy and cognitive status. Currently, theoretical integration of these features is difficult. Recent studies applying network

Modelling Perception with Artificial Neural Networks, ed. C. R. Tosh and G. D. Ruxton. Published by Cambridge University Press.

theory to this population have shown that the configuration of neural networks may be essential for our understanding of these multifactorial issues in neuro-oncology.

This chapter deals with the question whether we can use functional connectivity and network analysis to characterise the complex patterns of neural and behavioural consequences of brain tumours. Network theory has proven applicable to neural networks. Especially the 'small world' phenomenon, which assumes both local segregation and overall integration in a network, seems necessary for optimal brain functioning. Abnormal connectivity and dysfunctional network topology (i.e. less small world) in neuro-oncological patients may have a detrimental effect on cognition on the one hand, and on epileptic seizures on the other hand.

In this chapter, we focus on the impact of a brain neoplasm on both the functional status of the patient and on the integrity of the entire brain as a complex system, although this is not the main subject of this book. Instead, we hope to elucidate the relation between the computational side of neural networks and their functional significance (e.g. perception). This will therefore be a highly translational chapter which is also of great interest to those working with artificial neural networks, and it will provide some insights into the application of modelling in a clinical setting.

Section 8.1 will provide a general introduction into neuro-oncology, the subspecialism of neurology that deals with the diagnosis and treatment of brain tumours. Section 8.2 consists of a concise overview of network theory, followed by a discussion of the application of network theory to the brain in Section 8.3. Section 8.4 will summarise research using functional connectivity and network analysis in brain tumour patients. Sections 8.5 and 8.6 will handle network literature regarding epilepsy and cognition, respectively, and the research that has been executed concerning these symptoms in brain tumour patients. Finally, Section 8.7 contains some concluding remarks and future prospects.

8.1.1 Brain tumours

Primary brain tumours, arising from tissue in the brain, account for 2% of the incidence of all cancer types. However, the mortality rates of CNS cancer patients are multiple times higher than those of more frequently occurring types of cancer, such as breast and lung cancer. In addition to high mortality, most patients with intracranial tumours also experience devastating symptoms of their disease, including epileptic seizures and cognitive dysfunction. The annual incidence of primary brain tumours varies from 7 to 19 cases per 100 000 people (Levine *et al.*, 1993; Tatter *et al.*, 1996). Secondary brain tumours (i.e. metastases of tumours elsewhere in the body) are much more prevalent than primary tumours, but are beyond the scope of this chapter. Primary brain tumours (on which we focus in this chapter) can roughly be divided into low-grade and high-grade tumours. The majority of primary brain tumours arise from the supporting or glial tissue of the brain and are therefore called gliomas. Low-grade gliomas grow relatively slowly, while high-grade gliomas grow much faster.

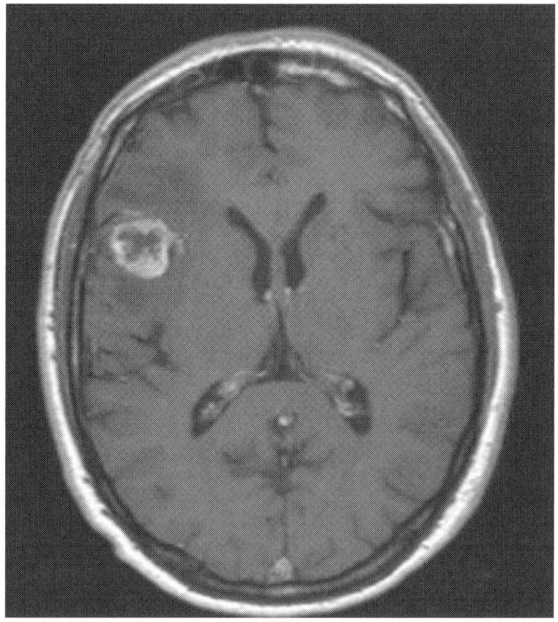

Figure 8.1. A T2-weighted MRI after administration of contrast in a patient with a right-sided tumour. Left and right side are switched due to radiological conventions.

Most high-grade glioma (HGG) patients present with increased intracranial pressure, resulting in headache or lowered consciousness and neurological deficits, such as hemiplegia (DeAngelis, 2001). The first symptom in the slowly growing low-grade glioma (LGG) is epilepsy in at least two-thirds of all patients (Wessels *et al.*, 2003). Other frequently observed symptoms in LGG patients are headache and fatigue, while neurological deficits are usually not as profound as in HGG. Both patient groups frequently suffer from cognitive deficits. The median survival rates of brain tumour patients vary from ten years in LGG to a discouraging median of one year in the case of the most malign HGG (Curran *et al.*, 1993; DeAngelis, 2001; Buckner, 2003).

8.1.2 Tumour-related epilepsy

Epilepsy is often the first sign of the presence of a brain tumour. Between 10 and 15% of adult patients presenting with epileptic seizures are diagnosed with a brain tumour as the underlying cause of the seizures (Forsgren, 1990; Bromfield, 2004). Conversely, most patients suffering from brain tumours develop epileptic seizures at some point during the course of their disease. This is the case in 85% of LGGs, while almost 50% of HGG patients will experience epileptic seizures at some point (Lote *et al.*, 1998). Generally, slow-growing tumours more frequently cause epilepsy at some point than tumours that progress very fast. Of course, this difference can in part be explained by the significantly longer survival time of LGG patients when compared with HGG. However, this does not explain why patients with LGG present with epileptic seizures as a first symptom so much more often (75% of patients) than HGG patients (30%; Villemure & de Tribolet, 1996).

The interval between two seizures, in which the patient does not experience seizures, is termed the 'interictal' period ('ictus' meaning seizure). The seizure occurs during the 'ictal' period, which is marked by specific epileptiform activity in the electroencephalogram (EEG) or magnetoencephalogram (MEG). After the seizure, confusion, drowsiness and headache may be present during the 'postictal' period. These different periods are frequently accompanied by characteristic changes in the EEG, which is used to diagnose and classify the seizure (see Figure 8.2).

The underlying cause of epileptic seizures in brain tumour patients has not yet been determined. It is thought that multiple factors contribute to the development and propagation of an epileptic seizure (epileptogenesis), such as the tumour itself and various consequences of the tumour on the surrounding tissue (for a review, see Beaumont & Whittle, 2000). However, all possible factors that are researched up to now do not suffice when trying to explain epileptogenesis. In particular the propagation of seizures (which is very difficult to predict) remains poorly understood when solely looking at biological factors. Tumour-related epilepsy remains a highly complex, multifaceted notion, which may only be understood when combining neurobiology with complex mathematics and graph theory. We will return to epilepsy and network analysis in Section 8.6.

8.1.3 Cognitive functioning in brain tumour patients

Most brain tumour patients experience cognitive deficits at some point during their disease. Severe neuropsychological impairments have been found in up to 89% of HGG patients (Imperato *et al.*, 1990; Klein *et al.*, 2001, 2003b). A comparable percentage of LGG patients display cognitive deficits (Hochberg & Slotnick, 1980; Taphoorn *et al.*, 1992, 1994a, 1994b; Reijneveld *et al.*, 2001; Klein *et al.*, 2002., 2003a; Klein & Heimans, 2004; Taphoorn & Klein, 2004).

The rate of tumour growth is related to the degree of cognitive dysfunction: a fast-growing tumour causes more profound cognitive deficits than a slow-growing tumour (Hom & Reitan, 1984; Anderson *et al.*, 1990). When comparing acute lesions with slowly growing tumours, it is clear that plasticity plays a major role in the resilience to cognitive deficits in brain tumour patients: brain tumour patients show impressive preservation of cognitive functioning when compared with acute lesions of the same size (Desmurget *et al.*, 2007). However, cognitive deficits that occur in brain tumour patients are surprisingly global, as they include memory disturbances, loss of concentration, planning and speech difficulties, and psychomotor slowness. These widespread disturbances can hardly be explained by the local damage caused by the tumour alone.

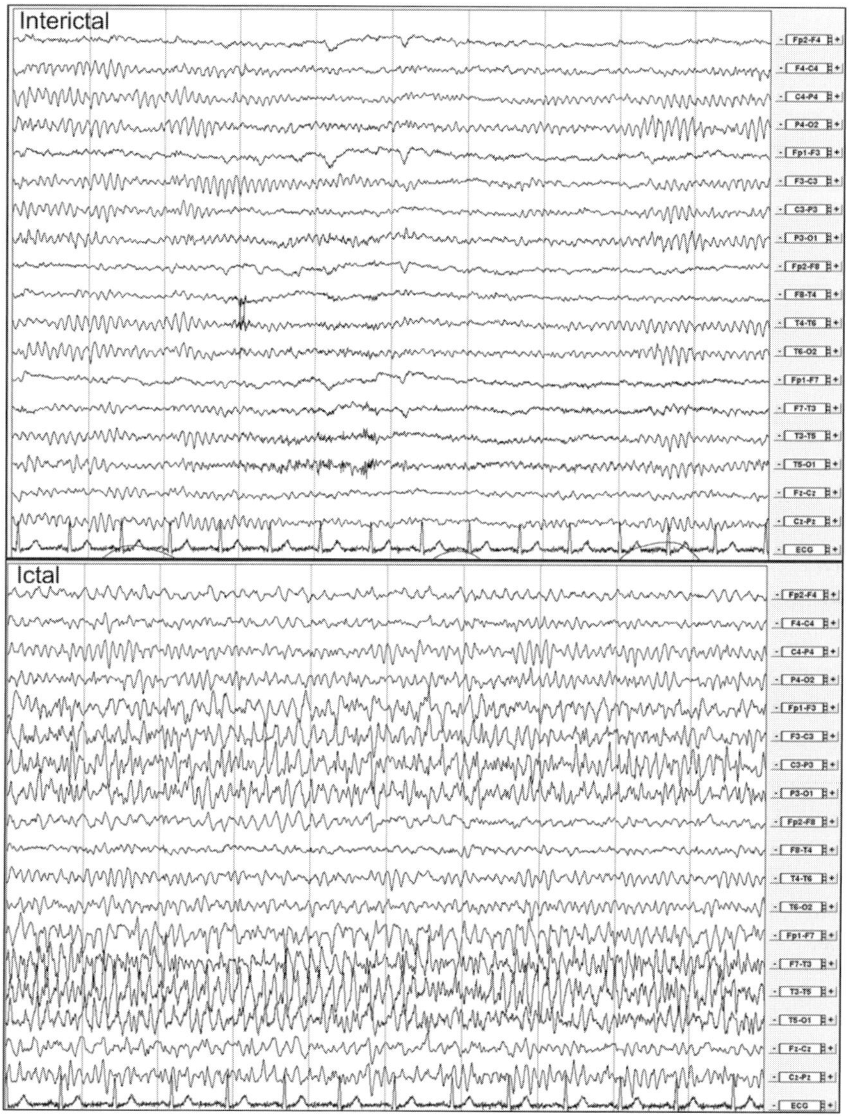

Figure 8.2. Interictal and ictal EEG recordings of a patient with temporal lobe epilepsy. Above: interictal normal EEG, predominantly showing brain activity in the alpha band, while the eyes are closed. Below: ictal EEG in the same patient, in which synchronous epileptiform peaks can be seen throughout the registration.

8.2 Network theory

Network or graph theory originates from two branches of science: mathematics and psychology. Combining these two bodies of thought has brought us numerous methods of analysing several types of networks by representing them in an abstract, theoretical figure called a 'graph'. Illustrated by Euler's problem of the bridges of Konigsberg in 1736 and the six degrees of separation of Milgram (Milgram, 1967), the quest basically comes down to discovering the optimal method of describing the biological and social networks that are everywhere around us. These networks usually combine two seemingly opposing concepts: integration and segregation. In many real-life networks, several highly specialised subsystems are very much integrated as a whole. These so-called 'small-world' characteristics thus include a locally clustered architecture, while even parts of the network that seem very remote from each other can actually be linked through only a few steps. The previously enigmatic co-existence of integration and segregation in a complex network has only recently been elucidated.

8.2.1 Small-world networks

Random networks were first described in the twentieth century (Solomonov & Rapoport, 1951; Erdos & Renyi, 1960) and seemed promising to model complex networks. However, these graphs did not meet up to the expectations of explaining previously mentioned small-world characteristics of real-life networks.

A major insight came with the seminal paper of Watts and Strogatz, who provided an elegant way of modelling small-world networks (Watts & Strogatz, 1998). They proposed a very simple model of a one-dimensional network (see Figure 8.3). Initially, each node or vertex in the network is only connected to its 'k' nearest neighbours (k being the degree of the network), representing a so-called 'regular' network. Next, with likelihood 'p', connections or edges are chosen at random and connected to another vertex, also chosen randomly. With increasing p, more and more edges become randomly reconnected and finally, for p = 1, the network is completely random. This highly comprehensible model allows investigation of all types of networks, ranging from regular to random.

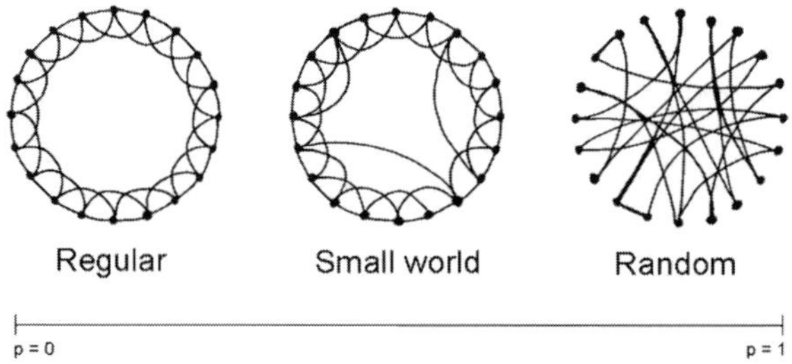

Figure 8.3. Regular, small-world and random networks. The regular network has both high clustering coefficient (C) and high path length (L), while the random network combines low C and low L. The intermediate small-world network can be achieved by relocating only a few long-distance connections from the regular network, which causes L to decrease drastically but preserves a high C. Adapted from Watts & Strogatz (1998).

The intermediate architecture between random and regular proved to be crucial to the solution of the small-world phenomenon. Two measures are of high importance when classifying a network on the continuum of regular to random: the clustering coefficient 'C', which is the likelihood that neighbours of a vertex are also connected, and the path length 'L', which is the average of the shortest distance between pairs of vertices counted in the number of edges. Regular networks are very clustered (high C) but it takes a lot of steps to get from one side of the graph to the other (high L). In contrast, random networks have a low C and low L. So, neither regular nor random networks explain the small-world phenomenon. However, when p is only slightly higher than 0 (with very few edges randomly rewired) the path length L drops sharply, while C hardly changes. In this manner, networks with a small fraction of randomly rewired connections combine both high clustering and a small path length, which corresponds to the small-world phenomenon. The authors demonstrated the existence of such small-world networks in the nervous system of *Caenorhabditis elegans*; a social network of actors; and the network of power plants in the USA. Furthermore, they showed that a small-world architecture might facilitate the spread of infection or information in networks (Watts & Strogatz, 1998).

As C and L are relative measures (dependent on the size of a network, level of functional connectivity etc.), their values are often normalised by dividing them by the C and L of a number of random networks with the same number of nodes and edges. The normalised C and L thus indicate how far the network's topology is from a completely random one. A single index of 'small-worldness', 'S' or 'σ' later was defined, which is the ratio between normalised values of C and L (Humphries & Gurney, 2008).

Another important measure in graph theory is the 'degree distribution'. The 'degree' or 'k' signifies the number of edges that are connected to a node. Thus, the degree distribution refers to the way edges are spread out across the network. Nodes with many edges connecting to them can be termed 'hubs', and are key nodes in the network. This degree distribution gives us a lot of information on the structure of the network.

The degree correlation refers to the association between the degrees of nodes. A high degree correlation indicates that most nodes have approximately the same number of edges (which is the case in random networks) and is called an assortative network, while low degree correlations means that degree varies across nodes (disassortative) and suggests the presence of hubs. Generally, social networks are assortative, while biological and technological networks tend to be disassortative (Newman, 2003).

The discovery of small-world networks initiated a widespread interest in complex networks and gave rise to many new theoretical and experimental studies. The following section provides some basic knowledge on several aspects of network properties. For more detailed information on mathematical backgrounds, see Stam & Reijneveld (2007).

8.2.2 Weighted versus unweighted graphs

In unweighted graphs, edges either exist or do not exist. Moreover, all edges in this binary graph have the same significance or value. In a weighted graph, weights are assigned to each of the edges in the network. Weights can be used to indicate the strength or effectiveness of connections, or the distance between vertices. Weighted graphs are more accurate models of real networks in many cases (Latora & Marchiori, 2001, 2003; Barrat *et al.*, 2004; Newman, 2004; Park *et al.*, 2004; Barthelemy *et al.*, 2005; Onnela *et al.*, 2005), but the great complexity of these graphs has made them difficult to analyse. Instead, weighted graphs are often subjected to a threshold, hereby converting them to unweighted graphs. Although it is fast and easy, this method has several disadvantages: (1) much of the information available in the weights is not used, (2) when the threshold is too high, some vertices may become disconnected from the graph, which poses problems with the computation of C and L, and (3) the choice of the threshold remains arbitrary. At this point, no final solution to this methodological issue has been postulated, and methods of analysis vary considerably between research groups.

When comparing networks of graphs with different levels of functional connectivity, this difference can influence the values of C and L. After all, when one graph has more edges than the other to begin with (because of high synchronisation), this difference will be reflected in varying network variable values. In order to control for differences in connectivity when comparing multiple graphs, we can deliberately ‘fix’ the average degree of these graphs by making sure that the threshold that is used ensures that these graphs are equal regarding the absolute number of connections. In this manner, we are certain that the graphs can be compared regarding network variables even when synchronisation differs. Thus, the actual degree of a network is inherent to a weighted graph, but it can be fixed when comparing unweighted graphs.

8.2.3 Synchronisation of networks

The description of real-life networks by using graph theory provides us with a basic understanding of the architecture of these networks, but does not establish how they achieve optimal functioning. To this end, correlations between structural characteristics and synchronisation dynamics of networks can be investigated (Motter *et al.*, 2006). Network synchronisation is particularly important in brain networks, since it is thought that the synchronisation of activity in distant brain areas plays an important role in brain functioning (Aertsen *et al.*, 1989). Initially, it was assumed that synchronisability was directly related to the average path length of a network, and that particularly small-world networks displayed a high level of synchronisability (Watts & Strogatz, 1998). Several research groups, however, have demonstrated that other factors also influenced synchronisability. In some cases, networks with higher average path lengths even synchronised more easily than those with lower path lengths (Barahona & Pecora, 2002; Hong *et al.*, 2002; Nishikawa *et al.*, 2003).

In order to better quantify the term synchronisability, several methods were used to define a measure and describe the 'optimal' network architecture for synchronisability (Nishikawa *et al.*, 2003; Donetti *et al.*, 2005; Zhou & Lipowsky, 2005; Boccaletti *et al.*, 2006; Motter *et al.*, 2006; Zhou *et al.*, 2006a). Currently, it is thought that in the case of weighted networks, random networks are most likely to synchronise, followed by (in order of decreasing synchronisability) small-world and regular networks (Chavez *et al.*, 2005, 2006). However, synchronisability does not depend solely on the type of the network: it is also influenced by the delay between couplings in the case of time series (Atay *et al.*, 2004), the number of links between two networks (Atay & Biyikoglu, 2005), the connectivity level between nodes (Lee, 2005; Zhou & Lipowsky, 2005) and the degree and heterogeneity of the intensity (i.e. the sum of the strengths of all inputs of a node, reflecting the weights of the degree as well as the links) of the nodes (van den Berg & van Leeuwen, 2004; Zhou & Kurths, 2006; Zhou *et al.*, 2006a).

Topological features of networks, therefore, seem to determine the extent to which networks facilitate synchronisation. The question remains whether graph theory and network synchronisation are up to the challenge of explaining the formidable intricacy of the brain.

8.3 Neural networks

The first important question when reviewing network topology in brain tumour patients is to what extent network theory is applicable to the brain. Synchronisation or 'functional connectivity' may prove to be fundamental for brain dynamics. The basic assumption of functional connectivity is that statistical interdependencies between time series of neuronal activity or related metabolic measures at separate areas in a neural network reflect functional interactions between these neurons and brain regions (Aertsen *et al.*, 1989). Multiple local networks are likely to be managed by long-distance patterns of functional connectivity to achieve higher brain functioning, such as planning, memory and executive functioning (Tononi & Edelman, 1998; Singer, 1999; Bressler, 2002; Reijneveld *et al.*, 2007). Functional connectivity between brain areas may be used to construct graphs of the brain. It is generally calculated based on some correlational measure, indicating the amount of synchrony between two measures of brain activity (see Figure 8.4 for an example in an MEG recording).

Another important consideration concerns the correlation between function and structure. Describing the brain's dynamics in terms of synchronisation between several brain areas raises the question whether the pattern of anatomical connectivity determines the functional connectivity pattern (Felleman & Van Essen, 1991; Guye *et al.*, 2008). In the next section, we will demonstrate the applicability of network theory to the brain and the correlations between these networks and brain anatomy.

As stated before, the two prerequisites of local segregation, referring to local specialisation in specific tasks, and integration, combining information from lower-level networks at a global level, are important in any real-life network, but they are crucial for optimal brain functioning (McIntosh, 2000; Sporns *et al.*, 2000a, 2000b). As said, the small-world

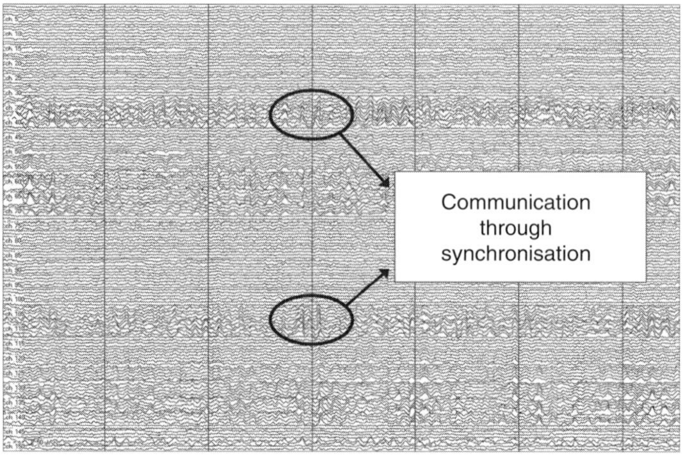

Figure 8.4. Functional connectivity in a 151-channel MEG recording.

network may be a highly adequate model of organisation in the brain, because it supports both segregated as well as integrated information processing (Sporns & Zwi, 2004).

8.3.1 Simulated neural network models

Modelling studies have shown that networks displaying small-world characteristics are more likely to meet the dual requirements of functional segregation and integration than those that do not (Sporns *et al.*, 2000a; Sporns & Tononi, 2002). This type of network architecture supports the development of a giant cluster or 'dynamic core', an integration system that emerges at the large-scale level from the cooperative interactions among widely distributed neuronal populations (Sporns *et al.*, 2000b; Sporns & Tononi, 2002; Sporns & Zwi, 2004). This giant cluster or dynamic core could speculatively be a potential substrate of higher cognitive functioning and consciousness. Furthermore, a small-world topology facilitates synchronisation between simulated neurons and also promotes fast responses in a neural model (Lago-Fernandez *et al.*, 2000; Masuda & Aihara, 2004; Roxin *et al.*, 2004). Other studies on properties of neural networks mainly demonstrated a correlation between shorter path lengths and optimal performance, while local properties such as the clustering coefficient were less important (Vragovic *et al.*, 2005; French & Gruenstein, 2006). The balance between excitation and inhibition in modelled synaptic dynamics also sufficed the requirements of small-world topology (van Vreeswijk & Sompolinsky, 1996; Zemanova *et al.*, 2006; Zhou *et al.*, 2006b; Honey & Sporns, 2008).

8.3.2 In vitro and in vivo experimental studies

Before ample confirmation of their theoretical ideas was achieved, Watts and Strogatz had already applied their small-world paradigm to a neuroscientific question. They found that the C and L of the nervous system of *Caenorhabditis elegans* were in agreement with the definition of a small-world configuration (Watts & Strogatz, 1998). Later, similar conclusions could be drawn from cortico-cortical connection data from macaques and cats (Hilgetag *et al.*, 2000; Sporns & Zwi, 2004; Kaiser & Hilgetag, 2006; Yu *et al.*, 2008).

A small-world pattern in the spread of (epileptic) activity in lesioned macaque cortex in vivo was found, suggesting a correlation between anatomical and functional connectivity patterns (Stephan *et al.*, 2000). The correspondence between functional and structural connectivity patterns has also been investigated by using computational models of macaque cortex (Honey *et al.*, 2007). Functional networks proved to overlap with the underlying structural networks. Others modelled the propagation of epileptic activity in a large-scale model of cat cortex. They concluded that association fibres and their connection strengths were useful predictors of global topographic activation patterns in the cerebral cortex, indicating a global structure-to-function relationship (Kotter & Sommer, 2000).

8.3.3 Studies in humans

Research regarding neural networks in humans has grown rapidly since the initial paper of Watts and Strogatz. Neural network analysis can be based on several types of measurements. Findings from these different types of data match globally, but comparing any two studies researching functional connectivity and networks in the brain is difficult because of methodological differences (Laufs, 2008). For instance, significant differences in small-worldness and degree distribution have been reported when comparing two different methods of subdividing the brain into regions (Wang *et al.*, 2008). In the following subsection, we will discuss the two main types of functional neural network measurements separately.

8.3.3.1 Functional magnetic resonance imaging

Functional connectivity has been applied to functional magnetic resonance imaging (fMRI) since 1995 (Biswal *et al.*, 1995), becoming increasingly popular in recent years (Auer, 2008). Applying graph theory to fMRI generally starts with blood oxygen level dependent (BOLD) time series of brain activity, after which a matrix of correlations between the time series is computed. In one of the first fMRI studies into connectivity, the matrix was converted to a graph by assigning edges to all supra-threshold correlations between activated brain areas. Various functional clusters in the form of subgraphs were demonstrated during a finger-tapping task (Dodel *et al.*, 2002). Another study investigated clustering coefficients, path lengths and degree distributions in relation to fMRI data. It was stated that any two fMRI voxels were functionally connected when their temporal correlation exceeded a predefined threshold. The reported functional brain networks indeed displayed small-world features (Eguiluz *et al.*, 2005).

A different approach was taken by the Cambridge group, who described several studies regarding fMRI BOLD time series during a resting state with eyes closed and no task

(Salvador *et al.*, 2005a, 2005b; Achard *et al.*, 2006; Achard & Bullmore, 2007). In the first study, BOLD time series were taken from 45 regions of interest in both hemispheres of 12 healthy subjects (Salvador *et al.*, 2005a). From these 90 time series, a matrix of partial correlations was obtained and thresholded. Graph analysis applied to these matrices suggested small-world topology of the resting state functional network. The authors noted that the anatomy did not always precisely predict functional relationships (Salvador *et al.*, 2005b). An extensive graph analysis of this data set displayed a single giant cluster of highly connected brain regions (Achard *et al.*, 2006), which might reflect the previously described dynamic core (Sporns *et al.*, 2000b; Sporns & Tononi, 2002; Sporns & Zwi, 2004).

In another study, MRI scans of 124 healthy subjects were obtained. A connection between two regions was assumed if they displayed statistically significant correlations in cortical thickness (He *et al.*, 2007). With this approach, the authors showed that the human brain network had small-world characteristics. Anatomical network analysis using MRI has been performed and demonstrated small-world features (Hagmann *et al.*, 2007; He *et al.*, 2007; Iturria-Medina *et al.*, 2008). Human brain-stem anatomical as well as functional networks also meet the criteria of being a small-world network (Humphries *et al.*, 2006), again pointing towards correspondence between functional and anatomical networks.

In addition to exploring network architecture in the brain itself, fMRI is becoming increasingly popular to compare network architectural differences between the healthy population and several patient groups. A recent study in patients with Alzheimer's disease (AD) and controls investigated small-world properties of the brain (Supekar *et al.*, 2008). The authors calculated a wavelet correlation matrix based on parcellation of the brain into 90 regions, and thresholded this matrix. They found healthy controls to display small-world characteristics (particularly in terms of clustering coefficient), while AD patients did so less evidently. When trying to use the clustering coefficient (normalised for degree) as a biomarker for AD, this yielded 72% sensitivity and 78% specificity. The application of network analysis in the diagnosis of certain brain disease may be a useful tool in the future. Furthermore, disrupted small-world topology was reported when comparing schizophrenic patients with healthy controls (Liu *et al.*, 2008). Again, patients in this study had significantly higher clustering coefficients than controls. These observations were based on graphs constructed from the partial correlations between 90 parcellated brain regions. Other studies have researched network-related fMRI changes in multiple sclerosis (Lowe *et al.*, 2002), attention deficit hyperactivity disorder (Cao *et al.*, 2006) and depression (Anand *et al.*, 2005). Although global patterns may be derived from these studies, more research is needed to clarify the precise characteristics of networks in the brain. At this moment, fMRI has not been used to assess functional connectivity and network topology in brain tumour patients.

8.3.3.2 Electroencephalography (EEG) and magnetoencephalography (MEG)

Electroencephalography and MEG register brain activity by measuring electrical and magnetic flow, respectively, within the brain, resulting in one time series per EEG electrode or MEG sensor. Magnetoencephalography has a much higher spatial resolution than EEG, because the skull and scalp do not distort the magnetic field patterns. Also, MEG does not require the use of a reference electrode as EEG does, rendering MEG analysis somewhat more straightforward than EEG. Disadvantages of both EEG and MEG are the inverse problem and volume conduction. The inverse problem refers to the non-uniqueness of a solution for the deduction of magnetic and electric currents from the measured time series. In other words: there is more than one way in which the measured signals can come about, but the most likely solution is calculated. Volume conduction refers to the fact that EEG or MEG signals picked up at different electrodes or sensors may originate from the same source. Analysis of EEG and MEG is usually performed in different frequency bands, which can roughly be divided into delta (0.5–4 Hz), theta (4–8 Hz), lower alpha (8–10 Hz), upper alpha (10–13 Hz), beta (13–30 Hz), lower gamma (30–45) and upper alpha bands (55–80 Hz).

The first application of graph analysis to MEG was published in 2004 (Stam, 2004). We studied correlations between time series of 126 MEG channels in five healthy subjects with the synchronisation likelihood (SL), a measure of synchronisation or functional connectivity (Stam & van Dijk, 2002; Montez *et al.*, 2006). The SL takes both nonlinear and linear coupling between time series into account, and varies between 0 (desynchronisation) and 1 (total synchronisation). The matrices of SL values between each sensor pair were converted to unweighted graphs by assuming an edge between pairs of channels with an SL above a threshold, which was varied. This analysis was performed for data filtered in different frequency bands. In the alpha band, graphs were close to regular networks. For delta, theta and gamma frequencies, the graphs showed small-world features.

Graph theoretical properties of MEG recordings in healthy subjects were studied more extensively in a recent paper (Bassett *et al.*, 2006). The authors applied graph analysis to MEG recordings in 22 healthy subjects during a no-task, eyes-open state, and a simple motor task (finger tapping). Pair-wise channel correlations were converted to an unweighted graph by use of a threshold, after which a range of graph theoretical measures was calculated. Small-world architectural features were found in the six major frequency bands, which were remarkably stable over different frequency bands as well as over experimental conditions.

A recent seminal paper has shown that functional connectivity is a useful tool to assess the functionality of brain tissue (Guggisberg *et al.*, 2008). The authors subjected 15 patients with circumscribed brain lesions (13 tumours), who were eligible for surgical removal of the lesion, to MEG registration before surgery took place. Imaginary coherence was used to compute synchronisation matrices containing all channels (Nolte *et al.*, 2004). Patients' functional deficits that were due to the brain regions surrounding their lesions were related to connectivity values of the same areas. The authors found connectivity to be lower in brain areas that were included in the lesion than in the intact tissue. Notably, connectivity in brain tumour patients was only decreased in the regions corresponding to the functional deficits but not in the entire area invaded by the tumour.

Thus, functional connectivity may be useful to distinguish between functional and dysfunctional tissue within a brain tumour. This important study shows that the level of functional connectivity in the brain is not merely a scientific classification, and that it may indeed be used for clinical decision making in (brain tumour) patients.

The general picture that emerges from these fMRI and neurophysiological data, is that neural networks display small-world characteristics. Other factors (e.g. anatomical structures, processing efficiency), however, may be responsible for deviation of neural networks from the optimal small-world configuration.

8.4 What is the effect of brain tumours on network properties?

Functional connectivity and network properties are considered to be physiological substrates for segregated and distributed information processing (Salvador *et al.* 2005a, 2005b; Achard *et al.*, 2006; Achard & Bullmore, 2007). Following this hypothesis, both purposeful (such as neurosurgery) and random attacks (e.g. lesions) would lead to changes in these parameters.

8.4.1 The lesioned brain

Global and local efficiency measures were applied to fMRI data in healthy young and old subjects (Achard & Bullmore, 2007). Efficiency is a measure of how optimally information is exchanged in a certain network, and is related to both clustering (local efficiency) and path length (global efficiency; Latora & Marchiori, 2001). This measure can be used to investigate small-world networks. Subjects were studied during a resting state no-task paradigm, either with placebo treatment or with sulpiride. Sulpiride is an antagonist of the dopamine D2 receptors in the brain and has sedating effects. The analysis was based upon wavelet correlation analysis of low-frequency correlations between BOLD time series of 90 regions of interest, followed by thresholding (Bassett *et al.*, 2006). The efficiency measures were related to a 'cost' factor, defined as the actual number of edges divided by the maximum number of edges possible in the graph. Local and global efficiency, normalised for cost, were shown to be lower in the older subjects than in the young group, and it was lower in the sulpiride condition as compared with the placebo condition. The effect of age on efficiency was stronger and involved more brain regions than the sulpiride effect.

Cortical connectivity of five patients with cervical spinal cord injury (SCI) and five healthy volunteers has been investigated through analysis of EEG recordings of 12 regions of interests (ROIs; De Vico Fallani *et al.*, 2007). A connection matrix containing connectivity values for each pair of ROIs was thresholded. In addition to significant differences compared with computed random networks of the same size in both groups, higher local (but not global) efficiency was found in the SCI group, suggesting a compensatory higher level of internal organisation in these patients. We will come back to this idea of compensatory mechanisms in the following sections.

A highly interesting modelling study has researched the effects of lesions in two models of oscillatory cortical neurons (Honey & Sporns, 2008). Lesions to the most heavily connected nodes (i.e. hubs) had the greatest effect on cortico-cortical interactions, although the architecture of the specific part of the cortex that was investigated influenced the extent of damage after targeted attack of a hub.

Unfortunately, the impact of natural lesions on functional connectivity and neural networks has not been further investigated. It would be highly interesting to explore this in for instance patients with cerebrovascular accidents, as the effects of acute brain lesions could be assessed in this patient group. A first step to further elucidate the affects of such acute lesions is a study that we recently conducted in patients who underwent the Wada test (Douw *et al.*, 2009). During the Wada test or intra-arterial amobarbital procedure (IAP), a sedative is injected into only one of the cerebral hemispheres in order to selectively suppress activity in that hemisphere. After injection, functioning of the non-anaesthetised hemisphere can temporarily be assessed by means of neuropsychological testing. This procedure is used in patients that are eligible for neurosurgery due to pharmaco-resistant epilepsy, in order to determine pre-operatively whether functioning in the hemisphere contralateral to the origin of the seizures is sufficient to ensure optimal functional status after the planned removal of the part of the hemisphere where seizure activity originates. During the procedure, an EEG is recorded in order to assess the duration of the effects of the sedative. We analysed functional connectivity patterns in the injected and non-injected hemisphere by means of the SL, mainly to determine whether reversibly 'shutting down' one hemisphere would induce changes in the contralateral hemisphere. Results showed that while synchronisation increased consistently in the injected hemisphere after injection, functional connectivity in the contralateral hemisphere showed a more complex pattern of change. Delta and theta band functional connectivity within the contralateral hemisphere seemed to decrease after injection, while the opposite was true in the beta band (e.g. SL increased). Connectivity between the two hemispheres decreased in the delta and theta bands, while it increased after injection in the beta band. We concluded that, in general, acute local functional disturbances of the brain can have significant effects on connectivity in more remote areas of the brain.

8.4.2 Brain tumour patients

The application of graph theory to a neuro-oncological population is rather novel, although changes in EEG coherence (a measure of connectivity) in these patients have already been reported as early as 1994 (Harmony *et al.*, 1994). Our own group reported on functional connectivity in brain tumour patients for the first time in 2006 (Bartolomei *et al.*, 2006a, 2006b). In these studies, we used the synchronisation likelihood to assess the degree of functional connectivity in MEG recordings of both brain tumour patients and healthy controls. Brain tumour patients displayed a different pattern of functional connectivity than healthy controls: they showed a general broadband (0.5–60 Hz) decrease of synchronisation. When looking at separate frequency bands, patients tended to show increased connectivity in the lower frequency bands (delta to alpha) compared to healthy controls, while lower SL was found in the beta and gamma bands. These differences in connectivity were not limited to regions surrounding the tumour, but were also present within the hemisphere contralateral to the tumour. Notably, patients' changes in synchronisation were more profound in the dominant left hemisphere, regardless of tumour lateralisation. These connectivity results were confirmed by a later study we conducted, using a more homogenous group of brain tumour patients, namely LGG patients ((Bosma *et al.*, 2008); see Figure 8.5). Again, LGG patients had increased connectivity compared to controls in the lower frequency bands, while the opposite was true in the higher frequencies. This interesting pattern of change was rather puzzling, in particular the widespread nature of the changes in the presence of a local lesion, prompting further research regarding these findings.

We subsequently used graph theory in the same brain tumour patients and controls (Bartolomei *et al.*, 2006a). Based on the SL, graph analysis was performed while using an individually adapted threshold regarding the number of edges per vertex, normalised for a fixed k of 10. Values of the clustering coefficient and path length were then divided by the computed values of 10 random graphs (Sporns & Zwi, 2004). Brain tumour patients were found to have decreased path lengths in the theta, beta and gamma bands when compared to healthy controls. Moreover, they showed less clustering than controls in the theta and gamma bands. These results imply that brain tumour patients have more random neural network topology than healthy persons do. As stated previously, a very low path length and thus random network structure might also be related to the tendency towards hypersynchronisation and epileptic seizures (see Section 8.5).

In a very recent study, we investigated neural networks in the previously described LGG patients and compared them to healthy controls (Bosma *et al.*, 2009). In this study, the phase lag index (PLI; Stam *et al.*, 2007b) was used to obtain synchronisation matrices. The PLI is a relatively novel method of calculating synchronisation between time series. In contrast to other methods of analysis, the PLI is scarcely influenced by volume conduction effects that occur in neurophysiological measurements. Graph analysis was applied to obtain the normalised C and L from a thresholded synchronisation matrix. Brain networks showed a small-world configuration in both patients and controls. However, there were significant differences between patients and controls: patients' C in the theta band was higher compared to controls, whereas the opposite was true for the beta band (e.g. clustering was lower in the patient population). Thus, patients' network topology seemed more small world than in controls in the theta band, while it was more random in the beta bands. This difference in the beta band was also reflected in the lower value of σ (as a measure of small worldness) in LGG patients.

The different findings between our first and second study into brain networks of brain tumour and LGG patients, respectively, indicate that much is still to be done in order to understand which factors account for the changes in connectivity that were found (see Figure 8.6). Possibly influential variables are the type of the tumour, treatment that has been applied for the tumour as well as the epilepsy, different methodologies (SL versus

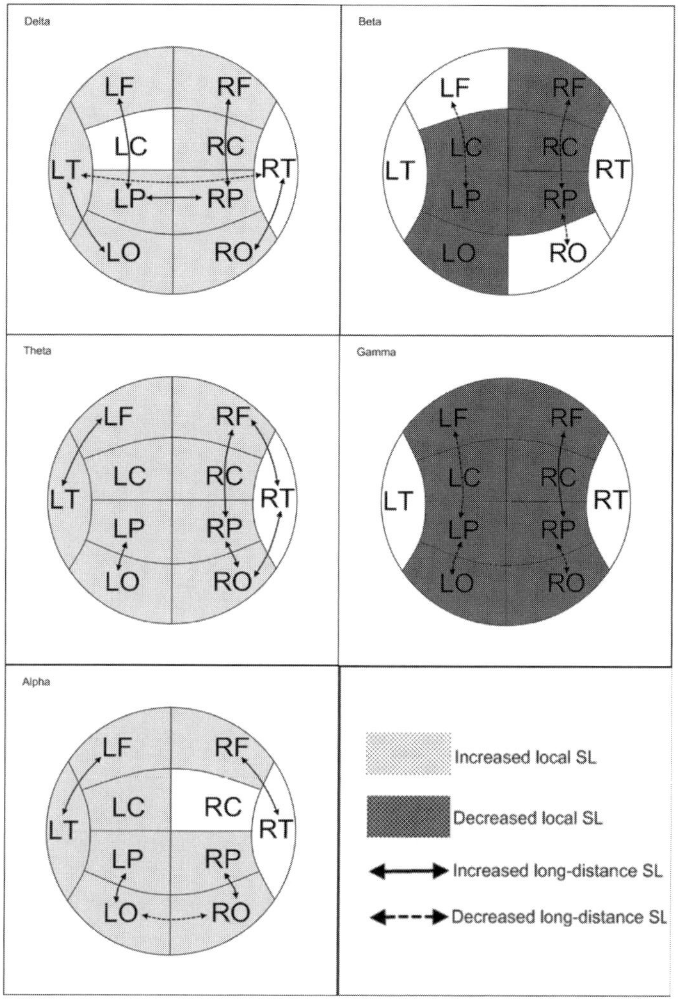

Figure 8.5. Summarised overview of significant findings regarding functional connectivity in brain tumour patients compared with healthy controls.
Note. Findings from patients with both right-sided and left-sided tumours (Bartolomei *et al.*, 2006a) are displayed in this figure. SL = synchronisation likelihood, LF = left frontal, LC = left central, LP = left parietal, LO = left occipital, LT = left temporal, RF = right frontal etc. Adapted from Bartolomei *et al.* (2006a) and Bosma *et al.* (2008).

PLI as measure of functional connectivity), and the presence of symptomatology (e.g. seizures, cognitive deficits). Furthermore, a compensatory mechanism may account for the unexpected small-world topology of patients in the theta band, although this notion is highly speculative. More thoughts on this compensatory mechanism are described in the following sections.

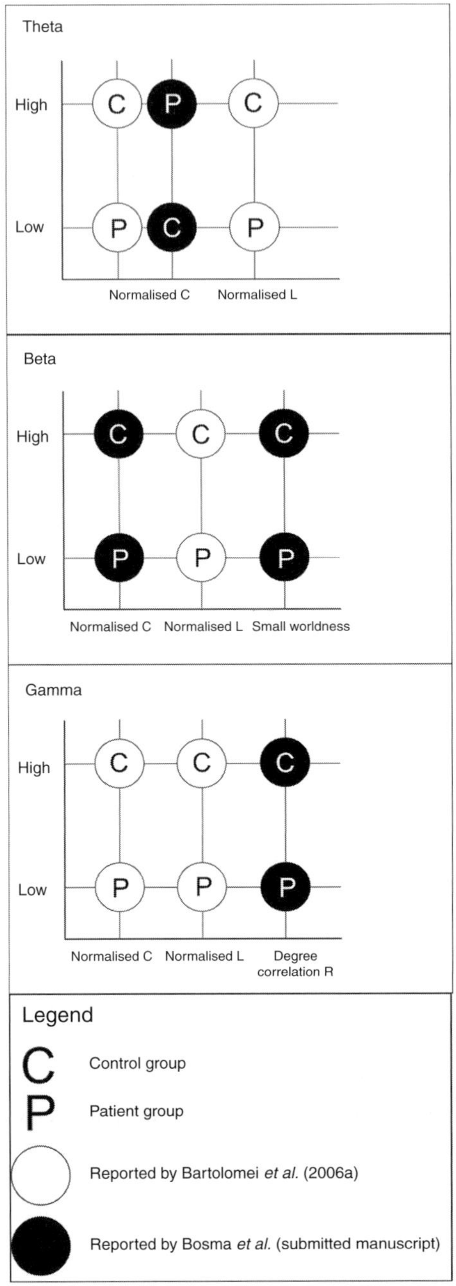

Figure 8.6. Summarised overview of significant findings regarding network analysis in brain tumour patients compared with healthy controls. 'High' indicates an increased value compared with the 'low' group regarding variables placed along the x-axis. C = clustering coefficient, L = path length, R = degree correlation. Derived from Bartolomei *et al.* (2006a) and Bosma *et al.* (submitted).

As brain tumour patients usually undergo treatment for their disease, the question arises whether this might influence functional connectivity and neural networks. We have investigated 15 brain tumour patients with MEG before and after resective surgery to address this issue (Douw *et al.*, 2008). Based on the PLI, brain tumour patients displayed decreased long-distance interhemispheric functional connectivity in the theta band after tumour resection. This finding was not influenced by several patient-, treatment- and tumour-related factors, and thus could be attributed to removal of the tumour. When comparing these results with earlier studies that report increased theta band functional connectivity (Bartolomei *et al.* 2006a, 2006b; Bosma *et al.*, 2008), we speculated that the postoperative decrease of interhemispheric theta synchronisation might be a tendency towards a more "normal" state of the theta band after tumour resection in brain tumour patients. This idea is further corroborated by the effect of connectivity changes on seizure frequency in this study. Indeed, patients with a large decrease of interhemispheric theta synchronisation more often were epilepsy-free after tumour resection, as compared to patients with a small decrease or an increase of synchronisation.

8.5. Cognition and network topology

It is tempting to hypothesise that topological changes leading to suboptimal network dynamics are correlated with suboptimal higher brain functioning, which was already suggested in 1995 (Bressler, 1995). Since then, evidence has accumulated that higher cognitive functions indeed require functional connectivity between multiple distinct neural networks (Lowe *et al.*, 1998; Meyer-Lindenberg *et al.*, 1998; Quigley *et al.*, 2001; Stam *et al.*, 2002a, 2002b; Micheloyannis *et al.*, 2003; Salvador *et al.*, 2005a), and that optimal functional connectivity requires optimal neural network architecture. It has been demonstrated that while at rest, the 'basic' brain network gives rise to constantly changing, weakly synchronised networks (Stam *et al.*, 2002b; Langheim *et al.*, 2006). This process of constantly creating and dissolving functional networks, also called 'fragile binding', is thought to underlie spontaneous information processing. The optimal brain dynamics are thought to be near the phase transition between low and high levels of synchronisation (Stam, 2005).

A clinical hint towards the relation between cognition and network topology was found in AD patients (Stam *et al.*, 2007a). A group of AD patients was compared to a non-demented control group. EEG recorded during an eyes-closed, no-task state and filtered in the beta band (13–30 Hz) was analysed with the SL, after which synchronisation matrices were converted to graphs. When C and L were computed as a function of threshold (same threshold for controls and patients), the path length was significantly higher in the AD group. When C and L were studied as a function of degree k (same k for both groups), the path length was higher in the AD group, but only for a small range of k (around 3). In both controls and patients, the graphs showed small-world features when C and L were compared to those of random control networks. A higher (more optimal) cognitive performance on the Mini Mental State Examination (MMSE) correlated with a higher C and smaller L, which is even more noteworthy as this measure of cognition is very crude and has relatively low differentiating power. One might hypothesise that the type of damage in AD, which is best described as 'random error' (Stam *et al.*, 2008), leads to a less optimal (i.e. less small-world like) network organisation.

8.5.1 Compensatory mechanisms

Compensatory mechanisms of synchronisation may play a role in cognitive functioning. We used the SL to compare functional connectivity in AD patients, mild cognitive impairment (MCI) patients, and healthy controls (Pijnenburg *et al.*, 2004). MCI is often considered as a preceding phase of AD, in which subtle cognitive deficits are present but the criteria for AD are not strictly met. Increased SL in the alpha band was related to poorer cognitive performance in the MCI patients, which could indicate a compensatory mechanism. However, this increase in connectivity was not present in patients suffering from full-blown AD, possibly because of the impossibility to compensate in this advanced stage of the disease. In another study, fMRI was performed to characterise temporal lobe functional connectivity (measured by correlation coefficients between time series) in epileptic patients and healthy controls (Bettus *et al.*, 2008a). The patients showed increased functional connectivity contralateral to the epileptic hemisphere. This increase in connectivity was related to poorer memory performance, again suggesting that functional connectivity and cognition are associated, and that some changes in connectivity may reflect compensatory mechanisms in patients with cognitive deficits.

Compensatory mechanisms may also be demonstrated when applying graph theory to the brain. Graph analysis has been performed during a finger-tapping task in a previously mentioned study (see Section 8.3; Bassett *et al.*, 2006). During the motor task, relatively small changes in network topology were observed, mainly consisting of the emergence of long-distance interactions between frontal and parietal areas in the beta and gamma bands. Micheloyannis and colleagues applied graph analysis to EEG recorded at rest and during a working memory test (Micheloyannis *et al.*, 2006b). Twenty healthy subjects with a few years of formal education and low IQs were compared to the same number of healthy subjects with university degrees and high IQs. SL matrices were converted to graphs by use of a threshold (while keeping the mean k of the nodes equal between both groups). No differences in network architecture were present between the groups during the resting state. However, during the working memory task, the networks of the group of lower educated subjects were closer to small world in the theta, alpha, beta and gamma bands. One might speculate that the controls with low education display a compensatory mechanism during the task, which is not needed by the highly educated controls. In a subsequent study, the 20 control subjects with higher education were compared to 20 patients with schizophrenia (stable disease, under drug treatment; Micheloyannis *et al.*, 2006a). During the working memory task, C was lower in the schizophrenia group compared to controls in the alpha, beta and gamma bands. Consequently, task-related networks in the schizophrenia group were less small world in terms of topology, and more random compared to controls. Combining these results with those of the first study, there seems to be a decrease of small-world features between controls with low education to controls with high education, and then from controls with high education to schizophrenic patients. Possibly, the controls with low education display a compensation mechanism, which involves a more small-world like brain network configuration in the lower frequency range during the task, which is not needed by the highly educated controls and which completely fails in the case of the schizophrenics.

8.5.2 Patients with brain tumours

We have investigated the relation between functional connectivity and cognitive functioning in LGG patients using MEG (Bosma *et al.*, 2008). As stated in Section 8.4.2, synchronisation (as measured by SL) in the lower frequency bands was increased when compared to healthy controls, while it was lower in the high frequencies. Regarding cognition, patients performed significantly poorer than controls in the domains of psychomotor functioning, attention, information processing speed and working memory. Within the patient group, increased short- and long-distance functional connectivity were associated with poorer cognitive functioning in the delta, theta and gamma band. We speculated that the observed cognitive deficits may be due to changes in functional connectivity due to the tumour and/or its treatment. The associations further suggest – again – that a compensatory mechanism might be present: LGG patients show increased lower band connectivity, but have a decreased higher frequency band SL. Our analysis proved that higher synchronisation in the lower frequency bands was related to poorer performance. Patients may need more effort and thus more synchronisation to be able to perform the neuropsychological tasks. It should be noted, however, that this speculative compensation mechanism does not optimise cognitive functioning to the level of healthy controls, since cognitive deficits were still present.

In our previously mentioned study, we assessed the association between neural network architecture and cognitive performance in LGG patients (Bosma *et al.*, 2009). As discussed in Section 8.4.2, patients had a higher clustering coefficient than healthy matched controls in the theta band, whereas a lower clustering coefficient and less small worldness were observed in the beta band. A lower degree correlation was found in the upper gamma band. Higher clustering coefficient, longer path length and lower degree correlations in delta and lower alpha band were associated with poorer cognitive performance. These deviations from small worldness in the strength and spatial organisation of brain networks may be responsible for cognitive dysfunction in LGG patients. The increased small worldness in LGG patients that is related to cognitive deficits may also reflect a compensatory mechanism in the lower frequency bands.

The general pattern that emerges from these studies is that networks of brain tumour patients are further remote from a small-world configuration in the lower frequency bands than those of healthy people, while the opposite is true in especially the higher frequency bands. These changes are related to poorer cognitive functioning, but the specifics of these phenomena are yet to be determined.

8.6. Neural networks and epilepsy

Synchronisation of neurons may be pivotal for optimal brain functioning, but it can also reflect abnormal dynamics related to epilepsy. From that point of view, epilepsy could be regarded as the price that we pay for our intelligence; we are living on a small edge between optimal intellectual functioning and epileptic seizures.

8.6.1 Simulated, in vitro and in vivo studies

Initial results regarding functional connectivity and graph analysis in epilepsy patients were achieved using computational models. Models of synchronisation and networks in the brain have linked increased randomness of the interictal network to epileptogenesis (Lopes da Silva *et al.*, 2003). In a brain-based model of epilepsy, an important region of the hippocampus (a mainly memory-related structure in the brain) showed seizure-like activity lasting for seconds, while other areas showed short bursts of activity (Netoff *et al.*, 2004). Small-world network models were then constructed for various types of neurons, in order to investigate the transition from seizures to bursting. Results indicated that the bursting behaviour may represent a dynamical state *beyond* the preceding seizures. This is an important finding, since similar bursting-like phenomena have also been observed in the scalp recorded EEGs of neurological patients, and their epileptic significance is still poorly understood (Brenner, 2002).

In vitro hippocampal rat neurons were injured with an exposure to glutamate (Srinivas *et al.*, 2007). After this injury, the neuronal network became hypersynchronous and fired bursts at high frequency, which was described as 'induced epileptic activity'. The authors then measured and characterised network properties, and showed that the clustering coefficient decreased after the injury. In other words, the neuronal network became more random as epileptic activity developed.

Others noted that in medial temporal lobe epilepsy (TLE), epileptogenesis is amongst others characterised by structural network remodelling and aberrant axonal sprouting (Percha *et al.*, 2005). A two-dimensional model of neurons was used to study the influence of network topology changes on seizure threshold. For increasing p (increasing randomness), they found a transition between a state of local to a state of global coherence. The authors speculated that neural networks might develop towards a critical regime between local and global synchronisation; seizures would result if pathology pushes the system beyond this critical state. A similar concept can be found in two other studies (Kozma *et al.*, 2005; Stam, 2005). The influence of temporal lobe architecture on seizures has also been addressed through a computational model of rat dentate gyrus with a small-world architecture (Dyhrfjeld-Johnsen *et al.*, 2007). A loss of long-distance cells in the temporal lobe had only little influence on global network connectivity, as long as a few of these long-distance connections were preserved. Simulations of the dynamics of this model showed that the ability of the network to synchronise also remained equal when some (apparently crucial) connections were removed.

We can conclude from these studies that increased network randomness may be a risk factor for hypersynchronisation and thus epileptic seizures to occur. Moreover, a critical point may be present, determining whether or not the network falls prey to runaway synchronisation.

8.6.2 Studies in humans

Several studies have investigated functional connectivity in patients with epilepsy, most of them by using EEG recordings. Pre-ictal desynchronisation has been reported, which was not only limited to the epileptic zone, but was even present in the contralateral hemisphere (Mormann *et al.*, 2003). Interictal recordings have shown both desynchronisation and hypersynchronisation (Le Van Quyen *et al.*, 2005; Schevon *et al.*, 2007; Aarabi *et al.*, 2008; Bettus *et al.* 2008a, 2008b), indicating the complexity of synchronisation patterns in the origin of epilepsy. Hypersynchronised epileptogenic zones are thought to be surrounded by isolating zones of hyposynchronisation (Le Van Quyen *et al.*, 2005). During a seizure, hypersynchronisation has been reported in most studies (Guye *et al.*, 2006; Schindler *et al.*, 2007a, 2007b; Aarabi *et al.*, 2008), which could promote seizure termination by driving neuronal networks into a refractory state (Schindler *et al.*, 2007a). These studies imply that epileptogenic zones might be identified by their synchronisation pattern. Synchronisation analysis has been useful when localising specific sites that participate in the initiation and propagation of seizures (Bartolomei *et al.*, 2008; Ortega *et al.*, 2008a), which is key to the optimal treatment of epilepsy.

Graph analysis may also prove highly valuable for epilepsy research, as evidence accumulates that seizure dynamics are related to the state of both interictal and ictal networks (Lytton, 2008). The first network analysis in human epilepsy was performed with six-channel EEG depth recordings in a single patient during an epileptic seizure (Wu *et al.*, 2006). A phase-coupling matrix of the EEG signals was thresholded to construct a network. During the seizure, a change in network configuration was detected in the direction of a small-world network: there was an increase in C and a decrease of L. Conversely, one might argue that the preceding interictal network was more random.

We have investigated seven patients during temporal lobe seizures recorded with intracranial depth electrodes (Ponten *et al.*, 2007). Matrices of pair-wise SL values were converted to graphs by use of a threshold with a fixed k of 15. During seizures, the normalised C increased in the delta, theta and alpha bands; the L also increased in these bands. In conclusion, a movement away from a random interictal towards a more ordered ictal network configuration was found, again suggesting that epilepsy is characterised by an interictal network with a pathological random structure. Such a random structure may have an even lower threshold for the spread of seizures than the normal small-world configuration. It seems as though the brain network of epilepsy patients becomes more and more random until the seizure 'resets' the brain, after which a more optimal (i.e. small-world like) architecture develops. In this viewpoint, the seizure is not the most pathological feature of an epilepsy syndrome: it is the interictal randomisation that is paramount to the disease. However, this hypothesis is still speculative: longitudinal measurements are needed to confirm this idea.

This hypothesis was further supported by another study using intracranial EEG recordings at seizure onset in relation to network topology (Kramer *et al.*, 2008). Coupling between electrodes changed at seizure onset, with an increase of small worldness during seizures. In another study, the neural network initially became more small-world, after which it returned to its pre-ictal random state before the seizure terminated (Schindler *et al.*, 2008). These results indicate that the network change indeed precedes the end of a seizure, and points toward the importance of understanding network changes when studying epilepsy. In a recent study, we performed network analysis with surface EEG recordings during absence seizures (Ponten *et al.*, 2009). Absence seizures are characterised by lowered consciousness and staring, after which the patient is amnesic for the seizure. As expected, we found an increase of synchronisation (SL and PLI) during the seizure when compared to the preictal state. Moreover, an increase of the normalised C and L (both weighted and unweighted) was found, indicating a more regular organisation during the absences, while the interictal network was more strongly small-world organised. Apparently, differences exist between temporal lobe seizures and absence seizures in terms of network topology, although both disorders are accompanied by a deviation from random network structure.

In other recent research at our department, we compared patients with short-term (i.e. shorter than 10 years) temporal lobe epilepsy with long-term (longer than 10 years) patients (van Dellen *et al.*, 2009). Intraoperative electrocorticography (ECoG) recordings were analysed regarding functional connectivity and network topology. During ECoG, so-called 'grids' containing 20 electrodes are placed directly on the cortex, allowing for direct measurement of brain activity without interference of the skin, scalp and skull. In this study, all grids were placed on the temporal cortex. We found that broadband functional connectivity was lower in patients suffering from long-term TLE when compared to short-term TLE patients. Furthermore, long-term TLE patients showed significantly less local clustering in the temporal cortex, while the path length in this area remained unaffected. There was less small worldness in the temporal cortical networks of long-term TLE patients when compared to short-term patients, suggesting that a less optimal network configuration occurs due to long-term TLE.

A highly interesting study also using ECoG data has investigated functional connectivity in epileptic patients with several measures of connectivity (Ortega *et al.*, 2008a). They found clusters of synchronised activity at specific areas of the lateral temporal cortex in most patients. Further analysis suggested that surgical removal of essential synchronisation clusters or critical nodes correlated with postoperative seizure control (Ortega *et al.*, 2008b). These studies suggest that assessment of synchronisation patterns and network architecture can aid neurosurgeons in achieving postoperative outcome by performing more 'tailored' epilepsy surgery, which is a very promising notion for the future of epilepsy and neuro-oncological surgery.

At this moment, no studies have been performed investigating the relation between neural networks and epilepsy in a neuro-oncological patient population. As previously mentioned, we found the possible normalisation of pathological theta band connectivity in brain tumour patients to be related to the outcome of epilepsy after tumour resection (Douw *et al.*, 2008). Network analysis in brain tumour patients seems to suggest that network randomisation might be a general result of brain damage. More random networks might have a lower threshold for seizures in these patients, facilitating epileptogenesis and propagation.

8.7 Conclusions and future prospects

In this chapter, we have demonstrated that modern network theory provides an excellent framework for the study of complex networks in the brain, and specifically brain tumour patients. It has been shown that brain tumour patients show alterations in functional connectivity patterns when compared to healthy subjects (Bartolomei *et al.* 2006a, 2006b; Bosma *et al.*, 2008). Pathologically increased connectivity is reported in the lower frequency bands, while in contrast the higher frequency bands show decreased synchronisation. Since increased lower frequency connectivity correlated with cognitive functioning, a compensatory mechanism may account for this pathological low-frequency increase.

Moreover, neuro-oncological patients are likely to have a more random resting state neural network than the healthy population in the higher frequency bands (Bartolomei *et al.*, 2006a), while networks may be more small-world-like in particularly the theta band (Bosma *et al.*, 2009). A randomly constructed network has been thought to be most vulnerable to hypersynchronisation (Chavez *et al.*, 2005, 2006). Brain tumour patients' increased network randomness may be related to epileptic seizures, as we showed for patients with non-tumour-related epilepsy (Ponten *et al.*, 2007). Furthermore, the changes in network topology are correlated with cognitive deficits, and possibly reflect some compensatory mechanisms (Bosma *et al.*, 2009).

In conclusion, research suggest that tumour-related epilepsy and cognitive deficits are both associated with changes in functional connectivity patterns resulting in more random brain networks (Reijneveld *et al.*, 2007; Stam & Reijneveld, 2007; Brogna *et al.*, 2008). We would like to speculate that brain tumours change the brain's network topology, hereby causing epileptic seizures and cognitive problems. In the beginning, tumour cells invade the healthy brain tissue, changing neurobiological and neuroimmunological features of the neurons on a very local scale. As the neoplastic cells multiply, a larger area of the brain displays altered functioning. Since the healthy brain is likely to be a small-world network, these mostly local dysfunctions gradually begin to impact other, more remote clusters of specialised action, eventually changing architectural characteristics throughout the brain (Honey & Sporns, 2008). Based on the compensatory mechanisms mentioned in several studies, we hypothesise that this pattern of network randomisation and regularisation goes relatively unnoticed for quite some time, as increased effort is applied to neutralise it. At some point, a threshold may be reached and a phase transition occurs, after which the network shifts and decreased efficiency starts to become expressed in symptomatology. Cognition may be characterised by global dysfunctioning, due to the pathologically random network topology that is now present across the brain. In addition, the randomness of the brain may cause hypersynchronisation and epileptic seizures. During the seizure, network topology goes from random interictally to more small-world-like at the end of the seizure, possibly terminating the seizure and 'resetting' the pathological status until a critical point is reached again. Defining the critical point or threshold after which the network undergoes a phase transition and becomes pathological may be crucial for our understanding of neuro-oncological disease and its consequences. Moreover, this hypothesis of network decompensation can also be applied to other neurological disorders. Thus, we believe that clarifying the process proposed above will not only elucidate the effects of a tumour on the surrounding brain tissue, but can also provide us with a more general paradigm of (optimal) brain functioning. Ideally, new treatments will be developed.

Although clues to solve the multifaceted problems in neuro-oncology are accumulating, important issues are to be resolved in future studies. First, when using EEG and MEG, volume conduction and the inverse problem are always of influence on the signals used for analysis. However, the impact of these factors on graph theoretical measures has not been determined at this point, while it may be of importance in the calculation of especially the clustering coefficient. Second, the ambiguous and unclear optimal methodology of network analysis is still an important complication. The best method of converting functional imaging data (either fMRI, EEG or MEG) to graphs that can be analysed by using network theory has not been defined yet. A third issue is that most network analyses are performed in signal space at the moment, while algorithms for source reconstruction have not been defined. Fourth, the threshold that is used to convert correlational matrices to an unweighted binary graph is rather arbitrarily chosen, while calculation of a whole range of thresholds causes an accumulation of type II errors because of multiple testing. Converting correlational graphs to weighted graphs may be a solution to this problem, but sparse methods are available to do this. Also, when using weighted graphs, differences in synchronisation between the groups that are compared can contaminate network analysis and produce inaccurate results. A fifth problem that frequently occurs when converting matrices of correlations to graphs is the fact that some of the nodes may become disconnected from the network; this presents difficulties in the calculation of clustering coefficients and path lengths.

A framework to address some of these problems has been proposed through defining the efficiency of the path between two vertices as the inverse of the shortest distance between the vertices (note that in weighted graphs the shortest path is not necessarily the path with the smallest number of edges; Latora & Marchiori, 2001, 2003; Vragovic *et al.*, 2005). Other definitions of C have been proposed for weighted networks (Barrat *et al.*,

2004; Onnela *et al.*, 2005). However, very few studies compare several methods of network analysis, which could improve measure selection in the future. New and better measures would be very helpful in our network quest. In order to find the optimal method of analysis, collaboration between neuroscientists, mathematicians and physicists may prove indispensible.

A number of conceptual issues for future studies also deserve mentioning. First of all, it remains unknown which network properties are the best predictors of cognitive deficits, and it is still to be defined what the exact correlation is between network properties and susceptibility to seizures. Longitudinal studies researching the evolution of epilepsy and its relation with network status are key to answering this question, since current studies are all based on cross-sectional investigations and cannot address the evolution of network topology as the epileptic disorder progresses and develops. Furthermore, a longitudinal approach would minimise inter-individual variation of our measurements. The combination of several measurement methods (EEG, MEG, fMRI, ECoG) may be essential for our complete understanding of this issue. Second, our hypothesis concerning network randomness-related epilepsy and cognitive dysfunction in brain tumour patients needs to be further explored. It is particularly difficult to separate these three notions, in order to later investigate their mutual associations. Moreover, diverse types of brain tumour may very well induce subtly different changes in network topology. Third, to place tumour patients in a broader view of brain disease, it would be interesting to sort out whether different

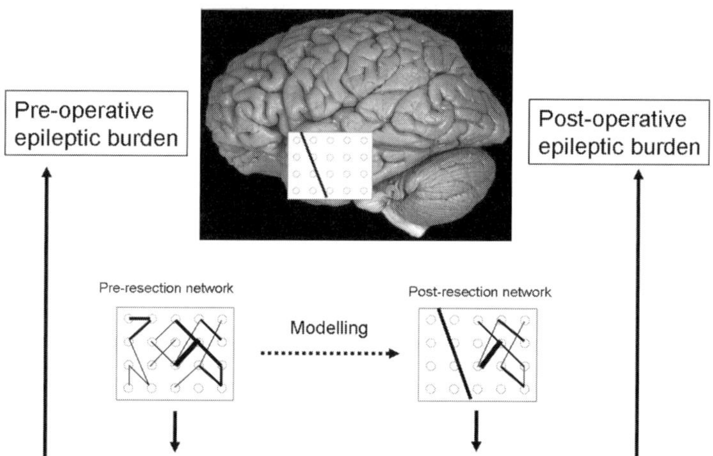

Figure 8.7. Prediction of the effect of surgical resection on network changes and seizure outcome. Schematic representation of a simulation model. Pre-, intra- and post-operative data on seizure burden and network characteristics are used to construct a model, and investigate how changes of network characteristics through different simulated interventions in the model correlate with the network changes induced by actual intervention and with clinical outcome.

pathological processes change network structure and function in different ways. It would be particularly interesting to know if different types of brain disease lead to either random error or targeted attack of brain networks, and when the threshold for phase transition is reached. This would enable us to predict when and how brain disease gives rise to clinical symptoms. Finally, the clinical application of network analysis is the ultimate goal of our efforts. In the case of brain tumour patients, it has already been shown that the functionality of brain tissue can be predicted by means of functional connectivity (Guggisberg *et al.*, 2008), and that network analysis can help determine whether regions of the brain are pathological (Ortega *et al.*, 2008a 2008b). While invasive imaging and stimulation are currently in use, it is hoped that future network-based strategies can minimise treatment burden for these patients. Moreover, treatment may also be planned based on network analysis. Better guided resection of the tumour and epilepsy-related areas may possibly be performed if the exact location of essential regions of the brain's network are known (see Figure 8.7). In the end, we hope to minimise burdensome cognitive deficits and help alleviate the devastating epileptic seizures these patients experience.

References

Aarabi, A., Wallois, F. *et al.* 2008. Does spatiotemporal synchronization of EEG change prior to absence seizures? *Brain Res* **1188**, 207–221.

Achard, S. & Bullmore, E. 2007. Efficiency and cost of economical brain functional networks. *PLoS Comput Biol* **3**(2), e17.

Achard, S., Salvador, R. *et al.* 2006. A resilient, low-frequency, small-world human brain functional network with highly connected association cortical hubs. *J Neurosci* **26**(1), 63–72.

Aertsen, A. M., Gerstein, G. L. *et al.* 1989. Dynamics of neuronal firing correlation: modulation of "effective connectivity". *J Neurophysiol* **61**(5), 900–917.

Anand, A., Li, Y. *et al.* 2005. Activity and connectivity of brain mood regulating circuit in depression: a functional magnetic resonance study. *Biol Psychiatry* **57**(10), 1079–1088.

Anderson, S. W., Damasio, H. *et al.* 1990. Neuropsychological impairments associated with lesions caused by tumor or stroke. *Arch Neurol* **47**(4), 397–405.

Atay, F. M. & Biyikoglu, T. 2005. Graph operations and synchronization of complex networks. *Phys Rev E Stat Nonlin Soft Matter Phys* **72**(1 Pt 2), Art. no. 016217.

Atay, F. M., Jost, J., *et al.* 2004. Delays, connection topology, and synchronization of coupled chaotic maps. *Phys Rev Lett* **92**(14), Art. no. 144101.

Auer, D. P. 2008. Spontaneous low-frequency blood oxygenation level-dependent fluctuations and functional connectivity analysis of the 'resting' brain. *Magn Reson Imaging* **26**, 1055–1064.

Barahona, M. & Pecora, L. M. 2002. Synchronization in small-world systems. *Phys Rev Lett* **89**(5), Art. no. 054101.

Barrat, A., Barthelemy, M. *et al.* 2004. The architecture of complex weighted networks. *Proc Natl Acad Sci USA* **101**(11), 3747–3752.

Barthelemy, M., Barrat, A. *et al.* 2005. Characterization and modelling of weighted networks. *Physica A* **346**, 34–43.

Bartolomei, F., Bosma, I. *et al.* 2006a. Disturbed functional connectivity in brain tumour patients: evaluation by graph analysis of synchronization matrices. *Clin Neurophysiol* **117**(9), 2039–2049.

Bartolomei, F., Bosma, I. *et al.* 2006b. How do brain tumors alter functional connectivity? A magnetoencephalography study. *Ann Neurol* **59**(1), 128–138.

Bartolomei, F., Wendling, F. *et al.* 2008. [The concept of an epileptogenic network in human partial epilepsies]. *Neurochirurgie* **54**(3), 174–184.

Bassett, D. S. & Bullmore, E. 2006. Small-world brain networks. *Neuroscientist* **12**(6), 512–523.

Bassett, D. S., Meyer-Lindenberg, A. *et al.* 2006. Adaptive reconfiguration of fractal small-world human brain functional networks. *Proc Natl Acad Sci USA* **103**(51), 19518–19523.

Beaumont, A. & Whittle, I. R. 2000. The pathogenesis of tumour associated epilepsy. *Acta Neurochir (Wien)* **142**(1), 1–15.

Bettus, G., Guedj, E. *et al.* 2008a. Decreased basal fMRI functional connectivity in epileptogenic networks and contralateral compensatory mechanisms. *Hum Brain Mapp.*

Bettus, G., Wendling, F., *et al.* 2008b. Enhanced EEG functional connectivity in mesial temporal lobe epilepsy. *Epilepsy Res.*

Biswal, B., Yetkin, F. Z. *et al.* 1995. Functional connectivity in the motor cortex of resting human brain using echo-planar MRI. *Magn Reson Med* **34**(4), 537–541.

Boccaletti, S., Latora, V. *et al.* 2006. Complex networks: structure and dynamics. *Phys Reports* **424**, 175–308.

Bosma, I., Douw, L. *et al.* 2008. Synchronized brain activity and neurocognitive function in patients with low-grade glioma: a magnetoencephalography study. *Neuro Oncol.*

Bosma, I., Reijneveld, J. C. *et al.* 2009. Disturbed functional brain networks and neurocognitive function in low-grade glioma patients: a graph theoretical analysis of resting-state MEG *Nonlinear Biomed Phys* **3**(1), 9.

Brenner, R. P. 2002. Is it status? *Epilepsia* **43 (Suppl. 3**), 103–113.

Bressler, S. 2002. Understanding cognition through large-scale cortical networks. *Current Directions in Psychological Science* **11**, 58–61.

Bressler, S. L. 1995. Large-scale cortical networks and cognition. *Brain Res Brain Res Rev* **20**(3), 288–304.

Brogna, C., Gil Robles, S. *et al.* 2008. Brain tumors and epilepsy. *Expert Rev Neurother* **8**(6), 941–955.

Bromfield, E. B. 2004. Epilepsy in patients with brain tumors and other cancers. *Rev Neurol Dis* **1(Suppl. 1**), S27–S33.

Buckner, J. C. 2003. Factors influencing survival in high-grade gliomas. *Semin Oncol* **30**(6 Suppl. 19), 10–4.

Cao, Q., Zang, Y. *et al.* 2006. Abnormal neural activity in children with attention deficit hyperactivity disorder: a resting-state functional magnetic resonance imaging study. *Neuroreport* **17**(10), 1033–1036.

Chavez, M., Hwang, D. U. *et al.* 2006. Synchronizing weighted complex networks. *Chaos* **16**(1), Art. no. 015106.

Chavez, M., Hwang, D. U. *et al.* 2005. Synchronization is enhanced in weighted complex networks. *Phys Rev Lett* **94**(21), Art. no. 218701.

Curran, W. J., Jr., Scott, C. B. *et al.* 1993. Recursive partitioning analysis of prognostic factors in three Radiation Therapy Oncology Group malignant glioma trials. *J Natl Cancer Inst* **85**(9), 704–710.

De Vico Fallani, F., Astolfi, L. *et al.* 2007. Cortical functional connectivity networks in normal and spinal cord injured patients: evaluation by graph analysis. *Hum Brain Mapp* **28**(12), 1334–1346.

DeAngelis, L. M. 2001. Brain tumors. *N Engl J Med* **344**(2), 114–123.

Desmurget, M., Bonnetblanc, F. *et al.* 2007. Contrasting acute and slow-growing lesions: a new door to brain plasticity. *Brain* **130**(Pt 4), 898–914.

Dodel, S., Hermann, J. M. *et al.* 2002. Functional connectivity by cross-correlation clustering. *Neurocomputing* **44–46**, 1065–1070.

Donetti, L., Hurtado, P. I. *et al.* 2005. Entangled networks, synchronization, and optimal network topology. *Phys Rev Lett* **95**(18), Art. no. 188701.

Douw, L., Baayen, H., *et al.* 2008. Treatment-related changes in functional connectivity in brain tumor patients: a magnetoencephalography study. *Exp Neurol* **212**, 285–290.

Douw, L., Baayen, J. C. *et al.* 2009. Functional connectivity in the brain before and during intra-arterial amobarbital injection (Wada test) *Neuroimage* **46**(3), 584–588.

Dyhrfjeld-Johnsen, J., Santhakumar, V. *et al.* 2007. Topological determinants of epileptogenesis in large-scale structural and functional models of the dentate gyrus derived from experimental data. *J Neurophysiol* **97**(2), 1566–1587.

Eguiluz, V. M., Chialvo, D. R. *et al.* 2005. Scale-free brain functional networks. *Phys Rev Lett* **94**(1), Art. no. 018102.

Erdos, P. & Renyi, A. 1960. On the evolution of random graphs. *Publications Mathemat Inst Hungarian Acad Sci* **12**, 17–61.

Felleman, D. J. & Van Essen, D. C. 1991. Distributed hierarchical processing in the primate cerebral cortex. *Cereb Cortex* **1**(1), 1–47.

Forsgren, L. 1990. Prospective incidence study and clinical characterization of seizures in newly referred adults. *Epilepsia* **31**(3), 292–301.

French, D. A. & Gruenstein, E. I. 2006. An integrate-and-fire model for synchronized bursting in a network of cultured cortical neurons. *J Comput Neurosci* **21**(3), 227–241.

Guggisberg, A. G., Honma, S. M. *et al.* 2008. Mapping functional connectivity in patients with brain lesions. *Ann Neurol* **63**, 193–203.

Guye, M., Bartolomei, F. *et al.* 2008. Imaging structural and functional connectivity: towards a unified definition of human brain organization? *Curr Opin Neurol* **24**(4), 393–403.

Guye, M., Regis, J. *et al.* 2006. The role of corticothalamic coupling in human temporal lobe epilepsy. *Brain* **129**(7), 1917–1928.

Hagmann, P., Kurant, M. *et al.* 2007. Mapping human whole-brain structural networks with diffusion MRI. *PLoS ONE* **2**(7), e597.

Harmony, T., Marosi, E. *et al.* 1994. EEG coherences in patients with brain lesions. *Int J Neurosci* **74**(1–4), 203–226.

He, Y., Chen, Z. J. *et al.* 2007. Small-world anatomical networks in the human brain revealed by cortical thickness from MRI. *Cereb Cortex* **17**(10), 2407–2419.

Hilgetag, C. C., Burns, G. A. *et al.* 2000. Anatomical connectivity defines the organization of clusters of cortical areas in the macaque monkey and the cat. *Phil Trans R Soc B* **355**(1393), 91–110.

Hochberg, F. H. & Slotnick, B. 1980. Neuropsychologic impairment in astrocytoma survivors. *Neurology* **30**(2), 172–177.

Hom, J. & Reitan, R. M. 1984. Neuropsychological correlates of rapidly vs. slowly growing intrinsic cerebral neoplasms. *J Clin Neuropsychol* **6**(3), 309–324.

Honey, C. J., Kotter, R. *et al.* 2007. Network structure of cerebral cortex shapes functional connectivity on multiple time scales. *Proc Natl Acad Sci USA* **104**(24), 10240–10245.

Honey, C. J. & Sporns, O. 2008. Dynamical consequences of lesions in cortical networks. *Hum Brain Mapp* **29**(7), 802–809.

Hong, H., Choi, M. Y. *et al.* 2002. Synchronization on small-world networks. *Phys Rev E Stat Nonlin Soft Matter Phys* **65**(2 Pt 2), Art. no. 026139.

Humphries, M. D. & Gurney, K. 2008. Network 'small-world-ness': a quantitative method for determining canonical network equivalence. *PLoS ONE* **3**(4), e0002051.

Humphries, M. D., Gurney, K. *et al.* 2006. The brainstem reticular formation is a small-world, not scale-free, network. *Proc R Soc B* **273**(1585), 503–511.

Imperato, J. P., Paleologos, N. A. *et al.* 1990. Effects of treatment on long-term survivors with malignant astrocytomas. *Ann Neurol* **28**(6), 818–822.

Iturria-Medina, Y., Sotero, R. C. *et al.* 2008. Studying the human brain anatomical network via diffusion-weighted MRI and Graph Theory. *Neuroimage* **40**(3), 1064–1076.

Kaiser, M. & Hilgetag, C. C. 2006. Nonoptimal component placement, but short processing paths, due to long-distance projections in neural systems. *PLoS Comput Biol* **2**(7), e95.

Klein, M., Engelberts, N. H. *et al.* 2003a. Epilepsy in low-grade gliomas: the impact on cognitive function and quality of life. *Ann Neurol* **54**(4), 514–520.

Klein, M. & Heimans, J. J. 2004. The measurement of cognitive functioning in low-grade glioma patients after radiotherapy. *J Clin Oncol* **22**(5), 966–967; author reply 967–968.

Klein, M., Heimans, J. J. *et al.* 2002. Effect of radiotherapy and other treatment-related factors on mid-term to long-term cognitive sequelae in low-grade gliomas: a comparative study. *Lancet* **360**(9343), 1361–1368.

Klein, M., Postma, T. J. *et al.* 2003b. The prognostic value of cognitive functioning in the survival of patients with high-grade glioma. *Neurology* **61**(12), 1796–1798.

Klein, M., Taphoorn, M. J. *et al.* 2001. Neurobehavioral status and health-related quality of life in newly diagnosed high-grade glioma patients. *J Clin Oncol* **19**(20), 4037–4047.

Kotter, R. & Sommer, F. T. 2000. Global relationship between anatomical connectivity and activity propagation in the cerebral cortex. *Phil Trans R Soc B* **355**(1393), 127–134.

Kozma, R., Puljic, M. *et al.* 2005. Phase transitions in the neuropercolation model of neural populations with mixed local and non-local interactions. *Biol Cybern* **92**(6), 367–379.

Kramer, M. A., Kolaczyk, E. D. *et al.* 2008. Emergent network topology at seizure onset in humans. *Epilepsy Res* **79**(2–3), 173–186.

Lago-Fernandez, L. F., Huerta, R. *et al.* 2000. Fast response and temporal coherent oscillations in small-world networks. *Phys Rev Lett* **84**(12), 2758–2761.

Langheim, F. J., Leuthold, A. C. *et al.* 2006. Synchronous dynamic brain networks revealed by magnetoencephalography. *Proc Natl Acad Sci USA* **103**(2), 455–459.

Latora, V. & Marchiori, M. 2001. Efficient behavior of small-world networks. *Phys Rev Lett* **87**(19), Art. no. 198701.

Latora, V. & Marchiori, M. 2003. Economic small-world behavior in weighted networks. *Eur Phys* **32**, 249–263.

Laufs, H. 2008. Endogenous brain oscillations and related networks detected by surface EEG-combined fMRI. *Hum Brain Mapp* **29**(7), 762–769.

Le Van Quyen, M., Soss, J. *et al.* 2005. Preictal state identification by synchronization changes in long-term intracranial EEG recordings. *Clin Neurophysiol* **116**(3), 559–568.

Lee, D. S. 2005. Synchronization transition in scale-free networks: clusters of synchrony. *Phys Rev E Stat Nonlin Soft Matter Phys* **72**(2 Pt 2), Art. no. 026208.

Levine, V. A., Leibel, S. A. *et al.* 1993. Neoplasms of the central nervous system. *Cancer: Principles and Practice of Oncology, 4th Edn.* (ed. V. T. DeVita, S. Hellman & S. A. Rosenberg), pp. 2022–2032. J.B. Lippincott.

Liu, Y., Liang, M. *et al.* 2008. Disrupted small-world networks in schizophrenia. *Brain* **131**(4), 945–961.

Lopes da Silva, F., Blanes, W. *et al.* 2003. Epilepsies as dynamical diseases of brain systems: basic models of the transition between normal and epileptic activity. *Epilepsia* **44(Suppl. 12**), 72–83.

Lote, K., Stenwig, A. E. *et al.* 1998. Prevalence and prognostic significance of epilepsy in patients with gliomas. *Eur J Cancer* **34**(1), 98–102.

Lowe, M. J., Mock, B. J. *et al.* 1998. Functional connectivity in single and multislice echoplanar imaging using resting-state fluctuations. *Neuroimage* **7**(2), 119–132.

Lowe, M. J., Phillips, M. D. *et al.* 2002. Multiple sclerosis: low-frequency temporal blood oxygen level-dependent fluctuations indicate reduced functional connectivity initial results. *Radiology* **224**(1), 184–192.

Lytton, W. W. 2008. Computer modelling of epilepsy. *Nat Rev Neurosci* **9**(8), 626–637.

Masuda, N. & Aihara, K. 2004. Global and local synchrony of coupled neurons in small-world networks. *Biol Cybern* **90**(4), 302–309.

McIntosh, A. R. 2000. Towards a network theory of cognition. *Neural Netw* **13**(8–9), 861–870.

Meyer-Lindenberg, A., Bauer, U. *et al.* 1998. The topography of non-linear cortical dynamics at rest, in mental calculation and moving shape perception. *Brain Topogr* **10**(4), 291–299.

Micheloyannis, S., Pachou, E. *et al.* 2006a. Small-world networks and disturbed functional connectivity in schizophrenia. *Schizophr Res* **87**(1–3), 60–66.

Micheloyannis, S., Pachou, E. *et al.* 2006b. Using graph theoretical analysis of multi channel EEG to evaluate the neural efficiency hypothesis. *Neurosci Lett* **402**(3), 273–277.

Micheloyannis, S., Vourkas, M. *et al.* 2003. Changes in linear and nonlinear EEG measures as a function of task complexity: evidence for local and distant signal synchronization. *Brain Topogr* **15**(4), 239–247.

Milgram, S. 1967. The small world problem. *Psychol Today* **2**, 60–67.

Montez, T., Linkenkaer-Hansen, K. *et al.* 2006. Synchronization likelihood with explicit time-frequency priors. *Neuroimage* **33**(4), 1117–1125.

Mormann, F., Kreuz, T. *et al.* 2003. Epileptic seizures are preceded by a decrease in synchronization. *Epilepsy Res* **53**(3), 173–185.

Motter, A. E., Mattias, M. A. *et al.* 2006. Dynamics on complex networks and applications. *Physica D* **224**, vii–viii.

Netoff, T. I., Clewley, R. *et al.* 2004. Epilepsy in small-world networks. *J Neurosci* **24**(37), 8075–8083.

Newman, M. E. 2003. The structure and function of complex networks. *SIAM Review* **45**, 167–256.

Newman, M. E. 2004. Analysis of weighted networks. *Phys Rev E Stat Nonlin Soft Matter Phys* **70**(5 Pt 2), Art. no. 056131.

Nishikawa, T., Motter, A. E. *et al.* 2003. Heterogeneity in oscillator networks: are smaller worlds easier to synchronize? *Phys Rev Lett* **91**(1), Art. no. 014101.

Nolte, G., Bai, O. *et al.* 2004. Identifying true brain interaction from EEG data using the imaginary part of coherency. *Clin Neurophysiol* **115**(10), 2292–2307.

Onnela, J. P., Saramaki, J. *et al.* 2005. Intensity and coherence of motifs in weighted complex networks. *Phys Rev E Stat Nonlin Soft Matter Phys* **71**(6 Pt 2), Art. no. 065103.

Ortega, G. J., Menendez de la Prida, L. *et al.* 2008a. Synchronization clusters of interictal activity in the lateral temporal cortex of epileptic patients: intraoperative electrocorticographic analysis. *Epilepsia* **49**(2), 269–280.

Ortega, G. J., Sola, R. G. *et al.* 2008b. Complex network analysis of human ECoG data. *Neurosci Lett* **447**(2–3), 129–133.

Park, K., Lai, Y. C. *et al.* 2004. Characterization of weighted complex networks. *Phys Rev E Stat Nonlin Soft Matter Phys* **70**(2 Pt 2), Art. no. 026109.

Percha, B., Dzakpasu, R. *et al.* 2005. Transition from local to global phase synchrony in small world neural network and its possible implications for epilepsy. *Phys Rev E Stat Nonlin Soft Matter Phys* **72**(3 Pt 1), Art. no. 031909.

Pijnenburg, Y. A., v d Made, Y. *et al.* 2004. EEG synchronization likelihood in mild cognitive impairment and Alzheimer's disease during a working memory task. *Clin Neurophysiol* **115**(6), 1332–1339.

Ponten, S. C., Bartolomei, F. *et al.* 2007. Small-world networks and epilepsy: graph theoretical analysis of intracerebrally recorded mesial temporal lobe seizures. *Clin Neurophysiol* **118**(4), 918–927.

Ponten, S. C., Douw, L. *et al.* 2009. Indications for network analysis during absence seizures: weighted and unweighted graph theoretical analyses *Exp Neurol* **217**(1), 197–204.

Quigley, M., Cordes, D. *et al.* 2001. Effect of focal and nonfocal cerebral lesions on functional connectivity studied with MR imaging. *Am J Neuroradiol* **22**(2), 294–300.

Reijneveld, J. C., Ponten, S. C. *et al.* 2007. The application of graph theoretical analysis to complex networks in the brain. *Clin Neurophysiol* **118**(11), 2317–2331.

Reijneveld, J. C., Sitskoorn, M. M. *et al.* 2001. Cognitive status and quality of life in patients with suspected versus proven low-grade gliomas. *Neurology* **56**(5), 618–623.

Roxin, A., Riecke, H. *et al.* 2004. Self-sustained activity in a small-world network of excitable neurons. *Phys Rev Lett* **92**(19), Art. no. 198101.

Salvador, R., Suckling, J. *et al.* 2005a. Neurophysiological architecture of functional magnetic resonance images of human brain. *Cereb Cortex* **15**(9), 1332–1342.

Salvador, R., Suckling, J. *et al.* 2005b. Undirected graphs of frequency-dependent functional connectivity in whole brain networks. *Phil Trans R Soc B* **360**(1457), 937–946.

Schevon, C. A., Cappell, J. *et al.* 2007. Cortical abnormalities in epilepsy revealed by local EEG synchrony. *Neuroimage* **35**(1), 140–148.

Schindler, K., Elger, C. E. *et al.* 2007a. Increasing synchronization may promote seizure termination: evidence from status epilepticus. *Clin Neurophysiol* **118**(9), 1955–1968.

Schindler, K., Leung, H. *et al.* 2007b. Assessing seizure dynamics by analysing the correlation structure of multichannel intracranial EEG. *Brain* **130**(Pt 1), 65–77.

Schindler, K. A., Bialonski, S. *et al.* 2008. Evolving functional network properties and synchronizability during human epileptic seizures. *Chaos* **18**, DOI:10.1063/1.2966112

Singer, W. 1999. Neuronal synchrony: a versatile code for the definition of relations? *Neuron* **24**(1), 49–65, 111–25.

Solomonov, R. & Rapoport, A. 1951. Connectivity of random nets. *Bull Math Biophys* **13**, 107–117.

Sporns, O. & Tononi, G. 2002. Classes of network connectivity and dynamics. *Complexity* **7**, 28–38.

Sporns, O., Tononi, G. *et al.* 2000a. Connectivity and complexity: the relationship between neuroanatomy and brain dynamics. *Neural Netw* **13**(8–9), 909–922.

Sporns, O., Tononi, G. *et al.* 2000b. Theoretical neuroanatomy: relating anatomical and functional connectivity in graphs and cortical connection matrices. *Cereb Cortex* **10**(2), 127–141.

Sporns, O. & Zwi, J. D. 2004. The small world of the cerebral cortex. *Neuroinformatics* **2**(2), 145–162.

Srinivas, K. V., Jain, R. *et al.* 2007. Small-world network topology of hippocampal neuronal network is lost, in an in vitro glutamate injury model of epilepsy. *Eur J Neurosci* **25**(11), 3276–3286.

Stam, C. J. 2004. Functional connectivity patterns of human magnetoencephalographic recordings: a 'small-world' network? *Neurosci Lett* **355**(1–2), 25–28.

Stam, C. J. 2005. Nonlinear dynamical analysis of EEG and MEG: review of an emerging field. *Clin Neurophysiol* **116**(10), 2266–2301.

Stam, C. J., Haan, W. d. *et al.* 2008. Graph theoretical analysis of magnetoencephalographic functional connectivity in Alzheimer's disease. *Brain.*

Stam, C. J., Jones, B. F. *et al.* 2007a. Small-world networks and functional connectivity in Alzheimer's disease. *Cereb Cortex* **17**(1), 92–99.

Stam, C. J., Nolte, G. *et al.* 2007b. Phase lag index: assessment of functional connectivity from multi channel EEG and MEG with diminished bias from common sources. *Hum Brain Mapp* **28**(11), 1178–1193.

Stam, C. J. & Reijneveld, J. C. 2007. Graph theoretical analysis of complex networks in the brain. *Nonlinear Biomed Phys* **1**(1), 3.

Stam, C. J., van Cappellen van Walsum, A. M. *et al.* 2002a. Variability of EEG synchronization during a working memory task in healthy subjects. *Int J Psychophysiol* **46**(1), 53–66.

Stam, C. J., van Cappellen van Walsum, A. M. *et al.* 2002b. Generalized synchronization of MEG recordings in Alzheimer's disease: evidence for involvement of the gamma band. *J Clin Neurophysiol* **19**(6), 562–574.

Stam, C. J. & van Dijk, B. W. 2002. Synchronization likelihood: an unbiased measure of generalized synchronization in multivariate data sets. *Physica D* **163**(3–4), 236–241.

Stephan, K. E., Hilgetag, C. C. *et al.* 2000. Computational analysis of functional connectivity between areas of primate cerebral cortex. *Phil Trans R Soc B* **355**(1393), 111–126.

Supekar, K., Menon, V. *et al.* 2008. Network analysis of intrinsic functional brain connectivity in Alzheimer's disease. *PLoS Comput Biol* **4**(6), e1000100.

Taphoorn, M. J., Heimans, J. J. *et al.* 1994a. Quality of life and neuropsychological functions in long-term low-grade glioma survivors. *Int J Radiat Oncol Biol Phys* **29**(5), 1201–1202.

Taphoorn, M. J., Heimans, J. J. *et al.* 1992. Assessment of quality of life in patients treated for low-grade glioma: a preliminary report. *J Neurol Neurosurg Psychiatry* **55**(5), 372–376.

Taphoorn, M. J. & Klein, M. 2004. Cognitive deficits in adult patients with brain tumours. *Lancet Neurol* **3**(3), 159–168.

Taphoorn, M. J., Schiphorst, A. K. *et al.* 1994b. Cognitive functions and quality of life in patients with low-grade gliomas: the impact of radiotherapy. *Ann Neurol* **36**(1), 48–54.

Tatter, S. B., Wilson, C. B. *et al.* 1996. Neuroepithelial tumors of the adult brain. In *Neurological Surgery*, 4th Edn. (ed. J. R. Youmans), Vol. 4: Tumors, pp. 2612–2684. W.B. Saunders.

Tononi, G. & Edelman, G. M. 1998. Consciousness and complexity. *Science* **282**(5395), 1846–1851.

van Dellen, E., Douw, L. *et al.* 2009. Long-term effects of temporal lobe epilepsy on local neural networks: a graph theoretical analysis of corticography recordings. *PLoS One* **4**(11), 8081.

van den Berg & van Leeuwen, C. 2004. Adaptive rewiring in chaotic networks renders small-world connectivity with consistent clusters. *Europhysics Letters* **65**, 459–464.

van Vreeswijk, C. & Sompolinsky, H. 1996. Chaos in neuronal networks with balanced excitatory and inhibitory activity. *Science* **274**(5293), 1724–1726.

Villemure, J. G. & de Tribolet, N. 1996. Epilepsy in patients with central nervous system tumors. *Curr Opin Neurol* **9**(6), 424–428.

Vragovic, I., Louis, E. *et al.* 2005. Efficiency of informational transfer in regular and complex networks. *Phys Rev E Stat Nonlin Soft Matter Phys* **71**(3 Pt 2A), Art. no. 036122.

Wang, J., Wang, L. *et al.* 2008. Parcellation-dependent small-world brain functional networks: a resting-state fMRI study. *Hum Brain Mapp.*

Watts, D. J. & Strogatz, S. H. 1998. Collective dynamics of 'small-world' networks. *Nature* **393**(6684), 440–442.

Wessels, P. H., Weber, W. E. *et al.* 2003. Supratentorial grade II astrocytoma: biological features and clinical course. *Lancet Neurol* **2**(7), 395–403.

Wu, H., Li, X. & Guan, X. 2006. Networking property during epileptic seizure with multi-channel EEG recordings. *Lecture Notes Comput Sci* **3976**, 573–578.

Yu, S., Huang, D. *et al.* 2008. A small world of neuronal synchrony. *Cereb Cortex.*

Zemanova, L., Zou, C. *et al.* 2006. Structural and functional clusters of complex brain networks. *Physica D* **224**, 202–212.

Zhou, C. & Kurths, J. 2006. Dynamical weights and enhanced synchronization in adaptive complex networks. *Phys Rev Lett* **96**(16), Art. no. 164102.

Zhou, C., Motter, A. E. *et al.* 2006a. Universality in the synchronization of weighted random networks. *Phys Rev Lett* **96**(3), Art. no. 034101.

Zhou, C., Zemanova, L. *et al.* 2006b. Hierarchical organization unveiled by functional connectivity in complex brain networks. *Phys Rev Lett* **97**(23), Art. no. 238103.

Zhou, H. & Lipowsky, R. 2005. Dynamic pattern evolution on scale-free networks. *Proc Natl Acad Sci USA* **102**(29), 10052–10057.

Part III

Artificial neural networks as models of perceptual processing in ecology and evolutionary biology

9

Evolutionary diversification of mating behaviour: using artificial neural networks to study reproductive character displacement and speciation

Karin S. Pfennig and Michael J. Ryan

9.1 Introduction

When species with similar sexual signals co-occur, selection may favour divergence of these signals to minimise either their interference or the risk of mis-mating between species, a process termed reproductive character displacement (Howard, 1993; Andersson, 1994; Servedio & Noor, 2003; Coyne & Orr, 2004; Pfennig & Pfennig, 2009). This selective process potentially results in mating behaviours that are not only divergent between species that co-occur but that are also divergent among conspecific populations that do and do not occur with heterospecifics or that co-occur with different heterospecifics (reviewed in Howard, 1993; Andersson, 1994; Gerhardt & Huber, 2002; Coyne & Orr, 2004; e.g., Noor, 1995; Saetre *et al.*, 1997; Pfennig, 2000; Gabor & Ryan, 2001; Höbel & Gerhardt, 2003).

An oft-used approach to assessing whether reproductive character displacement has occurred between species relies on behavioural experiments that evaluate mate preferences from populations that do and do not occur with heterospecifics (sympatry and allopatry, respectively). In such experiments, individuals are presented the signals of heterospecifics and/or conspecifics to assess whether allopatric individuals are more likely to mistakenly prefer heterospecifics than are sympatric individuals (reviewed in Howard, 1993). The expectation is that individuals from sympatry should preferentially avoid heterospecifics, whereas those in allopatry should fail to distinguish heterospecifics from conspecifics (presumably because, unlike sympatric individuals, they have not been under selection to do so). Such patterns of discrimination have been observed, and they provide some of the strongest examples of reproductive character displacement (reviewed in Howard, 1993).

A problem with this approach, however, is that allopatric females also may discriminate against heterospecifics depending on how heterospecific signals vary relative to preferred conspecific signals (Rodriguez *et al.*, 2004; Pfennig & Ryan, 2007). If heterospecific signals possess characters that are disfavoured by allopatric females, heterospecific signals may be selected against even if they have never been encountered. For

Modelling Perception with Artificial Neural Networks, ed. C. R. Tosh and G. D. Ruxton. Published by Cambridge University Press.

example, if allopatric females prefer conspecific signals above a certain threshold frequency, heterospecifics will be selected against if they produce signals below that threshold. Thus, allopatric females, like sympatric females, may discriminate against heterospecifics.

Finding that both sympatric and allopatric individuals can successfully discriminate against heterospecifics seemingly undermines support for character displacement. Yet, such a pattern should not necessarily rule out the possibility that character displacement has occurred. When selection to avoid heterospecific interactions is strong, mate preferences can still diverge in sympatry, thereby resulting in an *enhanced* ability to discriminate against heterospecifics in sympatry (e.g. Noor, 1995; Gabor & Ryan, 2001). Moreover, the process of mate recognition is probably not sensitive to conspecifics and heterospecifics as special categories. Instead, selection likely favours the expression of preferences for conspecific signals that minimise interspecific interference and maximise the likelihood of mating with the correct species. Thus, selection influencing decision criteria that decrease the chances of mating with heterospecifics could directly affect how females choose conspecifics (Gerhardt, 1994; Pfennig, 2000; Ryan & Getz, 2000; Hankison & Morris, 2003; Höbel & Gerhardt, 2003), not simply how well they discriminate conspecifics from heterospecifics (Ryan & Rand, 1993; Pfennig, 1998; Ryan & Getz, 2000).

Understanding whether and how reproductive character displacement proceeds is important for understanding how mating behaviours diversify and why species possess distinct mating signals. Indeed, reproductive character displacement can play a critical role in speciation: by favouring the evolution of traits that minimise the risk of mating between species, reproductive character displacement can *finalise* speciation by enhancing reproductive isolation between hybridising species (i.e. reinforcement; Servedio & Noor, 2003; Pfennig & Pfennig, 2009). Yet, reproductive character displacement can also potentially *initiate* speciation (Howard, 1993; Hoskin *et al.*, 2005; Pfennig & Ryan, 2006; Pfennig & Pfennig, 2009). If mating behaviours diverge between conspecific populations that do and do not occur with a given heterospecific, individuals may fail to accept conspecifics from the alternative population type as mates. If so, these conspecific populations may become reproductively isolated and ultimately undergo speciation as a result (Howard, 1993; Hoskin *et al.*, 2005; Pfennig & Ryan, 2006). Thus, reproductive character displacement sets the stage for future speciation events even as it operates to exaggerate differences between existing or incipient species (Figure 9.1).

Here, we provide an example of how artificial neural networks can be used to address these issues. Specifically, we have used artificial neural networks to evaluate how interactions with different heterospecifics affect the evolution of female preferences for conspecific signals. This work has shown that reproductive character displacement can lead to divergent preferences for the properties of signals females use to recognise conspecifics, without necessarily causing differences between populations in their discrimination against heterospecifics. Perhaps more critically, the work has also illustrated how reproductive character displacement may initiate speciation by contributing to reproductive isolation among populations that diverge in mating behaviour.

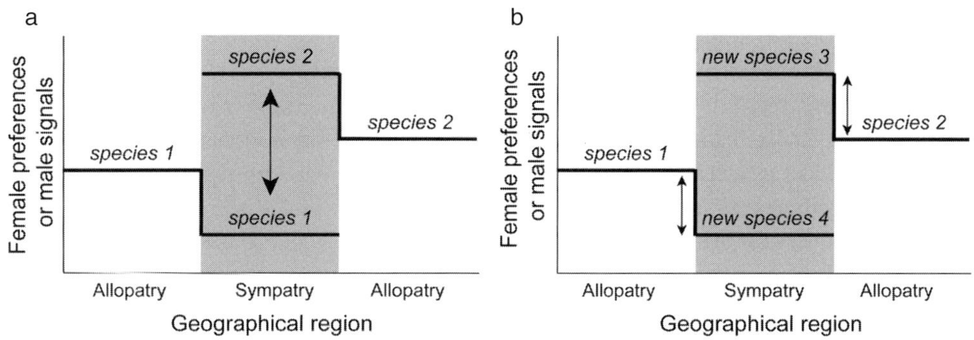

Figure 9.1. Reproductive character displacement can both finalise and initiate speciation. (a) Character displacement may finalise the speciation process by directly promoting the evolution of reproductive isolation between populations. When populations that have diverged in allopatry come together in only part of their geographical range (indicated by the shading), selection to minimise reproductive interference or hybridisation may exaggerate differences in mating behaviours (indicated by the doubled-headed arrow). Such divergence thereby enhances reproductive isolation between existing or incipient species. (b) Character displacement may also initiate speciation by indirectly promoting the evolution of reproductive isolation between conspecific populations. In particular, an indirect consequence of character displacement is that sympatric individuals will evolve different mate preferences and/or mate attraction signals than allopatric conspecifics. If male signals or female preferences diverge to the point that sympatric and allopatric individuals do not recognise each other as potential mates, reproductive isolation results. This process may eventually promote the formation of two new species (indicated here as 'new species 3' and 'new species 4'). Modified from Pfennig & Pfennig (2009) and Pfennig & Rice (2007).

9.2 Using artificial neural networks to study character displacement

Here, we describe our previously published work (Pffenig & Ryan, 2006, 2007) in which we used artificial neural networks to mimic the evolution of conspecific recognition in response to different heterospecific interactions. Artificial neural networks, also called connectionist models, consist of computational units ('neurons') that can stimulate or inhibit each other and are connected into networks. These interconnected units (networks) can simulate behaviour in response to an input and have been likened to the nervous system in function (Enquist & Ghirlanda, 2005).

Artificial neural network models are a potentially powerful tool for examining how mating behaviours diversify and the role of this diversification in speciation. Populations of networks can be generated that evolve mating behaviours under different selective contexts or that undergo different signalling interactions. Such models thereby allow for an understanding of how individual behaviours contribute to larger evolutionary patterns of diversification and speciation.

For example, neural network simulations have provided key insights into how both historical contingency and other species in the signalling environment influence how conspecific signals are recognised (Phelps & Ryan, 1998, 2000; Ryan & Getz, 2000;

Phelps *et al.*, 2001; Ryan *et al.* 2001). Indeed, Phelps & Ryan (1998, 2000) showed how the training of artificial neural networks could be used to mimic the past evolutionary history of frog calls to demonstrate how history influences recognition patterns of real female túngara frogs (*Physalaemus pustulosus*). Although they simulated a specific system, these studies came to the general conclusion that computational strategies used in mate recognition by current species are importantly influenced by the recognition strategies used by their ancestors. Such studies illustrate how artificial neural networks can be used as tools for better understanding evolutionary patterns and processes.

Artificial neural networks are particularly useful for investigating how mate recognition evolves among populations that co-occur with different heterospecifics. In natural systems, the presence of heterospecifics often covaries with changes in habitat or population evolutionary history. These factors can generate patterns that are consistent with, but that are not actually the result of, character displacement. Simulations with artificial neural networks provide a means for focusing on how signallers and receivers coevolve owing to heterospecific interactions in order to clarify the predictions that can be tested empirically. Indeed, as we note in our Discussion, simulations with artificial neural networks can identify how some empirical approaches may be overly conservative in their approach to character displacement.

9.3 The model

We mimicked a system in which males use pulsatile calls to attract females as mates (as occurs in many anuran and insect systems; Gerhardt & Huber, 2002). Although we simulated species recognition for acoustic signals, our results likely can be generalised to other sensory modalities. We based elements of our model on a naturally occurring spadefoot toad species, *Spea multiplicata*. As in many species, *S. multiplicata* occur with different species in different parts of their range in the southwestern region of the USA (Stebbins, 2003). In the eastern part of their range, for example, they co-occur with a congener, *S. bombifrons*. In the western part of their range, they occur with another spadefoot toad, *Scaphiopus couchii*. In still other populations they are the only spadefoot species present. These distributional patterns make *S. multiplicata* an excellent system for assessing how female behaviours evolve among disparate populations. We therefore used elements of this system to inform a model aimed at investigating how heterospecific interactions affect the evolution of female mate preferences among different populations. Because we did not model the spadefoot system explicitly, however, many features of our model differ markedly from the spadefoots' natural history. Our goal was not to mimic the spadefoot system per se, but to use this system to guide the modelling efforts described below.

We generated three population types consisting solely of networks belonging to the same species, 'species *A*'. Depending on the population type, the networks evolved conspecific recognition of advertisement signals of species *A* in the face of no heterospecific signals, or when faced with discrimination of signals from their own species versus signals from one of two heterospecific species. In particular, in one population type, networks were selected for the ability to discriminate representations of conspecific acoustic stimuli of 'species *A*' from white noise. The white noise stimulus controlled for the presence of a second stimulus and provided a means of assaying the networks' recognition of a conspecific signal. We refer to this population type as '*A*'. This population mimics the evolution of conspecific recognition in the absence of heterospecifics. In the second population type, 'species *A*' networks evolved to discriminate between conspecific stimuli of 'species *A*' and stimuli of a heterospecific, 'species *B*'. We refer to this population type as '*AB*'. Finally, in a third population type, networks evolved to discriminate between conspecific stimuli of 'species *A*', and stimuli from a second heterospecific, 'species *C*'. We refer to this population type as '*AC*'. For a list of definitions and usage of key terms, see Table 9.1.

We used the standard Elman network (Elman, 1990) available in the neural network toolbox in Matlab (Demuth & Beale, 1997). The network architecture consisted of a layer of 35 input neurons that received the stimulus (each neuron responded to a different frequency in the signal; see below for details of signal properties) and then fed this input forward to a single hidden layer of 23 neurons. Activity from this hidden layer was then fed forward to a single output neuron (see below). Elman networks are particularly effective at decoding stimuli that are temporally structured (e.g. acoustic stimuli) because the Elman architecture includes recurrent connections within the hidden layer so that the neurons of the hidden layer feed back onto themselves (Elman, 1990; Demuth & Beale, 1997; e.g. Phelps & Ryan, 1998, 2000; Ryan & Getz, 2000; Phelps *et al.*, 2001). This recurrence permits the processing of information in a current time-step contingent on the information from a preceding time-step. Evolutionary simulations using similarly structured networks have been shown to predict female preferences for both conspecific and heterospecific male calls in túngara frogs (Phelps & Ryan, 1998; Phelps & Ryan, 2000; Phelps *et al.*, 2001).

The activity of the input layer was not weighted, and was determined strictly by the stimulus input. The stimulus was input over the course of 190 time steps, where each time step corresponded to a column, analogous to a slice of time, in the signal matrix (see below for description). The activity of the hidden layer, $\mathbf{a}^1$, was determined using a hyberbolic tangent (tansig) transfer function that combined the activity and weights of connections from the input, the recurrent connections and a bias (notation here and below is that of Demuth & Beale, 1997):

$$\mathbf{a}^1(k) = \text{tansig}(\mathbf{IW}^{1,1}\mathbf{p} + \mathbf{LW}^{1,1}\mathbf{a}^1(k-1) + \mathbf{b}^1) \quad (1)$$

where $\mathbf{p}$ was a 35×1 vector from the input layer corresponding to the kth column from the signal matrix. $\mathbf{IW}^{1,1}$ was a 23×35 matrix, the elements of which constituted the weights of the connections between the input and hidden layer, $\mathbf{LW}^{1,1}$ was a 23×35 matrix that constituted the weights of the recurrent connections of the hidden layer neurons, and $\mathbf{b}^1$ was a 23×1 bias vector (Demuth & Beale, 1997). Biases enable networks to represent relationships between a signal and output more easily than networks without biases (Demuth & Beale, 1997). The sizes of the bias vectors corresponded

Table 9.1. *Definitions and usage of key terms used throughout the chapter.*

Term	Definition and usage
Conspecific	Of the same species. The calls of three species are used in the simulations: *A*, *B* and *C*. Species *A* is the focal species for all simulations.
Heterospecific	Of a different species. For example, the calls of *B* are heterospecific calls for *A*.
Population	A group of 100 networks that undergo selection, mutation and evolution in response to different discrimination tasks. Three types of populations were generated for our simulations: *A* populations, in which networks were presented conspecific calls of species *A* versus white noise; *AB* populations in which networks were presented conspecific calls of species *A* versus heterospecific calls of species *B*; and *AC* populations in which networks were presented conspecific calls of species *A* versus heterospecific calls of species *C*. Note that *A*, *AB* and *AC* populations are all conspecifics – they consist of networks of the same species (species *A*).
Replicate	A population that has undergone selection, mutation and evolution. Depending on the question that was being addressed, we generated either 20 or 30 replicates for each population type described above.
Sympatric	Of a population occurring with a given heterospecific species. Networks in the *AB* populations are sympatric with species *B* (but not species *C*); networks in the *AC* population are sympatric with species *C* (but not species *B*); networks in the *A* populations are not sympatric with any species.
Allopatric	Of a population that does not occur with a given heterospecific species. Networks in the *AB* populations are allopatric with species *C* (but not species *B*); networks in the *AC* population are allopatric with species *B* (but not species *C*); networks in the *A* populations are allopatric with both *B* and *C*.
Local	Of the same population type or replicate. In contrasting calls from different population types: calls from the *A* populations are local only for *A* populations; calls from the *AB* populations are local only for *AB* populations; and calls from the *AC* populations are local only for *AC* populations. In contrasting calls from different replicates *of the same population type*, local calls are those of a single population.
Foreign	Of a different population type or replicate. In contrasting calls from different population types: calls from the *A* populations are foreign to both *AB* and *AC* populations; calls from the *AB* populations are foreign to both *A* and *AC* populations; and calls from the *AC* populations are foreign to both *A* and *AB* populations. In contrasting calls from different replicates *of the same population type*, foreign calls are those of a different replicate.

to the number of neurons in the hidden and recurrent layers (Demuth & Beale, 1997). The biases were subject to mutation and so could evolve in our simulations (see below). The hyperbolic tangent transfer function limits the output from the hidden layer to values ranging from –1 to 1 (Demuth & Beale, 1997).

The activity of the output neuron, $\mathbf{a}^2$, was the result of a pure linear transfer function that combined the activity and connections to it with a bias:

$$\mathbf{a}^2(k) = \text{purelin}(\mathbf{LW}^{2,1}\mathbf{a}^1(k) + \mathbf{b}^2) \quad (2)$$

where $\mathbf{LW}^{2,1}$ was a 1×23 matrix that constituted the weights connecting the output neuron with the neurons of the hidden layer and $\mathbf{b}^2$ was a 23×1 bias vector. The pure linear transfer function calculated output by returning the value passed to it. Thus, there were no limits on output values.

The resulting output from each network was a vector of responses corresponding to each column in the signal matrix. We summed this vector to obtain a single scalar response measure to the entire signal matrix. Summing in this way was appropriate, as we had no a priori reason to weight the networks' responses to different time points in the signal differently. For further details and schematics of the network architecture see Demuth & Beale (1997) and Ryan & Getz (2000).

9.4 Simulating the evolution of conspecific recognition

We used a genetic algorithm to simulate the evolution of conspecific recognition. Networks underwent selection and mutation before being passed to the next generation. Our methods, which were similar to those of Ryan & Getz (2000), are described below.

For each population type, we created 100 networks consisting of the architecture described above. The matrix values used to specify each network were initially uniformly randomly generated with values constrained between –1 and 1. We then presented to each network a conspecific stimulus and either a noise stimulus or one of two different heterospecific stimuli (the particular stimuli depended on the population in which the network 'resided'; see above). We defined the fitness of a network as the difference between its response to the conspecific stimulus and its response to the heterospecific stimulus. This fitness function results in higher fitness for those networks that are better able to discriminate between conspecifics and heterospecifics (i.e. those that maximise their responses to conspecifics while minimising their responses to heterospecifics). In nature, females must typically discriminate among courting males of different species (e.g. in a frog chorus males of different species could be calling simultaneously), so selection likely operates to maximise the likelihood of choosing the correct species while minimising the likelihood of selecting the wrong species (Reeve, 1989; Wiley, 1994). Because fitness cannot be negative (e.g. a female cannot have fewer than no offspring from a mating), negative fitness values were truncated to zero.

Using these fitness measures, we randomly selected the networks that were passed to the next generation. In particular, we selected 100 networks at random with replacement (i.e. the same network could be chosen more than once) from those networks in the preceding generation. The likelihood that a network was represented in the next generation was proportional to its fitness: networks with higher fitness had a higher likelihood of being chosen for the next generation than did networks with lower fitness.

Following this selection process, all networks that were selected to pass to the next generation underwent mutation (except a single network with the highest fitness in the previous generation). Values for the weights and biases of each network were chosen for mutation with a probability of 0.001. For those values that were chosen for mutation, we then added a random value between –0.5 and 0.5 to the existing value in each matrix element. Any values that exceeded 1.0 or were less than –1.0 were truncated to 1.0 and –1.0 respectively. Limits were set in order to mimic real biological systems in which neural activity has limits. Moreover, setting such limits is likely to make our findings conservative in that divergence in network behaviour becomes less, rather than more, likely. Previous work varying the nature of this mutation regime suggests that alterations do not appear to affect the general outcome of the simulations.

We used this general approach to ask two questions regarding the effects of heterospecific interactions on the evolution of female preferences. First we asked: how does character displacement affect mate preferences for conspecifics and discrimination against heterospecifics? To answer this first question, we examined how network preferences for conspecifics' signals and their ability to discriminate between conspecifics and heterospecifics would evolve differently among populations that varied in the nature of the heterospecific interactions they encountered. In these simulations, only the networks were allowed to evolve. By doing so, we could isolate the effects of heterospecific interactions on the evolution mate preferences.

Following the simulations to address the first question, we posed a second question: can reproductive character displacement initiate speciation? To address this question, we ran a second set of simulations. In these simulations, we allowed conspecific signals to evolve so as to evaluate how divergent preferences might contribute to diversification of conspecific signals. Our goal was to assess whether this divergence would tend to generate reproductive isolation among conspecific populations.

We therefore ran two sets of simulations. Below, we describe the stimuli presented to the networks, our application of the genetic algorithm described above, and our methods for evaluating network preferences in each set of simulations.

9.5 How does character displacement affect preferences for conspecifics and discrimination against heterospecifics?

9.5.1 Stimuli sets

The networks were presented pulsatile calls mimicking those possessed by many anuran and insect species. The calls were presented in a 35 × 190 frequency by time matrix in which the cell values ranged from 0 to 1 and represented amplitude of the signal at a given frequency and time (analogous to a sonogram). We synthesised the calls using a program written in Matlab that generated each call by combining randomly chosen values (see below) of four parameters: call duration (the length of the call in terms of matrix columns); call dominant frequency (the frequency in the call with the greatest energy,

Table 9.2. *Mean (± SD) of call parameters for each species, measured in terms of matrix columns or rows. See text for description of how calls were generated. The values below for species* A *were used throughout the set of simulations in which only the networks evolved. In the simulations where male calls coevolved with network preferences, only the call parameters of* A, *but not* B *or* C, *were allowed to evolve. In these coevolutionary simulations, the values below for species* A *were used in the initial generation and are therefore the parameters of the 'ancestral* A' *calls. See Figure 9.4 for contrast of evolved* A *calls versus the ancestral* A *calls.*

	Species		
Call parameter	*A*	*B*	*C*
Call duration (cols.)	62.6 (7.9)	9.1 (0.7)	62.4 (5.0)
Inter-call interval (cols.)	72.0 (1.7)	64.8 (0.9)	87.6 (4.7)
Call pulse rate (pulses/col.)	0.05 (0.01)	0.42 (0.05)	0.34 (0.02)
Dominant frequency (rows)	15.6 (1.2)	18.5 (1.2)	18.4 (1.5)

measured in terms of matrix rows); pulse rate (measured as number of pulses per matrix column); and inter-call interval (the number of matrix columns between the last column of the first call and the first column of the second call). This last parameter is a measure of calling rate; greater inter-call intervals result in slower call rates, whereas smaller inter-call intervals result in faster call rates.

Each call presented to a network was generated by randomly choosing a parameter value from the appropriate distribution for the conspecific or heterospecific calls. The distributions used for these parameter values were those of three naturally co-occurring spadefoot toads (*S. multiplicata*, *S. bombifrons* and *Sc. couchii*) from southeastern Arizona, USA (Pfennig, 2000). Once these parameter values were chosen, the duration of the call was shortened to 13% of its original length and the inter-call interval was shortened to approximately 5% of its original value, so that the duration of the longest possible call sequence would fit within the matrix presented to the networks. Pulse rate values were not altered from those chosen from the natural distributions; we report measures of pulse rate herein in terms of columns of the stimulus matrix, which represent time. We multiplied this pulse rate by the shortened call duration to obtain the number of pulses that would make up each call. Pulse length therefore varied within and between species, and was dependent on the combined parameters of pulse rate and call duration. Dominant frequency was converted to row values of the matrix. The resulting distribution of the call parameters measured in terms of rows and columns of the matrix are given in Table 9.2.

Using the randomly chosen parameters, each call was synthesised by initially generating a single pulse. To do so, a value of 1 (the maximum value of amplitude in the signal matrix) was assigned in the row corresponding to the dominant frequency of the call at the column corresponding the onset of the call (the onset of the call in the call matrix was randomly determined). The values in the following columns then degraded from 1 exponentially, and the values in the adjacent rows degraded exponentially from the values in the columns. This pattern thereby created a triangular pulse. The pulse was then repeated as appropriate in subsequent columns and rows of the matrix to generate a single call with the appropriate duration and pulse rate. A gap of silence (where values within the columns were set to 0) equivalent to the inter-call interval followed the call, at the end of which we appended a single pulse to indicate the onset of a second call.

The white noise stimuli presented to networks in the *A* populations were generated by assigning uniform random values ranging from 0 to 1 in a matrix that was the same size as that of the male calls. Moreover, after generating the male calls as described above, we also added noise to calls to simulate communication in a noisy environment. We did this by adding uniform random values ranging from 0 to 1 to the elements in each call matrix; resulting values greater than 1 were truncated to 1. By adding noise to the call stimuli, we ensured that all populations experienced white noise and therefore any differences that arose would not be an artefact of the noise stimulus. The amplitude of all stimuli presented to the networks was standardised so that they were equal in total amplitude.

Although some individual call characters were similar between species *A* (the conspecific species) and at least one of the heterospecific species (Table 9.2), the multivariate means of the call parameters were significantly different among all three species based on a sample of 20 randomly generated calls for each species (Wilks' $F_{6,110} = 192.08$, $p < 0.001$). Indeed, a discriminant analysis showed that all calls could be reliably assigned to the correct species based on their characteristics, a pattern that differed significantly from random expectation (log-likelihood ratio $\chi^2_4 = 131.83$, $p < 0.001$). By using calls that could be discriminated statistically from one another based on a combination of the calls' characters, we created a situation in which the impact of heterospecific interactions on the evolution of mate preferences should have been minimal. If heterospecific calls are sufficiently different from conspecifics, females can possibly identify conspecifics based solely on the variation of conspecific calls rather than the variation of conspecific calls relative to that of heterospecific calls (Patterson, 1985).

9.5.2 Simulations, testing and analyses of networks' responses

The above stimuli were presented to the networks, and using the genetic algorithm described above, the selection and mutation process was repeated for 1000 generations. We then replicated the entire procedure 20 times for each population type. Both the mean population fitness and maximum fitness for all replicates reached a plateau prior to generation 1000.

Following the above simulations, we selected the single network with the highest fitness from the last generation in each population type from each of the 20 replicates. To determine the nature of selection on each call parameter by networks from the three different population types, we tested each network with a series of conspecific calls in which each call parameter was systematically varied while all the other call characters were held constant. In particular, for each conspecific call character we generated a series of calls in which each character took on values ranging from 3.5 standard deviations below the mean for that character to 4.0 standard deviations above the mean in 0.5 standard deviation intervals. All other call parameters were fixed at the mean values for those traits. Thus, for each of the four call characters we generated 15 variants. In addition to these call variants, we also presented the networks with a call in which all the call characters were set at the mean values for all four traits constituting a conspecific species *A* call (Table 9.2). Thus, the networks were presented a total of 61 different calls in this analysis.

We averaged the responses of the 20 networks from each population type to each of the call variants, and standardised these data so that they would be comparable across the different call parameters. We then regressed the network responses on the variation of each call character using a second-order polynomial regression (Sokal & Rohlf, 1995). If the second-order regression coefficient was not significant, that term was dropped from the model and a linear regression used. This analysis allowed us to determine the nature of selection on each call character exerted by the networks in each of the three populations (Falconer & Mackay, 1996; Conner & Hartl, 2004). Essentially, this analysis resulted in population level 'preference functions' for each call character in each population (Gerhardt, 1991; Wagner, 1998; Höbel & Gerhardt, 2003; Rodriguez *et al.*, 2004).

To evaluate whether networks in the different populations diverged in their preferences for male traits, we performed the following analysis. First, using standardised data, we regressed each network's response onto the systematic variation in each trait using second-order polynomial regression (Sokal & Rohlf, 1995). This gave us each network's preference function for each call character (Wagner, 1998). This analysis generated eight total regression coefficients (i.e. one first- and one second-order regression coefficient for each of four call characters) for each network in each population. We used principal component analysis (Sokal & Rohlf, 1995) to reduce the eight regression coefficients to a more manageable variable set. We then used MANOVA (Zar, 1984) to determine if the populations were significantly different in their values of these principal components. By doing so, we evaluated whether networks from the different populations differed significantly in their preference functions for, and therefore the pattern of selection they might exert on, conspecific male traits.

We next assayed whether the networks diverged in their ability to discriminate between conspecific and heterospecific calls. In one set of tests, we presented each network with a randomly generated conspecific call and a randomly chosen call of species *C*. In a second set of tests, we presented each network with a randomly generated conspecific call and a randomly generated call of species *B*. In each set of tests, we presented each network with 100 pairs of calls. In each pairing we scored a network as preferring a stimulus when it had a higher response to that stimulus. We then calculated the proportion of pairings in which the network showed preference for the conspecific stimulus. We used these individual measures to calculate population means. These means were compared among the populations with ANOVA and Tukey–Kramer HSD multiple comparisons tests to determine if the populations differed in their ability to discriminate between conspecific and heterospecific males. In each population we also tested whether the networks significantly preferred the conspecific male. To do so, we tested whether the population mean preference for or against conspecific calls was significantly different from 50%, which is the null expectation if the networks were random in their preference of conspecifics versus heterospecifics. In all analyses described above, the data met parametric assumptions.

9.6 Can reproductive character displacement initiate speciation?

9.6.1 *Stimuli sets*

To determine whether reproductive character displacement can initiate speciation, we generated call stimuli as described above. In this set of simulations, however, we allowed the conspecific male calls, but not the heterospecific calls, to evolve in our simulations. Heterospecific calls were not allowed to evolve, because there would be no reason to expect coevolution between preferences in one species and calls of another species. Our simulation assumes that the heterospecific calls are at an evolutionary equilibrium. Further work is required to understand how evolutionary dynamics in one species affects coevolutionary dynamics between preferences and sexual signals in another species.

To allow conspecific calls to evolve, at each generation, the 100 conspecific calls that had been presented to the 100 networks passed to the next generation were also passed to the next generation (i.e. the calls represented the sires and the networks represented the dams of the next generation's offspring). From these calls we obtained the mean and standard deviation for each call parameter. These new distributions were then used to generate the calls (as described above) in the subsequent generation. Thus, in each generation, calls were randomly generated from the distribution of calls of the 'sires' in the previous generation. Calls were not pooled across replicates. Each replicate represented an independent evolutionary simulation of both species recognition and signal evolution.

For each replicate, we calculated the mean call parameters of the 100 calls in the final generation. These means were combined into a single data set along with call parameters of 30 randomly generated calls for each of the ancestral *A* population, and *B* and *C* species. The randomly generated ancestral and heterospecific calls served as samples of these calls types.

We analysed these data using a principal component analysis, which generated two principal components that described the joint variation in the four parameters. Both principal components had eigenvalues greater than 1. The first explained 52.8% of the variation in the advertisement calls, whereas the second explained 26.1% of the variation. We used these principal component values to compare the calls among the *A*, *AB* and *AC* populations based on the combined variation in the four call parameters. Because the data did not meet parametric assumptions, we compared each principal component among pairs of populations using Wilcoxon rank sums tests. We used a Bonferroni corrected alpha level of 0.017 in these multiple comparisons (Sokal & Rohlf, 1995).

9.6.2 Simulations, testing and analyses of networks' responses

The above stimuli were presented to the networks, and using the genetic algorithm described above, the selection and mutation process was repeated for 200 generations. We then replicated the entire procedure 30 times for each population type. The mean population fitness and maximum fitness for all replicates reached a plateau by generation 200.

We selected the network with the highest fitness from every 8th generation up through to the last generation in each population type from each of the 30 replicates. We tested these networks for preferences of their own conspecific calls versus the heterospecific (or noise) stimulus with which they coevolved. More critically, we also assayed the responses of these networks to advertisement calls of their own population (*local calls*) versus those of the two alternative populations (*foreign conspecific calls*). In the tests described below, we used the male call distributions from the networks' own generation.

To test the networks' preferences for local calls versus the heterospecific (or noise) stimulus with which the networks coevolved, we presented each network with 100 pairs of a randomly generated call from its own population versus a randomly generated heterospecific or noise stimulus.

To test the networks' preferences for local calls versus foreign conspecific calls, we presented each network with two sets of calls. In one set, networks were presented local calls versus foreign conspecific calls from one of the alternative populations, and in the second set, networks were presented local calls versus foreign conspecific calls from the second alternative population (e.g. *A* networks were presented *A* vs. *AB* calls in one set and *A* vs. *AC* in a second set). Thus, we generated six possible pairings of local and foreign conspecific calls. For each set we presented 100 pairs of randomly generated local calls versus randomly generated foreign conspecific calls to each of the 30 networks in each population type.

In all tests of network preference, we calculated the difference in response between the local call and the alternative call. This raw measure of discrimination is analogous to the fitness measure used during the evolution of the networks. Because the magnitude of networks' discrimination differed not only across generations but also across independently evolved replicates and populations, we generated a relative measure of preference for local calls that was comparable among pairs of stimuli, generations, replicates and populations. We generated this relative preference measure as follows. After all simulations were completed, we obtained the highest discrimination score expressed by any network at any time within that network's own replicate for the pairings of local calls versus the heterospecific calls with which they coevolved (i.e. *B*, *C* or noise). We then divided a network's raw discrimination scores for a given call pair by this maximum value for its replicate. As with our fitness measure, negative values were truncated to 0. We thereby generated a relative preference score for local calls in each pairing that varied from 0 to 1. At values close to 0, networks expressed no discrimination. At values approaching 1, networks were expressing discrimination as strong as the highest level observed against heterospecifics (or noise) in that network's lineage. We therefore ascertained whether networks preferentially responded to local calls by comparing their average preference score in a given pair-wise test with the null expectation of 0.

Although we found that the calls evolved to be divergent among the *A*, *AB* and *AC* population types, there was variation in the call parameters that evolved among the different replicates of these population types (especially in the *AB* and *AC* populations; see Results and Figure 9.4). Such variation could result from stochasticity in the simulations or may represent alternative solutions to similar discrimination tasks. We examined how networks responded to these call variants from other replicates of their same population type and compared this to their responses toward foreign conspecific calls from other populations. By doing so, we could discern whether networks selected against foreign conspecific calls because they were from alternative population types not just alternative replicates.

To make this comparison, we generated an average call for each replicate using the mean values of all four call parameters for the given replicate. We then presented each network with the average call from its own replicate (the local call) versus each alternative replicate (the foreign replicate call) from its own population type. For example, a network from an *A* population was presented the average call for its population (the local call) and the average call of a different replicate *A* population (the foreign replicate call). Each network of the *A*, *AB* and *AC* populations was therefore presented 29 pairings of its own local call with calls from different replicates. Preferences were scored as above. From these preference scores, we generated a mean preference for local calls *within* a given population type that we then used as a null expectation against which to compare the networks' preferences for local calls versus calls from alternative population types. For example, the preference that networks from the *A* populations expressed for their own calls versus calls from the *AB* and *AC* populations was compared to the average preference that networks from *A* populations expressed for their own local calls versus those from other replicate *A* populations.

Finally, the networks might be more likely to discriminate against foreign calls as they become increasingly dissimilar from the local calls. If so, then preference for local calls should be negatively correlated with similarity between the local and foreign calls. To evaluate this possibility, we took the absolute difference between the principal component score of the average local call and the average foreign conspecific call presented to each network. We generated these values separately for both principal components. Because these data did not meet parametric assumptions, we used Spearman rank order correlation analysis to determine if the magnitude of difference between calls was associated with the average preference for local calls in a given pair type. These analyses utilised calls from across the independently evolved replicates, and so reflect patterns, if any, associated with reproductive character displacement rather than variation within a single lineage.

9.7 Results

9.7.1 How does character displacement affect preferences for conspecifics and discrimination against heterospecifics?

To answer this question, we focused only on network preferences; male calls were not allowed to coevolve. Our simulations revealed that artificial neural networks that did not encounter heterospecific calls or that interacted with different heterospecific calls diverged in their preferences for conspecific male call characters. In particular, we found that each population exerted a unique pattern of selection on the signal features that constituted conspecific advertisement calls (Figure 9.2)[1].

The eight regression coefficients measuring the networks' preference functions for the four call characters reduced to four principal components that each had an eigenvalue greater than 1.0. Together, the four principal components explained 82.3% of the variation in the regression coefficients. When we used MANOVA to compare these principal components among the three populations, we found a significant effect of population (Wilks' Lambda = 0.335, $F_{6,110} = 13.36$, $p < 0.001$). Contrasts of the populations revealed that the three populations were all significantly different from one another (*A* vs. *AB*: $F_{3,55} = 10.6$; *A* vs. *AC*: $F_{3,55} = 35.3$; *AB* vs. *AC*: $F_{3,55} = 7.7$; all contrasts are $p < 0.001$). These results indicate that the networks' preference functions for, and therefore the pattern of selection they might exert on, conspecific male traits differed among the three populations.

Although networks from the three population types differed in their preferences for conspecific calls, we found mixed evidence that sympatric and allopatric populations differed in their ability to discriminate against heterospecific calls. When we tested the networks for their preferences of conspecific versus heterospecific calls, we found they could potentially discriminate against a given heterospecific even when they had not evolved species recognition in the presence of that heterospecific species. Specifically, when given a choice of conspecific male calls versus the calls of species *C*, networks from the three populations differed in their ability to discriminate between conspecific and heterospecific male calls ($F_{2,57} = 83.37$, $p < 0.001$; Figure 9.3a). A Tukey–Kramer HSD test revealed that all populations were significantly different from one another at $p < 0.05$. Networks from the *AC* population showed the best discrimination against species *C* whereas networks from the *A* populations showed the worst discrimination against species *C* (Figure 9.3a). Indeed, networks from the *A* populations were random in their choices of conspecifics versus species *C* ($t_{19} = 1.5$, $p = 0.15$; Figure 9.3a). By contrast, networks from both the *AB* and *AC* populations preferred conspecific calls to those of species *C* (*AB* population: $t_{19} = 13.3$, $p < 0.001$; *AC* population: $t_{19} = 33.9$, $p < 0.001$; Figure 9.3a). Thus, networks from the *AB* population selected against heterospecific calls of species *C* even though they were allopatric with this species and had not evolved recognition in its presence.

[1] All of the results we present in this section (including figures) were previously published in Pfennig & Ryan (2007).

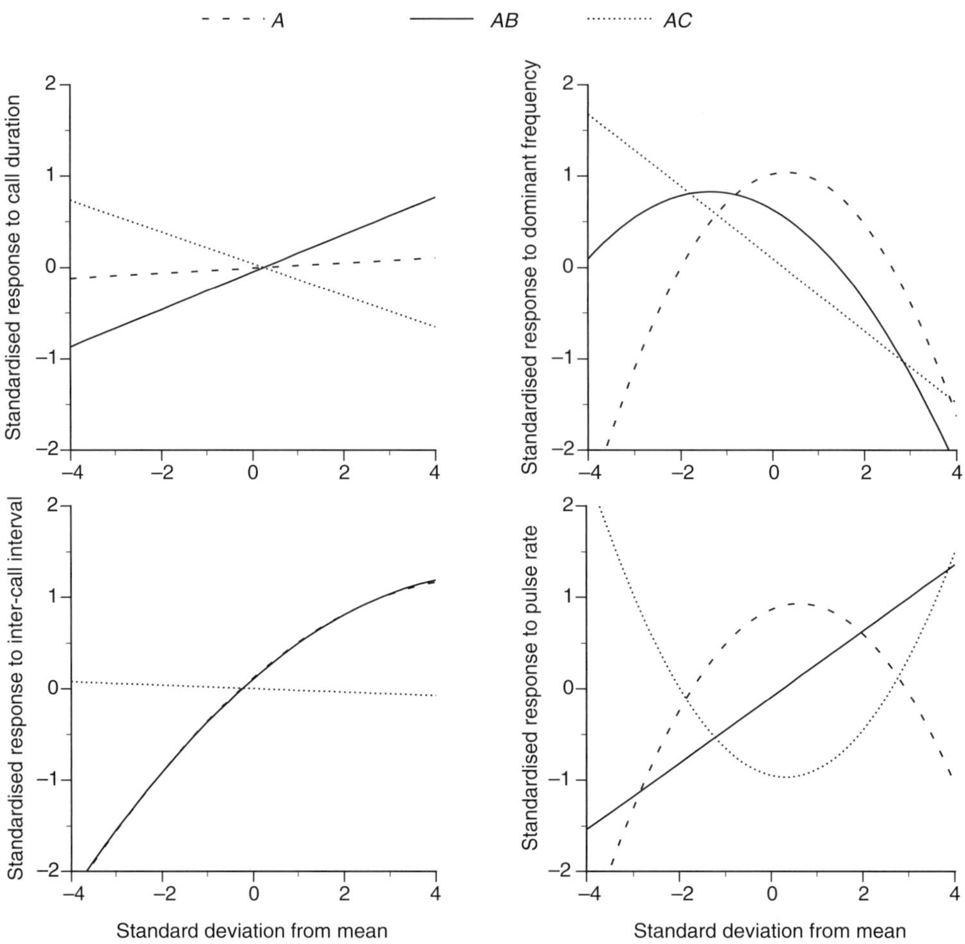

Figure 9.2. Preference functions showing the strength and nature of selection on each call character by networks of the three different population types. Preferences were measured on sets of calls in which each character was systematically varied while the other characters were held constant (see text for details). All relationships are statistically significant ($p < 0.05$), except the following: *AC* networks' selection on inter-call interval and call duration, *A* networks' selection on call duration and *AB* networks' selection on call duration (although for this relationship $p = 0.06$). For inter-call interval, the preference functions for *A* and *AB* overlap. Where the relationship shown is linear, the second-order regression coefficient was not significant and was dropped from the analysis.

When we presented the networks with calls from conspecifics versus those of species *B*, we found a significant difference in the populations' discrimination against heterospecifics ($F_{2,57} = 10.79$, $p < 0.001$; Figure 9.3b). A Tukey–Kramer HSD test revealed that only the *AB* and *AC* populations were significantly different at $p < 0.05$, however. The networks from the *AB* population were most effective at discrimination against species *B* whereas networks from the *AC* population were the least effective at discriminating against them

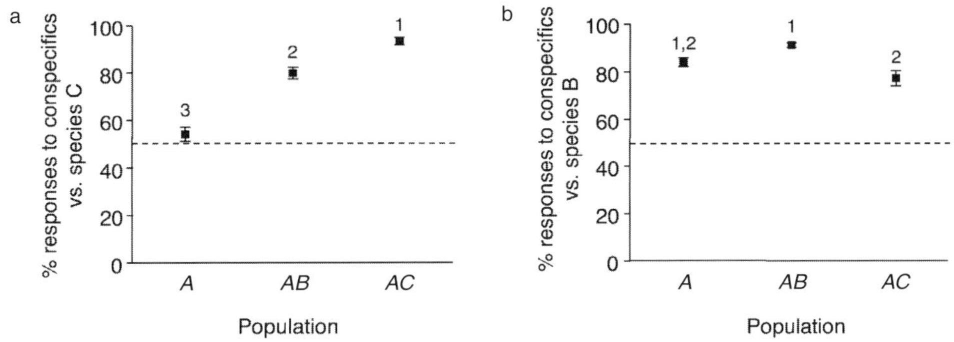

Figure 9.3. (a) Mean (± SE) per cent responses in which the conspecific signal was preferred in choice tests between conspecific calls of species *A* and heterospecific calls of species *C*; (b) Mean (± SE) per cent responses in which the conspecific signal was preferred in choice tests between conspecific calls of species *A* and heterospecific calls of species *B*. Dashed line shows 50% random expectation if networks within each population were indiscriminate in their choice of conspecifics versus heterospecifics. Within each panel, different numbers above each population mean indicate significant differences among the populations; populations that share the same number within each panel are not significantly different (see text for statistical analyses).

(Figure 9.3b). Although networks from the *A* population did not evolve conspecific recognition in the presence of species *B*, they did not differ from the *AB* networks in discrimination against species *B* (Figure 9.3b). Thus, although the networks showed divergence in their preference functions for conspecific male signals (Figure 9.2), this result did not necessarily translate into differences between sympatric and allopatric populations in discrimination of conspecific and heterospecific males. Moreover, despite differences between the *AB* and *AC* populations in their discrimination ability against species *B*, networks in all three populations significantly preferred the calls of conspecifics to those of species *B* (*A* population: $t_{19} = 23.0$, $p < 0.001$; *AB* population: $t_{19} = 36.2$, $p < 0.001$; *AC* population: $t_{19} = 8.6$, $p < 0.001$).

9.7.2 Can reproductive character displacement initiate speciation?

The divergence of signals among conspecific populations in response to heterospecifics could contribute to their reproductive isolation. We found that advertisement calls of all three populations evolved to be distinct from the ancestral call (Figure 9.4a) and from each other (Figure 9.4b). The principal component measures (PC 1 and PC 2) of the combined call parameters were both significantly different among the three populations (Wilcoxon normal approximation comparing PC 1 among population pairs: *A* vs. *AC*: $Z = -4.07$, $p < 0.0001$; *A* vs. *AB*: $Z = -6.11$, $p < 0.0001$; *AB* vs. *AC*: $Z = 5.45$, $p < 0.0001$; Wilcoxon normal approximation comparing PC 2 among population pairs: *A* vs. *AC*: $Z = -5.73$,

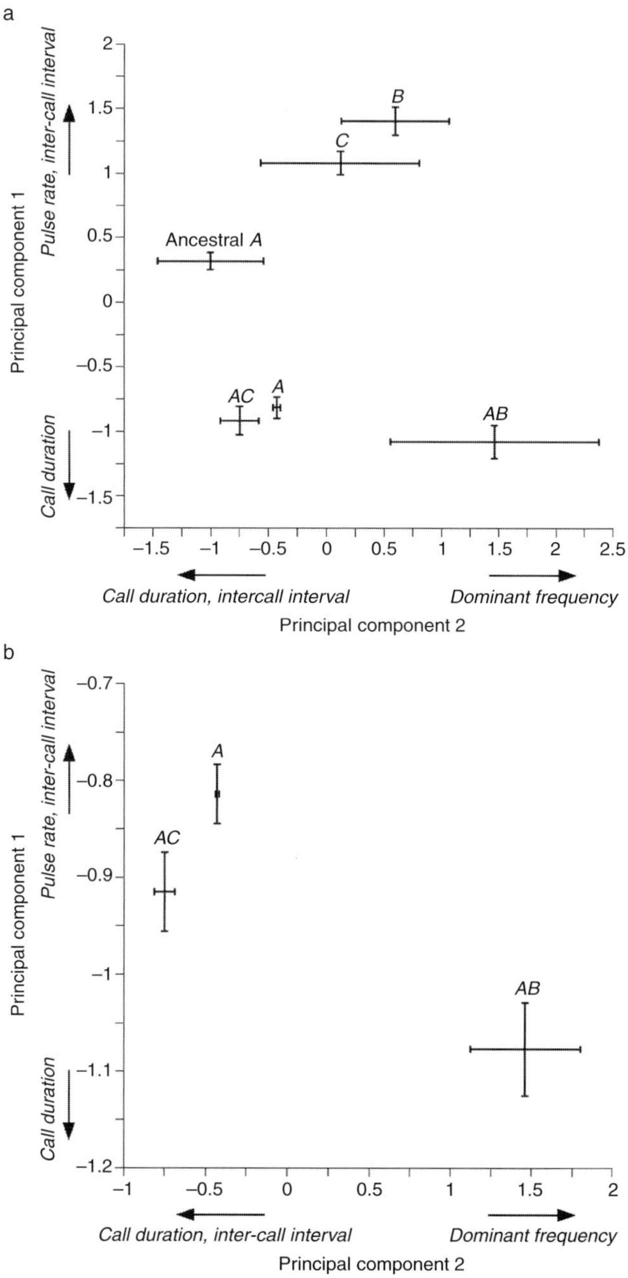

Figure 9.4. (a) Mean (± SD) for principal components that describe the combined variation in four call characters (dominant frequency, call duration, pulse rate and inter-call interval) for: the evolved calls from the three different conspecific populations (*A*, *AB* and *AC*), the heterospecific calls presented to the *AB* and *AC* populations during their evolution (*B* and *C*, respectively), and the initial conspecific call (ancestral *A*). Intersection of x-axis and y-axis standard deviation lines is the point of the mean for each. (b) Mean (± CI) for same principal components above comparing the evolved calls from *A*, *AB* and *AC* only. Non-overlapping confidence intervals indicate significant differences. Labels on axes indicate loading of call parameters on each principal component.

Table 9.3. *Network preferences for local calls versus foreign conspecific calls. The network's population is also the population of the local call. The preference for local calls was assessed in two ways. First, mean response to local calls is compared with the null expectation of 0 if networks respond equally strongly to local and foreign conspecific calls. Second, mean response to local calls is compared with a null expectation that is the mean response to independently evolved calls from replicates of the networks' own population type. This latter contrast controlled for the possibility that networks may have been generally selective against foreign conspecific calls, not just those that diverged due to character displacement. Means and confidence intervals of preference strength for local calls from these pairings are displayed in Figure 9.5.*

Network's population	Population of foreign call	t_{29} (p-value); H_0: local call preference = 0	t_{29} (p-value), H_0: local call preference = preference for calls from other replicates[a]
A	*AB*	21.48 (< 0.0001)[b]	17.96 (< 0.0001)[b]
	AC	9.46 (< 0.0001)[b]	6.70 (< 0.0001)[b]
AB	*A*	9.06 (< 0.0001)	3.24 (0.003)
	AC	10.72 (< 0.0001)	4.29 (0.0002)
AC	*A*	8.32 (< 0.0001)[b]	–1.04 (0.31)[b]
	AB	10.58 (< 0.0001)	6.71 (< 0.0001)

[a] Average response to replicates by networks in *A*: 0.027; *AB*: 0.310; *AC*: 0.203.

[b] Analysis used transformed data to meet parametric assumptions.

$p < 0.0001$; *A* vs. *AB*: $Z = 5.34$, $p < 0.0001$; *AB* vs. *AC*: $Z = -5.79$, $p < 0.0001$; $N = 30$ for each population).

We used two measures to determine if the networks preferred local calls versus foreign conspecific calls. First, we compared mean preference for local calls with the null expectation of 0 (see methods described above). Networks from the three populations significantly preferred local calls to foreign conspecific calls (Table 9.3; Figure 9.5).

Divergence of populations could potentially result from stochastic variation in preferences and calls among independently evolved lineages rather than due to character displacement per se. To control for this possibility, we also compared networks' mean preferences for local calls when presented with foreign conspecific calls with their mean preference for local calls when presented with independently evolved calls from alternative replicates of their own population type (see methods described above). We found that the *A* networks showed a weak preference for local calls versus calls from alternative replicates, whereas the *AB* and *AC* networks displayed relatively stronger preference for calls that evolved in their own replicate (Figure 9.5).

Although the networks discriminated against calls from alternative replicates of their own population type, they generally showed even stronger preferences for local calls when they were paired with foreign conspecific calls. In all but one pairing (*AC* networks presented with *AC* vs. *A* calls), the networks discriminated against foreign conspecific

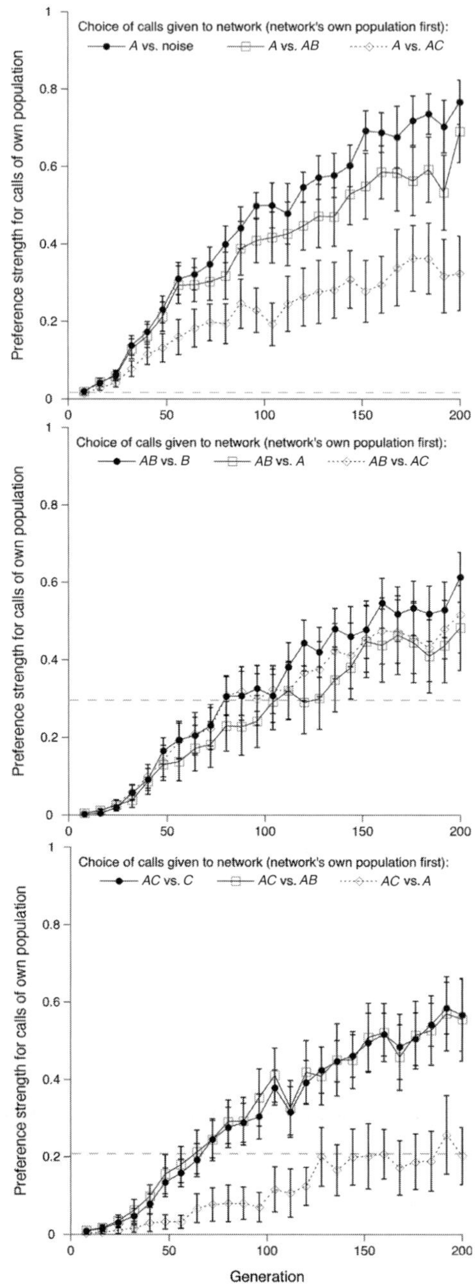

Figure 9.5. Mean (± 95% CI) preference for advertisement calls from networks' own population (local calls) versus calls of alternative populations (foreign conspecific calls) over time. Top panel shows results for *A* networks; middle panel shows results for the *AB* networks; and the bottom panel shows results for *AC* networks. Non-overlapping confidence intervals indicate significant differences. Dashed horizontal line shows null expectation derived from network preferences for local calls versus those from alternative replicates of same population type in final generation.

Table 9.4. *Correlation of preference strength for local calls with the absolute differences in principal component scores of local versus foreign conspecific calls (a measure of the similarity between the calls). Spearman rank order correlation coefficient reported because data did not meet parametric assumptions. The network's population is also the population of the local call. N = 30 for each correlation.*

Network's population	Population of foreign call	r_s (p-value) using difference in PC 1 scores	r_s (p-value) using difference in PC 2 scores
A	*AB*	−0.076 (0.69)	0.195 (0.30)
	AC	0.092 (0.63)	−0.366 (0.047)
AB	*A*	0.060 (0.75)	0.672 (< 0.0001)
	AC	0.246 (0.19)	0.393 (0.032)
AC	*A*	−0.345 (0.062)	0.528 (0.003)
	AB	0.126 (0.51)	0.339 (0.067)

calls significantly more strongly than they discriminated against calls from alternative replicates (Table 9.3; Figure 9.5).

The evolutionary trajectories of these preferences for local calls suggested that they arose in conjunction with the evolution of discrimination against the heterospecific calls (or noise) with which the networks coevolved (Figure 9.5). Indeed, in all of the populations, networks in the final generation exhibited similarly strong discrimination against at least one type of foreign conspecific call as they exerted against the heterospecific calls with which they coevolved (Figure 9.5).

This was exemplified in the *AB* population where we found no significant differences in preference for local conspecific calls among any of the three pairings the networks were presented (*AB* vs. *B* calls, *AB* vs. *A* calls and *AB* vs. *AC* calls; $F_{2,87} = 2.23$, $p = 0.11$; Figure 9.5). In the *A* population, we found a significant difference in preference for local calls among the three possible pairings the *A* networks faced ($F_{2,87} = 33.7$, $p < 0.0001$). The *A* networks, however, showed a similarly strong preference for locals calls when they were paired with *AB* calls as when local calls were paired with a white noise stimulus (as revealed by a Tukey–Kramer HSD test, $p > 0.05$; Figure 9.5). The *A* networks showed significantly lower preference for local calls in the pairing of *A* and *AC* calls than in the other call pairings with which they were tested (Tukey–Kramer HSD test, $p < 0.05$; Figure 9.5).

Similarly, in the *AC* population we found a significant difference in preference for local calls among the three possible pairings they faced ($F_{2,87} = 15.9$, $p < 0.0001$). Yet, the *AC* networks showed a similarly strong preference for local calls when they were paired with *AB* calls as when local calls were paired with heterospecific *C* calls (as revealed by a Tukey–Kramer HSD test, $p > 0.05$; Figure 9.5). The *AC* networks showed significantly lower preference for local calls in the pairing of *AC* and *A* calls than in the other call pairings with which they were tested (Tukey–Kramer HSD test, $p < 0.05$; Figure 9.5).

The above results emphasise that the networks sometimes discriminated against foreign conspecific calls as strongly as they did heterospecific (or noise) stimuli. These findings also indicate that the networks did not necessarily respond to foreign conspecific calls from different population types in the same way. In both the *A* and *AC* populations, the networks discriminated against the foreign *AB* call more strongly than they selected against each other (Figure 9.5). One explanation for this pattern is that because the *A* and *AC* calls were more similar (Figure 9.4), they were less likely to discriminate against each other than against the *AB* calls. We investigated whether the differences in how networks responded to foreign conspecific calls could be attributed to the level of similarity between the local calls and a given foreign conspecific call type. For variation described by PC1, we found no associations between preference for local calls and dissimilarity of local and foreign conspecific calls (Table 9.4). By contrast, we found that the greater the difference between local and foreign conspecific calls in PC2, the stronger the preference for local calls in four of the six pairings (Table 9.4).

9.8 Discussion

Using artificial neural network models, we simulated the evolution of conspecific recognition in the presence of different heterospecifics. We found that selection to avoid mating with heterospecifics can generate divergent mate preferences for aspects of conspecific signals among different conspecific populations. Moreover, these divergent preferences can ultimately promote diversification of mating signals among conspecific populations. Critically, we found that this divergence in preferences and signals can promote reproductive isolation among conspecific populations that differ in the nature of heterospecifc interactions they experience. Thus, reproductive character displacement can lead not only to enhanced differentiation of mating behaviour between species, it can also potentially initiate speciation events among populations that vary in the heterospecifics they encounter.

Many studies of reproductive character displacement assay whether females that are sympatric with a particular species of heterospecific are more likely to reject heterospecific mates than are allopatric females (reviewed in Howard, 1993). Similarly, such population differences in the ability to discriminate conspecifics from heterospecifics has been viewed as a critical prediction of reinforcement – the hypothesis that natural selection against hybridisation will promote divergent mating behaviours between hybridising species within sympatry but not allopatry (Howard, 1993; Noor, 1995; Servedio & Noor, 2003; Coyne & Orr, 2004). Yet, empirical studies are mixed as to whether they find support for this pattern (reviewed in Howard, 1993). The results of our model suggest that character displacement in female preferences for conspecific male calls does not necessarily result in differences between sympatric and allopatric populations in the ability to discriminate against heterospecifics. Moreover, because our results benefited from large sample sizes that would be unrealistic in many natural systems, such differences as those we did observe would be difficult to detect empirically. Thus, failure to find differences in the ability to discriminate against heterospecifics between populations of sympatry and allopatry in a natural system should not necessarily result in rejection of the hypothesis that reproductive character displacement or reinforcement has occurred. Instead, reproductive character displacement might best be detected by observing differences between sympatry and allopatry in mate preferences for aspects of conspecific signals rather than by searching for differences in discrimination against heterospecifics.

Our results indicate that whether females discriminate against heterospecifics in sympatry or allopatry with a given heterospecific likely depends on female preferences for aspects of conspecific male traits and the trait distribution of the heterospecific signals relative to those preferences. For example, preference for signals with lower dominant frequency by *AC* networks would result in them discriminating against not only species *C*, the heterospecific with which they coevolved, but also species *B*, with which they had no interactions during their evolutionary history (contrast Figure 9.2 with Figure 9.4a). Thus, females' risk of mating with heterospecifics may strongly depend on the nature of female preferences for conspecifics rather than whether the females occur in sympatry or allopatry per se (Ryan *et al.*, 2003; Rodriguez *et al.*, 2004).

Not all preferences that evolve necessarily contribute to successful species recognition, however. In the case of pulse rate, for example, the preferences for higher pulse rate expressed by both the *AB* and *AC* networks could possibly put them at risk of mating with heterospecifics if this trait were important to mate choice. Females in natural systems do not weight all traits equally (Gerhardt, 1994), and the fact that both the *AB* and *AC* networks strongly discriminated against both heterospecifics (Figure 9.3) suggests that pulse rate was not heavily weighted by the networks in their responses to the male calls. Why the networks evolved the preferences for pulse rate that they did remains unclear. One explanation is that the evolution of preferences for other characters could have had a pleiotropic effect on the evolution of pulse rate preferences. The degree to which heterospecific interactions generate diversity in mate preferences through pleiotropic effects rather than due to direct selection on traits that enhance discrimination remains an open question.

That the nature of heterospecific interactions alters preferences for conspecific signals has implications beyond the effects on species recognition. Our findings indicate that divergent preferences, in turn, drive the diversification of male signals among populations. This coevolution of female preferences and male signals thereby promotes assortative mating within – and reproductive isolation among – conspecific populations.

Although we observed divergence in male signals, call evolution was not strictly caused by differences in heterospecific interactions among the different populations. All three populations diverged dramatically from the ancestral call type (Figure 9.4a). Such evolution may have occurred if, for example, certain call characters were more easily discriminated against the noisy background that all three populations experienced. Yet, despite their similar evolution relative to the ancestral calls, the calls of the different populations also diverged from one another (Figure 9.4b). Interestingly, the calls that evolved in the populations that discriminated against heterospecifics were more variable (especially in *AB*) than those in the allopatric population, *A* (Figure 9.4). Why this was so is unclear. One explanation is that there were few optimal calls for discriminating against noise alone, but many alternative call solutions for discriminating against a given heterospecific. Generally, such variation could further promote diversification among populations.

Perhaps most critically, we found that networks preferred calls of their own population to those from alternative conspecific populations (Table 9.3; Figure 9.5). Indeed, in some cases, the networks discriminated against foreign conspecific calls and heterospecific calls similarly. These results suggest that character displacement in mating behaviours such as male signals (arising from selection to avoid heterospecifics in sympatry but not allopatry) can simultaneously promote assortative mating within sympatric and allopatric conspecific populations. In a natural system, this pattern of mate choice could generate reproductive isolation, and ultimately initiate speciation, among conspecific populations (e.g. Hoskin *et al.*, 2005).

Although networks from all three populations in our study tended to prefer local calls, networks from a given population did not necessarily show the same level of discrimination against different types of foreign conspecific calls (Figure 9.5). Such a finding indicates that the evolution of discrimination against heterospecifics does not necessarily result in the rejection (or equal treatment) of all foreign calls. Discrimination against foreign conspecific calls tended to be weaker when local and foreign conspecific calls were more similar (Table 9.4). Indeed, calls from the *A* and *AC* populations were the most similar (Figure 9.5) and networks from both populations were less discriminating against calls from the alternate population than they were against calls from the *AB* population. Similarly, when tested for their preferences of local calls versus calls from alternative replicates of their same population type, the preference for local calls was weakest in the *A* population (Figure 9.5; see also Table 9.3), which exhibited very low variation in calls across replicates (i.e. local calls and foreign conspecific calls were all similar; Figure 9.4). By contrast, the preference for local calls versus calls from alternative replicates was highest in the *AB* population (Figure 9.5; see also Table 9.3), which exhibited higher variation in calls across replicates (Figure 9.4). These results suggest that different types of heterospecific interactions may more likely contribute to reproductive isolation if they promote the evolution of opposing signal characters among conspecific populations. Thus, the particular mating behaviours that evolve in response to heterospecifics may determine whether populations become reproductively isolated.

One feature of our simulations that undoubtedly promoted the diversification of mating behaviours among the conspecific populations was the close coevolution between signals and receivers. Such a pattern of coevolution often occurs between males and females (Andersson, 1994). If, however, signal evolution (or receiver perception) is under direct countervailing selective pressures (e.g. from predators or energetic or physiological limitations) or affected indirectly by the evolution of correlated characters, divergence among populations may in turn be limited. Predicting the circumstances under which reproductive character displacement may promote the evolutionary diversification of mating behaviours, and possibly speciation, among conspecific populations may therefore require a comprehensive understanding of the selective and correlated factors that determine the evolution of mating behaviours within and among populations.

One factor not included in our model that can dramatically affect the degree to which populations diverge is gene flow. In our model, the populations were evolving in isolation, which facilitated their divergence. Gene flow among populations can reduce the likelihood of divergence, however, by introducing trait and preference alleles from one population into others (Barton & Hewitt, 1989; Kelly & Noor, 1996; Servedio & Kirkpatrick, 1997; Barton, 2001). If migration rates are sufficiently high and if alleles introduced via gene flow spread in a population, differences among conspecific populations for mating behaviours could disappear. Yet, although gene flow typically reduces divergence, it need not eliminate divergence especially if selection is strong (Liou & Price, 1994; Kelly & Noor, 1996; Kirkpatrick & Servedio, 1999). Moreover, our findings suggest that once populations begin to diverge in mating behaviours, migrant males or females would be at a selective disadvantage because they would be less likely to mate than resident individuals (Table 9.3; Figure 9.5). Consequently, as long as gene flow does not eliminate initial differentiation of mating behaviours among populations, their divergence could counteract the effects of gene flow and thereby further enhance the likelihood that populations become reproductively isolated.

Reproductive character displacement is generally viewed as a result of reinforcement and the final stages of speciation (Dobzhansky, 1940; Howard, 1993; Coyne & Orr, 2004) or a consequence of interactions that accentuate existing species boundaries (Butlin, 1987). Our results suggest that reproductive character displacement can also initiate speciation. Such a process has been described, for example, in the green-eyed tree-frog, *Litoria genimaculata* (Hoskin *et al.*, 2005). Because most species co-occur with heterospecifics and may even occur with different heterospecifics in different parts of their range, these results further suggest that reproductive character displacement could potentially initiate 'speciation cascades' – multiple speciation events across a given species' range. Yet, whether reproductive character displacement generates diversity in this way remains an open question. Discovering the role that reproductive interactions plays between species in rapid evolutionary diversification is therefore potentially critical for assessing how mate choice contributes to the speciation process.

9.9 Conclusions

Neural network models offer a valuable tool for examining how mating behaviours may diverge between conspecific populations experiencing unique selective environments. If extrapolated to natural systems, our findings suggest that interactions with heterospecifics can cause female preferences for conspecific male characters to diverge among populations co-occurring with different species. The finding that the populations diverged in mate preferences in response to selection to avoid heterospecific matings suggests that such interactions may facilitate the evolutionary diversification of both female mate preferences and male sexual signals. Ultimately, such a process could initiate reproductive isolation and speciation among the different populations if females from a given population fail to recognise males from different populations as acceptable mates (Howard, 1993; Hoskin *et al.*, 2005; Pfennig & Ryan, 2006). Thus, although reproductive character displacement is generally thought to occur following secondary contact of populations that already constitute two species, the process of reproductive character displacement itself could trigger further speciation events.

Acknowledgements

This work was previously published in Pfennig & Ryan (2006, 2007). We are grateful to D. Pfennig, H. Farris, P. Hurd, C. Smith, C. Tosh, M. Noor, M. Servedio, A. Welch, A. Rice, R. Martin, G. Harper, J. Nelson, Y. Zhang, T. Feldman, R. Bartlett and the M. Ryan and M. Kirkpatrick lab groups for discussion and comments on this work. This work was funded by postdoctoral fellowships and grants to K.P. from the National Science Foundation and the National Institutes of Health funded SPIRE program at the University of North Carolina, Chapel Hill.

References

Andersson, M. 1994 *Sexual Selection*. Princeton University Press.

Barton, N. H. 2001. The role of hybridization in evolution. *Mol Ecol* **10**, 551–568.

Barton, N. H. & Hewitt, G. M. 1989. Adaptation, speciation and hybrid zones. *Nature* **341**, 497–503.

Butlin, R. 1987. Speciation by reinforcement. *Trends Ecol Evol* **2**, 8–13.

Conner, J. K. & Hartl, D. L. 2004. *A Primer of Ecological Genetics*. Sinauer Associates, Inc.

Coyne, J. A. & Orr, H. A. 2004. *Speciation*. Sinauer Associates, Inc.

Demuth, H. & Beale, M. 1997. *Neural Network Toolbox*. The Math Works, Inc.

Dobzhansky, T. 1940. Speciation as a stage in evolutionary divergence. *Am Nat* **74**, 312–321.

Elman, J. L. 1990. Finding structure in time. *Cogn Sci* **14**, 179–211.

Enquist, M. & Ghirlanda, S. 2005. *Neural Networks and Animal Behavior*. Princeton University Press.

Falconer, D. S. & Mackay, T. F. C. 1996. *Introduction to Quantitative Genetics*. 4th edn. Longman Group Ltd.

Gabor, C. R. & Ryan, M. J. 2001. Geographical variation in reproductive character displacement in mate choice by male sailfin mollies. *Proc R Soc B* **268**, 1063–1070.

Gerhardt, H. C. 1991. Female mate choice in treefrogs – static and dynamic acoustic criteria. *Anim Behav* **42**, 615–635.

Gerhardt, H. C. 1994. Reproductive character displacement of female mate choice in the gray treefrog, *Hyla chrysoscelis*. *Anim Behav* **47**, 959–969.

Gerhardt, H. C. & Huber, F. 2002. *Acoustic Communication in Insects and Anurans: Common Problems and Diverse Solutions*. University of Chicago Press.

Hankison, S. J. & Morris, M. R. 2003. Avoiding a compromise between sexual selection and species recognition: female swordtail fish assess multiple species-specific cues. *Behav Ecol* **14**, 282–287.

Höbel, G. & Gerhardt, H. C. 2003. Reproductive character displacement in the acoustic communication system of green tree frogs (*Hyla cinerea*). *Evolution* **57**, 894–904.

Hoskin, C. J., Higgie, M., McDonald, K. R. & Moritz, C. 2005. Reinforcement drives rapid allopatric speciation. *Nature* **437**, 1353–1356.

Howard, D. J. 1993. Reinforcement: origin, dynamics, and fate of an evolutionary hypothesis. In *Hybrid Zones and the Evolutionary Process* (ed. R. G. Harrison), pp. 46–69. Oxford University Press.

Kelly, J. K. & Noor, M. A. F. 1996. Speciation by reinforcement: a model derived from studies of *Drosophila. Genetics* **143**, 1485–1497.

Kirkpatrick, M. & Servedio, M. R. 1999. The reinforcement of mating preferences on an island. *Genetics* **151**, 865–884.

Liou, L. W. & Price, T. D. 1994. Speciation by reinforcement of premating isolation. *Evolution* **48**, 1451–1459.

Noor, M. A. 1995. Speciation driven by natural selection in *Drosophila. Nature* **375**, 674–675.

Patterson, H. E. H. 1985. The recognition concept of species. In *Species and Speciation, Transvaal Museum Monograph No. 4* (ed. E. Vrba), pp. 21–29. Transvaal Museum.

Pfennig D. W. & Rice, A. M. 2007. An experimental test of character displacement's role in promoting postmating isolation between conspecific populations in contrasting competitive environments. *Evolution* **61**, 2433–2443.

Pfennig, K. S. 1998. The evolution of mate choice and the potential for conflict between species and mate-quality recognition. *Proc R Soc B* **265**, 1743–1748.

Pfennig, K. S. 2000. Female spadefoot toads compromise on mate quality to ensure conspecific matings. *Behav Ecol* **11**, 220–227.

Pfennig, K. S. & Pfennig, D. W. 2009. Character displacement: ecological and reproductive responses to a common evolutionary problem. *Quart Rev Biol* **84**, 253–276.

Pfennig, K. S. & Ryan, M. J. 2006. Reproductive character displacement generates reproductive isolation among conspecific populations: an artificial neural network study. *Proc R Soc B* **273**, 1361–1368.

Pfennig, K. S. & Ryan, M. J. 2007. Character displacement and the evolution of mate choice: an artificial neural network approach. *Phil Trans R Soc B* **36**, 411–419.

Phelps, S. M. & Ryan, M. J. 1998. Neural networks predict response biases of female túngara frogs. *Proc R Soc B* **265**, 279–285.

Phelps, S. M. & Ryan, M. J. 2000. History influences signal recognition: neural network models of túngara frogs. *Proc R Soc B* **267**, 1633–1639.

Phelps, S. M., Ryan, M. J. & Rand, A. S. 2001. Vestigial preference functions in neural networks and túngara frogs. *Proc Natl Acad Sci USA* **98**, 13161–13166.

Reeve, H. K. 1989. The evolution of conspecific acceptance thresholds. *Am Nat* **133**, 407–435.

Rodriguez, R. L., Sullivan, L. E. & Cocroft, R. B. 2004. Vibrational communication and reproductive isolation in the *Enchenopa binotata* species complex of treehoppers (Hemiptera: Membracidae). *Evolution* **58**, 571–578.

Ryan, M. J. & Getz, W. 2000. Signal decoding and receiver evolution – an analysis using an artificial neural network. *Brain Behav Evol* **56**, 45–62.

Ryan, M. J., Phelps, S. M. & Rand, A. S. 2001. How evolutionary history shapes recognition mechanisms. *Trends Cogn Sci* **5**, 143–148.

Ryan, M. J. & Rand, A. S. 1993. Species recognition and sexual selection as a unitary problem in animal communication. *Evolution* **47**, 647–657.

Ryan, M. J., Rand, W., Hurd, P. L., Phelps, S. M. & Rand, A. S. 2003. Generalization in response to mate recognition signals. *Am Nat* **161**, 380–394.

Saetre, G. P., Moum, T., Bures, S. *et al.* 1997. A sexually selected character displacement in flycatchers reinforces premating isolation. *Nature* **387**, 589–592.

Servedio, M. R. & Kirkpatrick, M. 1997. The effects of gene flow on reinforcement. *Evolution* **51**, 1764–1772.

Servedio, M. R. & Noor, M. A. F. 2003. The role of reinforcement in speciation: theory and data. *Ann Rev Ecol Evol System* **34**, 339–364.

Sokal, R. R. & Rohlf, F. J. 1995. *Biometry*. W. H. Freeman and Co.

Stebbins, R. C. 2003. *A Field Guide to Western Reptiles and Amphibians*. Houghton Mifflin Company.

Wagner, W. E. 1998. Measuring female mating preferences. *Anim Behav* **55**, 1029–1042.

Wiley, R. H. 1994. Errors, exaggeration, and deception in animal communication. In *Behavioral Mechanisms in Ecology* (ed. L. Real), pp. 157–189. University of Chicago Press.

Zar, J. H. 1984. *Biostatistical Analysis*. Prentice-Hall.

10

Applying artificial neural networks to the study of prey colouration

Sami Merilaita

10.1 Introduction

In this chapter I will examine the use of artificial neural networks in the study of prey colouration as an adaptation against predation. Prey colouration provides numerous spectacular examples of adaptation (e.g. Cott, 1940; Edmunds, 1974; Ruxton *et al.*, 2004). These include prey colour patterns used to disguise and make their bearers difficult to detect as well as brilliant colourations and patterns that prey may use to deter a predator. As a consequence, prey colouration has been a source of inspiration for biologists since the earliest days of evolutionary biology (e.g. Wallace, 1889).

The anti-predation function of prey colouration is evidently a consequence of natural selection imposed by predation. More specifically, it is the predators' way of processing visual information that determines the best possible appearance of the colouration of a prey for a given anti-predation function and under given conditions. Because predators' ability to process visual information has such a central role in the study of prey colouration, it follows that we need models that enable us to capture the essential features of such information processing.

An artificial neural network can be described as a data processing system consisting of a large number of simple, highly interconnected processing elements (artificial neurons) in an architecture inspired by biological nerve systems (Tsoukalas & Uhrig, 1997). Artificial neural networks provide a technique that has been applied in various disciplines of science and engineering for tasks such as pattern recognition, categorisation and decision making, as well as a modelling tool in neural biology (e.g. Bishop, 1995; Haykin, 1999). The structural and functional similarities between artificial neural networks and biological neural systems are an often-mentioned fact that has also drawn the attention of behavioural and evolutionary ecologists to this modelling technique. These similarities include the structure that constitutes a network of simple processing units (neurons), processing of data in a parallel mode and behaviours that correspond to memory, learning and generalisation (e.g. Enquist & Arak, 1998; Ghirlanda & Enquist, 1998). These features of artificial neural networks make them an appealing tool for modelling biological information processing.

Modelling Perception with Artificial Neural Networks, ed. C. R. Tosh and G. D. Ruxton. Published by Cambridge University Press.

This chapter on the use of neural network models for the study of prey colouration has been organised the following way. First, I will briefly review the function of prey colouration as an adaptation against predation. Then, I will in general terms describe simple artificial neural network models and how they can be applied for the study of prey colouration, followed by a more detailed description of two examples of studies applying neural networks. Finally, I will discuss the suitability of such models for the study of prey colouration. I hope this will help to introduce researchers of prey colouration to neural networks as well as researchers using neural networks to the study of prey colouration.

10.2 Prey colouration as an adaptation against predation

10.2.1 Concealment

Prey animals make use of colours and patterns in manifold ways to decrease the risk of predation. The two main functions of anti-predator colouration are concealment and signalling to predators. The most obvious way of concealment, or camouflage, is background matching, the use of colours and patterns that match the colour and patterns in the visual environment of the animal (Cott, 1940; Endler, 1978; Ruxton *et al.*, 2004; Merilaita & Lind, 2005). Such a match between a prey and its background in the eyes of the predators will minimise the amount of the information that the predators can use for detection. At first glance this may appear simple to achieve but there are some complications to it (Merilaita *et al.*, 1999, 2001; Houston *et al.*, 2007). Practically all habitats vary spatially and, therefore, a colour pattern that matches one patch or microhabitat does not necessarily provide an equally good match at another microhabitat. Background matching will in such cases be maximised either by a colouration that maximises background matching in that microhabitat only, in which the risk of becoming detected by a predator is highest, or by a colouration that is a compromise between the requirements of several different microhabitats. Which of these two strategies provides the best overall match, being thus the optimal strategy, depends on how the various microhabitats differ from each other, on the visual system of the predators and on which colours and patterns the prey can produce. In general, considering how common background matching appears to be among prey, questions related to optimisation of colours and patterns for background matching have not received attention they deserve, possibly because superficially background matching may appear a simple problem.

Background matching can make the body surface less easy to detect, but there are other traits, such as the shape and outline of the body or some specific body parts, that may give away even a background-matching prey. Therefore, additional principles of concealment may be necessary. Disruptive colouration breaks up the shape of the body or some body parts (Thayer, 1909; Cott, 1940; Merilaita, 1998; Cuthill *et al.*, 2005; Fraser *et al.*, 2007). Disruptive colouration uses a set of markings that creates the appearance of false edges and boundaries and hinders the detection or recognition of an object's, or part of an object's, true outline and shape (Stevens & Merilaita, 2009). This has been suggested to deceive the edge detection mechanism of the eye (Stevens & Cuthill, 2006).

Another suggested principle of concealment is distractive marking (Thayer, 1909). Distractive markings have sometimes been confused with disruptive colouration, but these two appear to be mechanistically different, targeting different processes in predator vision and therefore subjected to different selection pressures on their appearances (Stevens & Merilaita, 2009). Distractive markings have been proposed to be markings that contrast the rest of the prey colouration and therefore attract the attention of the predator to themselves and away from other characteristics that would be more likely to give away the prey (Thayer, 1909). Thus, although visible, the markings themselves are supposed to be meaningless to the predator and contain little useful information for detection or recognition of the prey. There now exists some empirical evidence for the efficacy of distractive markings (Dimitrova *et al.*, 2009).

Self-shadow concealment can be achieved by a colouration that is darker on those areas where light normally falls (typically the dorsal side) and lighter on the opposite site. Such colouration compensates for luminance differences on the body caused by directional light. This makes prey detection more difficult if it conceals a conspicuous shape or if it conceals cues of 3D shape of the prey (Thayer, 1909; Ruxton *et al.*, 2004; Rowland *et al.*, 2007a).

All the principles above can collectively be called cryptic colouration, because they more or less aim to hamper detection of the prey. Masquerade is a form of camouflage that is distinct from cryptic colouration, because it is based on hampering of recognition. An animal relying on masquerade has colours, patterns and morphology that resemble an uninteresting object, for example a twig or a leaf. Although cryptic colouration, too, may influence recognition, the defence has failed if the prey is detected. This is not the case with a prey that relies on masquerade.

Many studies have investigated optimisation of prey colouration in immobile prey. However, for many species it may be important to have an effective camouflage when they are moving. Furthermore, some animals may use their colours and patterns to make it difficult to detect their speed or trajectory (Stevens, 2007). These topics have not yet received very much attention, but would be potential subjects for modelling studies (see Borst, 2007).

To summarise, in order to reduce the risk of becoming detected by predators, a prey may use its colour pattern to decrease the amount of information available for detection through resemblance to background. However, because it is seldom possible to completely eliminate the availability of such information, it may be beneficial for the prey to also use colour pattern to provide predators with misleading information about its appearance. Although research on visual camouflage has increased massively, our surprisingly scarce knowledge about these questions suggests that there still are plenty of opportunities for future research in this area.

10.2.2 Signalling

Colouration may also be employed to deter predators. Some prey use colouration to, honestly or deceitfully, signal predators that they possess a defence that makes them an unprofitable or dangerous catch. Prey colouration that is associated with such unprofitability or risk and decreases the risk of attack, because predators tend to avoid prey with such colouration, is called an aposematic warning signal (Wallace, 1889; Cott, 1940; Edmunds, 1974; Ruxton *et al.*, 2004). Although the avoidance response may be innate, in many cases it appears to be at least partly learned. This means that one important aspect of aposematic signals should be that they are easy to remember (Ruxton *et al.*, 2004). Another aspect of aposematic colouration, also related to learning, is that predators' avoidance response towards aposematic prey strengthens with increasing abundance of the aposematic prey type (e.g. Lindström *et al.*, 2001). This is also related to an evolutionary dilemma that has been intensely studied: if predators effectively learn to avoid aposematic prey only when they are common enough, then how have the initially rare, aposematic mutants been able to reach such threshold levels of abundance in the first place?

Although the colouration of aposematic prey has often been described as conspicuous, it is not necessarily conspicuousness that is under selection. It is possible that aposematic prey is selected for being easily distinguishable from edible prey and conspicuousness is a by-product of this selection (Wallace, 1889; Sherratt & Beatty, 2003; Merilaita & Ruxton, 2007). This also implies that aposematic signals need not be maximally conspicuous to result in minimal risk of attack. For example, as study on the swallowtail butterfly larva suggests, the colouration may actually combine a warning function at a short distance and a concealing function at a longer distance (Tullberg *et al.*, 2005). Thus, one of the main questions regarding aposematism is how detectability, distinctiveness and memorability should be adjusted in combination with the secondary defence to produce the optimal warning colour pattern.

An aposematic prey may offer another prey an opportunity to benefit from its defence if the latter can mimic its appearance (Cott, 1940; Edmunds, 1974; Ruxton *et al.*, 2004). Such mimicry has traditionally been divided into Batesian and Müllerian mimicry. In Batesian mimicry an undefended prey has evolved a similar appearance to an aposematic prey, resulting in incorrect categorisation and reduced predation risk for the mimetic species. On the other hand, the model (aposematic) species may suffer from dilution of its defence, especially if the avoidance is based on a learned response. Thus, while the mimetic species is under selection for increased resemblance, the model species is expected to be under selection for distinctiveness from the mimetic and other less-defended prey species.

In Müllerian mimicry there are two or more prey species with similar appearances and the predator experiences them all as unprofitable (Cott, 1940; Edmunds, 1974; Ruxton *et al.*, 2004). In contrast to Batesian mimicry, in Müllerian mimicry the relationship among prey species cannot necessarily be described as purely parasitic, because all of them may gain some benefit from the common resemblance (Rowland *et al.*, 2007b). This is because a large number of prey with similar appearance can be expected to increase the efficacy of avoidance learning by predators and to decrease the predation risk by naïve predators *per* individual prey. Interestingly, a recent experiment suggests that qualitative variation in defence between Müllerian mimics is important: different defences may strengthen the avoidance response and hence increase the mutualistic benefit from Müllerian mimicry (Skelhorn & Rowe, 2005). These factors are likely to affect the evolutionary dynamics of the appearances of the mimetic species.

In addition to honest or deceitful warning of predators, anti-predator colouration may have other post-detection functions. These include intimidating eyespots, typically a pair of sets of relatively large concentric rings, for example on the wings of several Lepidoptera, which may be suddenly revealed when a predator is at a very close distance. Some researchers believe that predators find them intimidating because they experience them as the eyes of a potentially dangerous enemy (e.g. Vallin *et al.*, 2005), whereas others argue that the effect is simply based on the salience of the pattern (Stevens *et al.*, 2008). Also, it has been suggested that some other eyespots, such as the smaller, lateral eyespots at the wing margins of some butterflies, may function to direct an attack to a less vulnerable area of the body and increase the chance of the prey escaping and surviving an attack (Wourms & Wasserman, 1985; Ruxton *et al.*, 2004).

Also in the case of anti-predation signals there are plenty of unanswered questions. For example, how is the appearance of the signals (or prey colouration) optimised for the cognitive system of predators to maximise the desired response? How is this optimisation affected by factors such as presence and level of a secondary defence, the appearance of the background, the appearance of other prey types and their levels of secondary defences or the presence of multiple predator species?

10.2.3 Prey colouration in models

Generally, most questions about the evolution of anti-predation colouration deal with optimisation of the appearance of colouration of a given prey in relation to its environment. Prey colouration is optimised with respect to the predators' information about that and other potential prey items, and their ability to detect, distinguish and remember that prey. Basically, natural selection on anti-predation colouration can be thought of as adjustment of visual information (honest or deceitful), emitted by a prey and received by its predators, in an attempt to control the response of the predators.

When modelling approaches for the study of prey colouration are considered, three essential points can be concluded from the review above. First, the colouration of a prey is a pattern, and important information may be lost if it is reduced to a simple variable, such as 'conspicuousness'. Second, natural selection on anti-predator colouration is caused by processing of visual information by the predators. This also means that the abilities as well as the biases and limitation of the cognitive systems (vision and information processing) of predators have a central role in the natural selection on prey colouration (cf. Endler, 1992; Guilford, 1992; Arak & Enquist, 1993; Dukas, 1998; Enquist & Arak, 1998).

Third, because of the multiple factors (various prey types, their appearances, qualities as prey and abundances) at play, evolution of prey colouration may often involve complex interactions.

10.3 Neural networks and simulation of evolution of anti-predator colouration

10.3.1 Artificial neural networks

In behavioural and evolutionary ecology, artificial neural networks have been used to simulate a predator in studies about anti-predation colouration and a receiver in studies about signalling between and within species. The most often used network design in these studies has been a feedforward network with three layers (e.g. Arak & Enquist, 1993; Enquist & Arak, 1993, 1994, 1998; Holmgren & Enquist, 1999; Merilaita, 2003; Kenward *et al.*, 2004; Tosh *et al.*, 2007). In a feedforward network the signals traverse to one direction only, from the input layer towards the output layer. Each layer consists of a set of neurons, which are connected to neurons on the adjacent layers. A connection-specific 'weight' value is associated to each connection. Each neuron consists of a transfer (activation) function and a neuron-specific 'bias' value. When input data (usually values between –1 and 1) are projected on the input layer, signals will pass through the connections and neurons and reach the output layer. Because an artificial neural network consists of a network of connections, a single neuron typically receives signals from multiple input cells or neurons and may also forward its signal to more than one neuron. Signals coming to a neuron are first multiplied by the weights of the connections, and the products are then summed up together with the bias value. The resulting sum is then fed into the transfer function, and the output of the function is used as an input to the connections leading to the neurons on the next layer. Threshold functions are usually used as transfer functions.

In order to achieve the desired output values for given input values, a neural network has to be trained. Thus, training in this context means adjustment of weight and bias values of a network to produce a mapping between input data and output values. There are a number of various training methods based on the back-propagation algorithm. Back-propagation is based on a series of iterative calculations to decrease the errors in network weights. During the training passes the errors are first propagated backwards through the network in order to evaluate the derivatives of error function with respect to the weights, and these derivatives are then used to compute the adjustment to be made to the weights (Bishop, 1995; Haykin, 1999). This also means that a back-propagation algorithm requires that the transfer functions are differentiable.

Usually, the training data set only includes a subset of all possible values of input data. Yet, an appropriately trained network can be expected to produce correct outputs even for input data not included in the training data set, a feature called generalisation. This ability to generalise is interesting because it is one of the features that artificial neural networks share with biological neural networks (Enquist & Arak, 1998; Ghirlanda & Enquist, 1998).

Another way to train a network is training through evolution (e.g. Enquist & Arak, 1993, 1994; Holmgren & Enquist, 1999, Kenward *et al.*, 2004). This means that in a population of networks random variation in weight values are produced, and the best performing networks (judged by a fitness function) will be used to form the next generation of networks. Thus, it is comparable to an evolutionary process with genes coding for the weight values. In general, especially when used in disciplines other than evolutionary biology, such a method of searching for optimal solutions in a way that resembles biological evolution is often referred to as the genetic algorithm (e.g. Mitchell, 1996). Neural network training through evolution may locate the neighbourhood of an optimal solution quicker than back-propagation methods due to its global search strategy, but once in the neighbourhood of the optimal solution, it tends to converge to the optimal solution slower than back-propagation methods, because its convergence is controlled by mutation operations (Tsoukalas & Uhrig, 1997). Typically, models with evolving prey that have used training of the network (i.e. the predator) through evolution have produced slower responses in relation to prey evolution than do models that have used a back-propagation algorithm, but evidently, control over the rate of training of the network in relation to prey evolution is fully in the hands of the researcher. Anyway, the choice of training method may depend on whether the training simulates processes corresponding to evolutionary or learned change in the predator's response, or how desirable stochastic events are in the training process.

Input data for an artificial neural network can be described as points in a multidimensional hyperspace. In the present context these points might represent images, such as colour patterns of one or more species of prey or samples of the visual background. The number of cells in the input layer corresponds to the number of the dimensions. Most studies relevant in the context of prey colouration have used a feedforward neural network with threshold functions as transfer functions and one output neuron. Such networks can categorise the input data into two classes, such as 'detected' and 'not detected' in the case of cryptic colouration or 'attacked' and 'not attacked' in the case of warning colouration. Such classification can be compared with drawing a 'decision boundary' that separates input data points to the different categories. Thus, training a neural network adjusts the decision boundary. Insufficient training will result in inaccuracy, whereas too much training will result in 'over-fitting' of the decision boundary and decreased generalisation (i.e. the ability to correctly classify input data not included in the training data set). Importantly, the optimal network design (i.e. the number of layers and the number of neurons in the layers as well as how the neurons are connected to each other) depends on the complexity of the decision boundary and the dimensionality of the input vector.

Conversely, neural network performance depends on both the network design and training. This is in many ways analogous to fitting a polynomial curve to a data set (e.g. Bishop, 1995). Such a nonlinear curve not only describes the data, but it can also be used to predict output values for input values not included in the data set. However, to obtain a curve that satisfactorily fits the data one has to find the right number of terms in the polynomial as well as the right parameter values for the terms. Similarly, an artificial neural network satisfactorily performs a given mapping task (such as a categorisation task) if it fits the training data set and, moreover, can be used to predict output values also for input values not included in the training data. To obtain such network performance one has to choose an appropriate number of layers and neurons (network design) as well as find appropriate weight values for the connections between the neurons (training).

The design and training of a network have to be optimised for the modelling task in question. A network that has not enough neurons and layers or has not been trained well enough will not produce a correct mapping between the input and output values in the training data. On the other hand, if the network design is too complex or the network has been trained too much, then there is a risk that the network will over-fit the training data: the network will represent well the specific aspects of the training data (including any noise) at the expense of representing poorly the more general or systematic pattern (e.g. Bishop, 1995). In other words, the generalisation ability of the network will be poor. Consequently, the aim usually is to use the simplest design and shortest training that are sufficient for the task at hand. For small networks used for relatively simple tasks it is possible to gain an idea of the optimal architecture by deduction. Generally, information about the appropriateness of network design and training can be gained empirically by investigating network performance (which can be measured as mean of squared errors between observed and desired outputs) and how it varies when network design or number of iterations of the training algorithm used is changed. Note that if another data set (called validation or test data set) is used instead of the training data set for measuring performance then the measures of performance are less likely to be confounded by over-fitting, a method called cross-validation (Bishop, 1995, Haykin, 1999). More detailed and deeper discussion about various methods for optimisation of network design and training can be found elsewhere (e.g. Bishop, 1995, Haykin, 1999). However, when it comes to optimisation of design and training of neural networks in evolutionary models, one may have to consider some additional points that are typically not relevant for static, non-biological applications. In practice, the evolutionary process may affect training. For example, input data for the network are likely to change with proceeding evolution and, therefore, the training conditions may also change. Furthermore, it is important to bear in mind that the choice of design and training constitute a part of the assumptions of the model. Maximised or unconstrained predator performance may not always be the most realistic assumption, and constraints in information processing may have evolutionary importance (see the Introduction).

10.3.2 Prey colouration evolution

Evolutionary optimisation of prey colouration can be studied by letting prey colouration evolve through selection imposed by the neural network (e.g. Holmgren & Enquist, 1999; Merilaita, 2003; Merilaita & Ruxton, 2007; Merilaita & Tullberg, 2005). This can be done by simulating natural selection on a population of prey with genetically coded colouration (cf. genetic algorithm; e.g. Mitchell, 1996). The aim of this approach is to find optimal or good solutions as well as to produce adaptive responses (i.e. find new solutions) if the conditions are changing. Here the prey population corresponds to a set of candidate solutions. Prey colouration is described by chromosomes consisting of sets of alleles. It has been shown that a relatively small population size, about 30 individuals (20–100), usually works well, but this also depends on the rate of mutations and recombination events (Mitchell, 1996).

The prey population is subjected to selection by the neural network. Thus, the output of the neural network for a given colouration phenotype is used to determine the predation risk, which in turn can be used to determine the fitness of an individual with that phenotype. A fraction of the population with lowest fitness is removed and replaced by the offspring of the remaining individuals. To produce genetic variation that is necessary for evolution, some mutations and recombination events take place when the prey reproduces. Usually, the prey species can be assumed to be a haploid hermaphrodite if the study focuses on optimisation of colouration and not on the effects of specific genetic mechanisms. In a system that combines a continually trained neural network and a virtual prey species all the parties adapt to the responses of the other parties, resulting in evolution through counter-adaptive steps (and possibly co-adaptive steps among different prey species). Obviously, simulations like this have to be replicated because of the stochasticity involved in the system.

10.3.3 Neural networks in studies on prey colouration evolution

Neural network models have been applied to the study of various aspects of predation. For example, Tosh *et al.* (2007) used a neural network to investigate how a mixed-species grouping of prey could confuse the predator and thus benefit the prey. In their model prey colouration did not evolve, but instead the constitution of prey groups varied. The inputs to the neural network were projections of artificial prey groups represented simply by vectors of numbers with numerical values indicating the visual intensity of a prey item. So far there are, however, only few studies that have employed neural networks particularly to study prey colouration. In contrast, there are several neural network studies concerning the evolution of signalling. Some of these have studied questions that are relevant in the context of anti-predator signals. For example, Enquist & Arak (1993, 1994, 1998) studied the effect of biases in receivers' recognition mechanism on selection on different aspects of signal form, such as signal symmetry or amplification of a signal. In a recent study Kenward *et al.* (2004) investigated the evolution of repetitive patterns in visual signals.

Holmgren & Enquist (1999) used a neural network model to explicitly study prey colouration. They simulated a Batesian mimicry system using neural networks as predators and studied the evolutionary dynamics between the model and the mimic. Holmgren & Enquist (1999) used as a predator a feedforward network with nine input cells, five neurons in a hidden layer and one output neuron. The network was trained by evolution. Colourations of both the defended prey species (the model) and the undefended

Batesian mimic evolved and were represented by nine-cell vectors. The predator population consisted of 1000 individuals, the population of models consisted of 1000 individuals and the population of mimics consisted of 500 individuals. Each simulation was run for 50 000 generations. The study conducted by Holmgren & Enquist (1999) suggests that the appearance of the model and the mimic are constantly changing, and monotonically increasing response gradients cause the appearance of the model and the mimic to change in the same direction.

Merilaita & Ruxton (2007) modelled the evolution of aposematic warning colouration with respect to conspicuousness and distinctiveness. They used one neural network to model probability of prey detection as a function of visual deviance from background, and another neural network to model the dependency of probability that a predator attacks a prey with a given appearance on the abundance of individuals with that appearance that are unpalatable. The probability of survival of each prey appearance type was based on the outputs of the two neural networks. There were two prey species, the focal species that had an evolving appearance, and the reference species with a fixed appearance. In the various scenarios investigated in the study, either one of the prey species or both were unpalatable. The study suggests that prey conspicuousness may result from selection for distinctiveness, but that selection for distinctiveness does not result in maximisation of conspicuousness. Further, it suggests that selection for distinctiveness may even select for camouflage, depending on the appearance of other prey in the local environment and the relative benefits of camouflage and warning colouration.

Bain *et al.* (2007) used measurements of 17 different biometric variables to summarise the appearances of mimetic hoverflies, wasps, which were potential models of the hoverflies, and non-mimetic flies. The aim of the study was to identify those variables that trained pigeons had used to distinguish the hoverflies from the wasps in a previous, empirical study (Dittrich *et al.*, 1993). Bain *et al.* (2007) used a genetic algorithm both to choose the design of a feedforward neural network (the input variables and the number or neurons) and to train the network. The aim of this process was to find the simplest neural network model that would produce an output matching the response of the pigeons. The authors concluded that the input variables used by the optimised neural network most likely represented those features that were also used by the pigeons in the discrimination task. The most important predictors of pigeon behaviour included the number, colour and contrast of stripes and the antennal length. Note that in this study the neural network was in the first place used as a model to fit the data (cf. a statistical model) from the experimental study conducted by Dittrich *et al.* (1993) rather than to model perception.

Merilaita (2003) studied the effect of visual complexity of habitat on natural selection on cryptic colouration, and Merilaita & Tullberg (2005) studied the evolutionary choice of defence strategy between crypsis and aposematism. These two studies address quite different questions about prey colouration. Yet, to answer these questions both studies used artificial neural networks to simulate predation on populations of evolving, virtual prey. Below, I will describe these two studies in more detail to enlighten the use of this method.

10.4 Examples of studies

10.4.1 Prey camouflage and visual background complexity

Merilaita (2003) used a three-layer feedforward network to study the effect of visual background complexity on the evolution of cryptic colouration. Somewhat surprisingly, this appears to be the only neural network study which focuses on the evolution of cryptic colouration so far. In this study the predator's processing of visual information was assumed to be based on visual samples (vectors with eight cells) of its environment, which it classified either as background or prey. In the evolving prey this resulted in selection for colouration that the predator would incorrectly classify as background.

In the model two factors were varied. The first factor was the level of visual complexity of the background, meaning the number of different visual elements that occurred in the habitat. Here visual elements were defined as the components of colour patterns and can be thought of as colours or features such as stripes or spots. In this study, visual complexity refers to variation at a scale smaller than the size of the visual samples and it should not be confused with large-scale heterogeneity, such as differences between microhabitats. The complex habitat consisted of four different visual elements and the simple habitat consisted of three visual elements. The eight-cell samples of the backgrounds were created using simple rules. First, one of the visual elements was chosen as the 'basal element'. A sample of the complex habitat always contained two cells of the basal element and two cells of each of three remaining visual elements. In the simple habitat the samples always consisted of two cells of the basal element and three cells of each of the other two elements. Because the order of the cells was not determined, there were 2520 different possible samples of the complex habitat, but only 560 different possible samples of the simple habitat.

The second factor that was varied was the presence of a constraint in the colour pattern evolution of the prey. Either the prey was able to produce all the visual elements found in the habitat (unconstrained evolution), or it was able to produce all except one of the elements found in the habitat (constrained evolution). The initial prey colouration in each simulation run was constituted by eight cells of the basal element.

The feedforward network had eight input cells, five neurons in the hidden layer and one output neuron. Logistic sigmoid (i.e. smooth threshold) functions were used as transfer functions. The task of the neural network was to categorise a visual sample either as prey or background. The network outputs varied from 0 to 1, and the correct output for prey was 1 and for background it was 0. A back-propagation algorithm was used in the training. Before prey evolution, the network was first trained to categorise between the initial prey colouration and the background until the mean square error was decreased to 2×10^{-6}, a small value that was achieved in every run of the simulation. Then, while the prey was evolving, increment training was used, such that the network was presented with a training data set once every prey generation, and the weights and biases of the network were adjusted after each presentation of a vector in the training set. The training set consisted of a sub-sample of the prey population and an equally large sample of randomly chosen background samples. This training procedure was used because it was important to ensure that neural network performance was adjusted at the same rate in all the four combinations of habitat complexity (simple or complex) and type of evolution (unconstrained or constrained).

Fitness of an individual prey was determined by the output of the neural network for that colour pattern phenotype, such that the prey for which the output deviated most from one had the highest fitness. This caused prey colouration to evolve towards increased crypsis. As a consequence, the prey became less susceptible to predation (i.e. network outputs for prey decreased) and its resemblance to background increased during the course of evolution in all four combinations of habitat complexity (simple or complex) and type of evolution (unconstrained or constrained). The prey was allowed to evolve until either the prey susceptibility to predation was below a threshold value (i.e. the mean output for the most cryptic 25% of the prey population was lower than 0.001) or 200 prey generations had passed.

The main result from this study suggests that it was more problematic for the prey to evolve efficient crypsis in the simple habitat than in the complex habitat (Figure 10.1). Accordingly, in the complex habitat the detectability of the prey decreased below a threshold value in every run of the simulation both under unconstrained and constrained evolution of colouration. In contrast, in the simple habitat this was the case only under unconstrained evolution, whereas under constrained evolution the prey was unable to reach the threshold value in 34% of the simulation runs. In other words, the evolutionary constraint had a much more severe impact on predation susceptibility in the simple than in the complex habitat. This suggests that for a prey colour pattern to achieve a given probability of escaping detection (i.e. a given level of crypsis) more will be required from it, for example in terms of background matching, in a visually simple than in a visually complex habitat.

This result has several implications for the study of prey colouration. For one thing, it suggests that estimates of degree of crypsis based on similarity between colour pattern and habitat of prey (e.g. Endler, 1984) are not comparable between habitats that differ in complexity. Moreover, the result suggests that prey should make use of constraints in predators' processing of visual information. This is interesting, because the importance of predators' sensitivity to different wavelengths of light (colour vision) has received more attention, whereas the role of processing of visual information after it has passed the retina has not received much attention. The reason for this bias may be that more is known about colour vision than about processing of visual information after it has passed the retina. Hence, this bias should encourage the application of neural network models in behavioural and evolutionary ecology, as they can be used to study the evolutionary consequences of information processing, such as processing of visual information by predators.

The rationale behind the study and thus a central assumption of the model is the simple fact that the brain has a limited capacity to process visual information (e.g. Dukas, 1998), a limitation that is demonstrated by the trade-off between search rate and detection rate in predators (Gendron & Staddon, 1983; Gendron, 1986). Therefore, if the processing capacity (determined by neural network design and training) is constant, an increase in visual complexity (diversity of visual information) makes the detection task more difficult

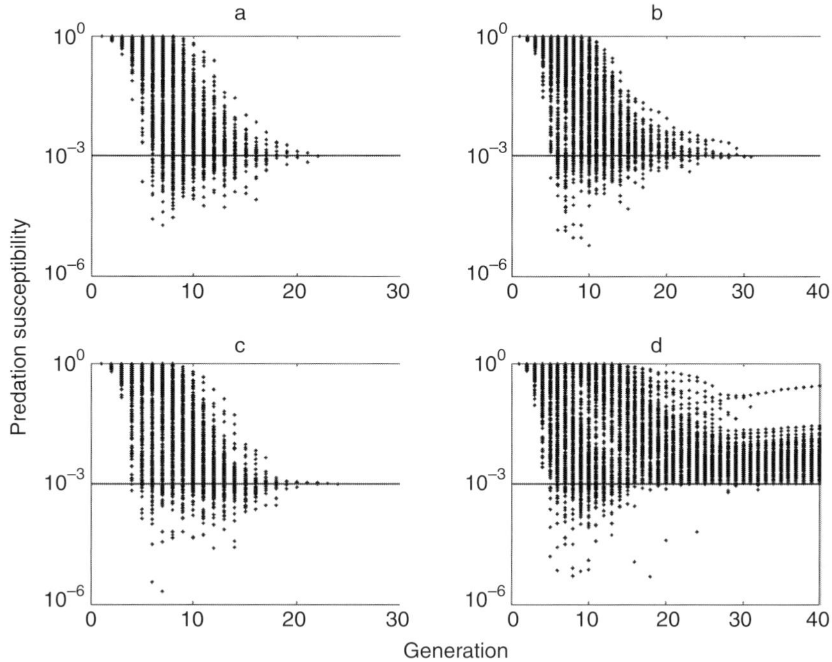

Figure 10.1. The predation susceptibility of the prey per generation, measured as the mean output of the neural network for the most cryptic 25% of the prey population: (a) unconstrained evolution in the complex habitat; (b) constrained evolution in the simple habitat; (c) unconstrained evolution in the complex habitat; (d) unconstrained evolution in the simple habitat. A simulation run (N = 500 in each category) was stopped when the predation susceptibility threshold value of 10^{-3} was reached. In (d) 34% of the runs of the simulation did not reach the threshold in 200 generations. One dot may represent multiple points. From Merilaita (2003) with permission.

and a prey even with constrained crypsis will be less likely to become detected. In spite of its simple rationale this model allowed conclusions that were novel in the study of cryptic colouration. There were at least two benefits from the use of an artificial neural network in this study. First, it enabled a feasible way to study selection on a 'colour' pattern and to draw conclusions about crypsis. With another modelling technique it would be more problematic to study these two features separately without confusing them. Second, the similarity in information processing between predators and artificial neural networks, in this case the constrained processing capacity and its consequences, was valuable.

10.4.2 The problems of crypsis and aposematism and the choice of defence strategy

Several previous studies on aposematism have emphasised how difficult it is for aposematism to evolve. An aposematic prey type has to be common enough for the predator to efficiently learn to associate its warning colouration with the secondary defence that the aposematic prey carries. Therefore, it has been argued that the evolution of aposematism is paradoxical as a rare aposematic mutant with a conspicuous aposematic colouration is unlikely to invade a cryptic prey population. For some reason this argument has been based on a more or less explicit assumption that the prey population is initially cryptically coloured and, furthermore, that this crypsis is highly efficient. However, there is no reason to expect crypsis to be an unproblematic adaptation either. As already touched upon, a general problem for cryptic prey is that visual heterogeneity of environment often makes it difficult to achieve through an invariable colouration a low risk of detection in every part of the habitat the prey uses (Edmunds, 1974; Merilaita *et al.*, 1999, 2001; Ruxton *et al.*, 2004; Houston *et al.*, 2007). Similarly, prey mobility may constraint crypsis because mobility as such facilitates detection (Cott, 1940; Edmunds, 1974; Ruxton *et al.*, 2004). Thus, it is reasonable to assume that constraints on crypsis are rather common.

Merilaita & Tullberg (2005) addressed these questions by studying the evolutionary choice of an optimal defence strategy, particularly between crypsis and anti-predator signalling. The first part of the study consisted of an evolutionary simulation model, in which crypsis was either constrained or not. More specifically, the prey either lived in a visually homogeneous habitat that allowed the evolution of a high degree of crypsis or in a heterogeneous habitat, consisting of two equally common but visually very different microhabitats, which imposed a constraint on the evolution of crypsis. There were two species of prey, Prey 1 and Prey 2, which differed only in one respect. Prey 1 was always edible, whereas Prey 2 might become inedible. Thus, initially Prey 2 was edible, but due to mutation the secondary defence was likely to arise and spread in the population sooner or later. Both the species had an evolving colour pattern. In the beginning of each simulation run both the prey species had the same, randomly chosen colour pattern. Thus the study concentrated on optimisation of prey colour pattern and choice of defence strategy when a constraint on crypsis caused by habitat heterogeneity was either present or absent. The prey was considered to have reached a cryptic optimum if the prey colour pattern matched the homogeneous habitat or matched either of the microhabitats in the heterogeneous habitat. The prey was considered to have reached an aposematic optimum if the prey was inedible, it had evolved a colour pattern that deviated from the habitat (or the microhabitats) and this colour pattern was not invaded by another during 30 successive prey generations.

Prey colour patterns and samples of the habitats were described by four-cell vectors. Each cell of a prey colour pattern vector was occupied by a colouration element denoted by 1, 2 or 3. The homogeneous habitat was uniform, consisting of one type of element only. This element was randomly chosen in the beginning of each simulation run to be either 1 or 3. In the heterogeneous habitat one of the microhabitats consisted uniformly of the element 1 and the other microhabitat uniformly of the element 3. The prey used both the microhabitats with equal probabilities in the heterogeneous habitat. Because habitat heterogeneity was assumed to constrain crypsis, the model parameters were chosen so that colouration adapted to one of the microhabitats would yield higher crypsis than any colouration that compromised the requirements of both the microhabitats.

In this model prey fitness was assumed to depend on the following factors: Prey crypsis increased and, thus, the probability of becoming detected by predators decreased with increasing resemblance between prey colour pattern and the background. In the heterogeneous habitat the two microhabitats were equally common and thus prey crypsis there was given by the average crypsis in the two microhabitats. Prey 2 could also benefit from aposematism. Accordingly, an increase in the proportion of inedible individuals among individuals with a given colour pattern decreased the risk of becoming attacked by a predator for all individuals with that colour pattern. Thus, for Prey 1 fitness was given by the probability of avoiding detection and for Prey 2 fitness was determined by that of the two strategies (crypsis or aposematism) that yielded the better protection against predation.

To simulate predation according to the assumptions of the model, two different neural networks were used. The first network was a radial basis function network. The radial basis function network gave the probability of detection for each prey colour pattern, based on the assumption that probability of detection is determined solely by the resemblance between the background and prey colour pattern. Radial basis function networks are used in pattern recognition, and they consist of an input layer, a hidden layer and an output layer. The transfer function of a radial basis function is typically a Gaussian function. Its output is determined by the Euclidian distance between the input vector and a template vector (e.g. Bishop, 1995; Tsoukalas & Uhrig, 1997; Haykin, 1999; Theodoridis & Koutroumbas, 1999). As the name implies, such radial basis function produces radially symmetric activations that decrease from 1 to 0 with increasing Euclidian distance from the template vector. Merilaita & Tullberg (2005) applied a very simplistic variant of such a network. It had one radial basis unit only, and a vector representing the background was used as the template vector. Consequently, the network transferred the amount of difference between prey colour pattern and background to hypothetical detection probability values.

A number of experimental studies on aposematism have demonstrated a tendency of predators to learn to avoid prey that they experience as unpleasant or dangerous by its appearance (reviewed in Ruxton *et al.*, 2004). The second neural network of the model in Merilaita & Tullberg (2005) was used to simulate such an avoidance response. It was a feedforward network with four input cells, eight neurons in the first hidden layer, four cells in the second hidden layer and one output neuron. The model assumed that an increase in a prey population in the relative frequency of inedible individuals with a given colouration increases the avoidance response and decreases the probability of attack towards that colour pattern. This required that the output of this network had to correspond to an attack probability and hence vary continuously from 0 to 1, instead of simply representing a binomial attack decision. Therefore, a typical feedforward network with one hidden layer and one output neuron with a threshold function as a transfer function was not suitable for the task, but it had to be modified. Thus, the output neuron of such a network was replaced with a second hidden layer and an output layer with one neuron that exceptionally had a linear transfer function (all the other transfer functions were smooth (logistic sigmoid) threshold functions). In other words, instead of a single output neuron assigning the summed signals from the first hidden layer in either of two decision classes, the multiple neurons on the second hidden layer enabled multiple outputs. These outputs were then summed and fed into the output neuron, which with the help of the linear transfer function transformed the signal to a value of attack probability. Consequently, training of the network created a relationship between colour pattern phenotypes and attack probabilities. The training data set included the colour pattern of each prey individual (Prey 1 and 2) and the proportions of inedible individuals within each colour pattern phenotype were used as training target values. The training was based on a back-propagation algorithm with an adaptive learning rate (e.g. Demuth & Beale, 2000), and the network was trained during each Prey 2 generation until the mean square error of the output was less than 0.001 or until the training data set had been presented 150 times.

As might be expected, Prey 1, which could not become inedible, always evolved to a cryptic optimum. When evolution of camouflage was not constrained by habitat heterogeneity, Prey 1 evolved in each of the 1000 replicates of the simulation a colour pattern that matched the background. Also in all the 1000 replicates of the simulation in the heterogeneous habitat Prey 1 evolved colouration matching either of the microhabitats (in 22 of the runs the population was polymorphic and consisted of individuals with either of the background matching colourations).

Because of the favourable conditions (e.g. high mutation rate for the secondary defence) Prey 2 evolved aposematism a number of times in both the habitats. However, the main result from the study was that in the homogeneous habitat Prey 2 evolved aposematism only 543 times out of 1000 compared with 915 times in the heterogeneous habitat. This suggests that constraints on camouflage (in this case due to habitat heterogeneity) favour the evolution of aposematism. In the second part of the study Merilaita & Tullberg (2005) back up this conclusion with empirical evidence. The evidence comes from a comparison of the commonness of crypsis and anti-predator signalling among day-active lepidopteran taxa, in which camouflage is constrained by mobility, and in night-active taxa, which rest during the day and therefore are not mobile when susceptible to visual predation. The comparison showed a significant association between day-activity and anti-predator signalling. Merilaita & Tullberg (2005) concluded that while focusing on the costs of aposematic or mimetic colouration, previous studies on the evolution of anti-predator signals have neglected the possible costs and constraints for the alternative strategy, crypsis, and that this may provide an explanation for why the evolution of anti-predator signals has been argued to be paradoxical.

10.5 Conclusions

As the brief review on prey colouration as an anti-predator adaptation indicated, the numerous questions related to evolution of prey colouration are attracting more and more attention. When considering a modelling approach for the study of prey colouration, there are some features, which are essential. First, in some cases it is possible to use models, in which colour patterns are represented by simplistic variables, such as ‘detectability’. A simple model is preferable if it meets the requirements set by the question studied.

However, natural selection acts on phenotypic traits such as colour pattern, instead of (directly) acting on conceptual traits such as predation risk or detectability. Thus, although such simplistic variables are suitable for modelling some questions, in other cases additional or more correct insights can be obtained by using a more realistic representation of colour patterns. Therefore, it may often be important to capture the multi-dimensional nature of colour patterns as variables. It is evident that artificial neural networks are well-suited for processing such variables, as for example their use in pattern recognition indicates (Bishop, 1995; Theodoridis & Koutroumbas, 1999). Moreover, neural networks provide a relatively simple method to deal with patterns.

The second essential feature is the central role of visual information processing of predators in natural selection on prey colouration. Therefore, the capacity to reproduce a desired aspect of information processing by a predator or to imitate a desired response of a predator to visual information with the required level of biological reality is important when one considers modelling approaches for the study of prey colouration. Neural networks are not just capable of information processing and responding to patterns, but the structural and functional similarities between artificial neural networks and biological neural systems may be helpful when the processing of visual information by predators is modelled. This is because these similarities may enable artificial neural networks to be used to model information processing and decision making in a biologically plausible fashion (e.g. Enquist & Arak, 1998; Ghirlanda & Enquist, 1998; Enquist *et al.*, 2002).

The neural networks used in studies of prey colouration and signalling have been rather simple. However, complexity as such is not a goal in modelling. On the contrary, simple models are generally easier to interpret and less likely to suffer from confounding, irrelevant factors or effects. Accordingly, when it comes to neural network models, comprehending the behaviour of a model, which is important, is easier if the model is simple than if it is complex. Thus simplicity ought to be preferred as far as it does not constrain the function of the model. Also, it is important to realise here that a neural network model is used for reproducing a given aspect of a predator's information processing or response to visual information. This means that it is wrong to interpret the aim to be something more, such as representing the whole visual system or the brain. For example, in the first neural network model (Merilaita, 2003) described above, studying crypsis and background complexity, the interesting aspect of predator behaviour was the limited capacity of information processing and its effect on a predators' ability to detect prey and eventually on natural selection on crypsis. In the study about evolutionary choice of defence strategy between aposematism and crypsis (Merilaita & Tullberg, 2005), the behavioural outcome (i.e. response to prey colouration) was more important than the process, through which it was achieved.

The third essential feature in modelling prey colouration is that it may involve dynamic and complex interactions. Neural network studies of prey colouration have used simulation of evolution to study prey colouration (Holmgren & Enquist, 1999; Merilaita, 2003; Merilaita & Tullberg, 2005; Merilaita & Ruxton, 2007). Evolutionary simulations are well-suited for studying adaptation and optimal phenotypes under dynamic and complex interactions, such as successive counter-adaptive or co-adaptive responses over multiple evolutionary steps or cases where fitness of a phenotype varies due to multiple, dynamic factors involved (see also Peck, 2004). For example, in the study about evolutionary choice of defence strategy between aposematism and crypsis (Merilaita & Tullberg, 2005), predation risk of a prey was affected by the appearance of the prey in relation to the background, the characteristics of the background, the presence or absence of the secondary defence, its appearance in relation to other prey and the abundance of prey with similar colouration phenotype. In the study by Holmgren & Enquist (1999) the focus was on the dynamics of selection on prey appearance in Batesian mimicry systems. Also, these dynamics are complex because change in any of the three involved factors (predator behaviour, appearance of the mimic and appearance of the model) are likely to impose a change in the other two factors. Generally, artificial neural networks are well suited as components of evolutionary simulation models because they can produce adaptive responses through training and they can respond to novel stimuli by generalising information from familiar stimuli.

An artificial neural network may exhibit various interesting behaviours. However, it is of paramount importance to bear in mind that, although there are some similarities between artificial neural networks and biological neural systems, not all behaviours of artificial neural networks are necessarily biologically relevant. In other words, these similarities seldom suffice to validate a model. This applies to modelling in general; empirical data are needed to confirm the validity of the assumptions or the results of a model. In the study about crypsis and background complexity (Merilaita, 2003) the model was based on a known fact that the brain has a limited capacity to process visual information simultaneously (Dukas, 1998). Further, psychological experiments using humans as subjects show that the difficulty of a visual detection task increases with increasing complexity of background patterns and colours, lending empirical support for the results of that study (Gordon, 1968; Farmer & Taylor, 1980). The study about evolution of defence towards aposematism or crypsis (Merilaita & Tullberg, 2005) used neural networks to imitate the decrease in detection probability with increasing resemblance between prey and their background as well as the ability of predators to learn to avoid defended prey by their appearance. Further, empirical support for the result of the model was provided by the phylogenetic comparison in the second part of the study (Merilaita & Tullberg, 2005).

To summarise, a capacity to deal with patterns as well as to reproduce aspects of biological information processing or behavioural responses to visual information are primary qualities of a modelling approach well suited for the study of prey colouration. Also, capability to deal with dynamic interactions is an advantage. Neural network models satisfy these requirements and may therefore provide an appropriate tool for many questions in the study of prey colouration. Although there are only a few studies so far that have applied neural networks to study prey colouration, the studies described above show that neural network models can help us to gain novel insights and promote the understanding of the specific appearances of prey colouration.

Acknowledgements

I thank Magnus Enquist, Charlotta Kvarnemo, Ian Henshaw and Colin Tosh for valuable comments on earlier versions of this manuscript. This study was supported by the Swedish Research Council and the Academy of Finland.

References

Arak, A. & Enquist, M. 1993. Hidden preferences and the evolution of signals. *Phil Trans R. Soc B* **265**, 1059–1064.

Bain, R. S., Rashed, A., Cowper, V. J., Gilbert, F. S. & Sherratt, T. N. 2007. The key mimetic features of hoverflies through avian eyes. *Proc R Soc B* **274**, 1949–1954.

Bishop, C. M. 1995. *Neural Networks for Pattern Recognition*. Oxford University Press.

Borst, A. 2007. Correlation versus gradient type motion detectors: the pros and cons. *Phil Trans R Soc B* **362**, 369–374.

Cott, H. B. 1940. *Adaptive Coloration in Animals*. Methuen.

Cuthill, I. C., Stevens, M., Sheppard, J. *et al.* 2005. Disruptive coloration and background pattern matching. *Nature* **434**, 72–74.

Demuth, H. & Beale, M. 2000. *Neural Network Toolbox for Use with Matlab, Version 4*. The MathWorks Inc.

Dimitrova, M., Stobbe, N., Schaefer, H. M. & Merilaita, S. 2009. Concealed by conspicuousness: distractive prey markings and backgrounds. *Proc R Soc B* **276**, 1905–1910.

Dittrich, W., Gilbert, F., Green, P., McGregor, P. & Grewcock, D. 1993. Imperfect mimicry: a pigeon's perspective. *Proc R Soc B* **251**, 195–200.

Dukas, R. 1998. Constraints on information processing and their effects on behavior. In *Cognitive Ecology: the Evolutionary Ecology of Information Processing and Decision Making* (ed. R. Dukas), pp. 89–128. University of Chicago Press.

Edmunds, M. 1974. *Defence in Animals*. Longman.

Endler, J. A. 1978. A predator's view of animal color patterns. *Evol Biol* **11**, 319–364.

Endler, J. A. 1984. Progressive background matching in moths, and a quantitative measure of crypsis. *Biol J Linn Soc* **22**, 187–231.

Endler, J. A. 1992. Signals, signal conditions and the direction of evolution. *Am Nat* **139**, S125–S153.

Enquist, M. & Arak, A. 1993. Selection of exaggerated male traits by female aesthetic senses. *Nature* **361**, 446–448.

Enquist, M. & Arak, A. 1994. Symmetry, beauty and evolution. *Nature* **372**, 169–172.

Enquist, M. & Arak, A. 1998. Neural representation and the evolution of signal form. In *Cognitive Ecology: The Evolutionary Ecology of Information Processing and Decision making* (ed. R. Dukas), pp. 21–87. University of Chicago Press.

Enquist, M., Arak, A., Ghirlanda, S. & Wachtmeister, C.-A. 2002. Spectacular phenomena and limits to rationality in genetic and cultural evolution. *Phil Trans R Soc B* **357**, 1585–1594.

Farmer, E. W. & Taylor, R. M. 1980. Visual search through color displays: effects of target-background similarity and background uniformity. *Percept Psychophys* **27**, 267–272.

Fraser, S., Callahan, A., Klassem, D. & Sherratt, T. N. 2007. Empirical tests of the role of disruptive coloration in reducing detectability. *Proc R Soc B* **274**, 1325–1331.

Gendron, R. P. 1986. Searching for cryptic prey: evidence for optimal search rates and the formation of search images in quail. *Anim Behav* **34**, 898–912.

Gendron, R. P. & Staddon, J. E. R. 1983. Searching for cryptic prey: the effect of search rate. *Am Nat* **121**, 172–186.

Ghirlanda, S. & Enquist, M. 1998. Artificial neural networks as models of stimulus control. *Anim Behav* **56**, 1383–1389.

Guilford, T. 1992. Predator psychology and the evolution of prey coloration. In *Natural Enemies: The Population Biology of Predators, Parasites and Diseases* (ed. M. J. Crawley), pp. 377–394. Blackwell.

Gordon, I. E. 1968. Interactions between items in visual search. *J Exp Psychol* **76**, 248–355.

Haykin, S. 1999. *Neural Networks: A Comprehensive Foundation*. 2nd edn. Prentice-Hall.

Holmgren, N. M. A. & Enquist, M. 1999. Dynamics of mimicry evolution. *Biol J Linn Soc* **66**, 145–158.

Houston, A. I., Stevens, M. & Cuthill, I. C. 2007. Animal camouflage: compromise or specialize in a 2 patch-type environment. *Behav Ecol* **18**, 769–775.

Kenward, B., Wachtmeister, C.-A., Ghirlanda, S. & Enquist, M. 2004. Spots and stripes: the evolution of repetition in visual signal form. *J Theor Biol* **230**, 407–419.

Lindström, L., Alatalo, R. V., Lyytinen, A. & Mappes, J. 2001. Strong antiapostatic selection against novel rare aposematic prey. *Proc Natl Acad Sci USA* **98**, 9181–9184.

Merilaita, S. 1998. Crypsis through disruptive coloration in an isopod. *Proc R Soc B* **265**, 1059–1064.

Merilaita, S. 2003. Visual background complexity facilitates the evolution of camouflage. *Evolution* **57**, 1248–1254.

Merilaita, S. & Lind, J. 2005. Background-matching and disruptive coloration, and the evolution of cryptic coloration. *Proc R Soc B* **272**, 665–670.

Merilaita, S., Lyytinen, A. & Mappes, J. 2001. Selection for cryptic coloration in a visually heterogeneous habitat. *Proc R Soc B* **268**, 1925–1929.

Merilaita, S. & Ruxton, G. D. 2007. Aposematic signals and the relationship between conspicuousness and distinctiveness. *J Theor Biol* **245**, 268–277.

Merilaita, S. & Tullberg, B. S. 2005. Constrained camouflage facilitates the evolution of conspicuous warning coloration. *Evolution* **59**, 38–45.

Merilaita, S., Tuomi, J. & Jormalainen, V. 1999. Optimisation of cryptic coloration in heterogeneous habitats. *Biol J Linn Soc* **67**, 151–161.

Mitchell, M. 1996. *An Introduction to Genetic Algorithms*. MIT Press.

Peck, S. L. 2004. Simulation as experiment: a philosophical reassessment for biological modeling. *Trends Ecol Evol* **19**, 530–534.

Rowland, H. M., Speed, M. P., Ruxton, G. D. *et al.* 2007a. Countershading enhances cryptic protection: an experiment with wild birds and artificial prey. *Anim Behav* **74**, 1249–1258.

Rowland, H. M., Ihalainen, E., Lindström, L., Mappes, J. & Speed, M. P. 2007b. Co-mimics have a mutualistic relationship despite unequal defence levels. *Nature* **448**, 64–66.

Ruxton, G. D., Sherratt T. M. & Speed M. P. 2004. *Avoiding Attack: The Evolutionary Ecology of Crypsis, Warning Signals and Mimicry*. Oxford University Press.

Sherratt, T. N. & Beatty, C. D. 2003. The evolution of warning signals as reliable indicators of prey defense. *Am Nat* **162**, 377–389.

Skelhorn, J. & Rowe, C. 2005. Tasting the difference: do multiple defence chemicals interact in Müllerian mimicry? *Proc R Soc B* **272**, 339–345.

Stevens, M. 2007. Predator perception and the interrelation between different forms of protective coloration. *Proc R Soc B* **274**, 1457–1464.

Stevens, M. & Cuthill, I. C. 2006. Disruptive coloration, crypsis and edge detection in early visual processing. *Proc R Soc B* **273**, 2141–2147.

Stevens, M., Hardman, C. J. & Stubbins, C. L. 2008. Conspicuousness, not eye mimicry, makes "eyespots" effective antipredator signals. *Behav Ecol* **19**, 525–531.

Stevens, M. & Merilaita, S. 2009. Defining disruptive coloration and distinguishing its functions. *Phil Trans R Soc B* **364**, 423–427.

Tosh, C. R., Jackson, A. L. & Ruxton, G. D. 2007. Individuals from different-looking animal species may group together to confuse shared predators: simulations with artificial neural networks. *Proc R Soc B* **274**, 827–832.

Thayer, G. H. 1909 *Concealing Coloration in the Animal Kingdom*. Macmillan.

Theodoridis, S. & Koutroumbas, K. 1999. *Pattern Recognition*. Academic Press.

Tsoukalas, L. H. & Uhrig, R. E. 1997. *Fuzzy and Neural Approaches in Engineering*. Wiley.

Tullberg, B. S., Merilaita, S. & Wiklund, C. 2005. Aposematic and crypsis combined as a result of distance dependence: functional versatility of the colour pattern in the swallowtail butterfly larva. *Proc R Soc B* **272**, 1315–1321.

Vallin, A., Jakobsson, S., Lind, J. & Wiklund, C. 2005. Prey survival by predator intimidation: an experimental study of peacock butterfly defence against blue tits. *Proc R Soc B* **272**, 1203–1207.

Wallace, A. R. 1889. *Darwinism*. Macmillan.

Wourms, M. K. & Wasserman, F. E. 1985. Butterfly wing markings are more advantageous during handling than during initial strike of an avian predator. *Evolution* **39,** 845–851.

11

Artificial neural networks in models of specialisation, guild evolution and sympatric speciation

Noél M. A. Holmgren, Niclas Norrström and Wayne M. Getz

11.1 Introduction

The existence of sympatric speciation has been a contentious issue because empirical support was scarce and the underlying theoretical mechanisms were not as fully understood as we might like (e.g. Futuyma & Mayer, 1980; Rundle & Nosil, 2005). The view on sympatric speciation is currently changing, however. Recent theories demonstrate how ecological adaptations can drive speciation (Dieckmann *et al.*, 2004; Doebeli *et al.*, 2005). In concert with theoretical development, empirical evidence corroborating this view is accumulating (Barluenga *et al.*, 2006; Panova *et al.*, 2006; Savolainen *et al.*, 2006). An obstacle for sympatric speciation is the exchange of alleles between lineages and the homogenising effect of recombination in sexual reproduction (Felsenstein, 1981; Rice & Salt, 1988). The current view on sympatric speciation is therefore that disruptive selection for evolutionary divergence has to be correlated with assortative mating and reproductive isolation (Felsenstein, 1981; Rundle & Nosil, 2005). This can be through linkage between ecological genes and mating genes, or a pleiotropic effect of ecological genes on mating behaviour. Orr & Smith (1998) make the distinction between extrinsic and intrinsic barriers to gene flow. Extrinsic factors are physical barriers in the environment that prevent encounters between individuals. Intrinsic factors are genetic traits that increase pre- or post-zygotic reproductive isolation. They define sympatric speciation as 'the evolution of intrinsic barriers to gene flow in the absence of extrinsic barriers'.

Host races have been defined as populations of a species that are partly reproductively isolated from one another as a direct consequence of adaptation to different hosts (Abrahamson *et al.*, 2001). Host races in phytophagous insects are believed to be precursors to full species, an idea that goes back to 1864 (Walsh, 1864), these being the first candidate examples for sympatric speciation. Specialisation on hosts is a prerequisite for host races to be reproductively isolated, and if the isolation evolves as a correlated character to specialisation, it may lead to sympatric speciation (Rice & Salt, 1990).

Resource competition between phenotypes with fixed and limited diet breadth is the driving force of current theories of sympatric speciation (Dieckmann & Doebeli, 1999;

Modelling Perception with Artificial Neural Networks, ed. C. R. Tosh and G. D. Ruxton. Published by Cambridge University Press.

Geritz & Kisdi, 2000). Among insects, diet breadth does not seem to be limited. Laboratory studies show that larvae feed and grow equally well on plants other than those chosen by the female to oviposit on (Dethier, 1947; Ballabeni & Rahier, 2000). It is still unknown why females are so restricted in diet breadth, and why specialists are more common than generalists in insects (Jermy, 1984; Jaenike, 1990). It has been proposed as a major enigma in the evolution of insects (Futuyma, 1991). Other theoretical explanations for host specificity in insects include avoidance of interspecific competition or predation, reduction of parasitism, and increased probability of mate finding, but the empirical support is often circumstantial at best (Futuyma & Moreno, 1988). Recently, host specificity in insects has been suggested to be the result of limitations in brain function, more specifically in the recognition systems that process information (here host signals) for effective recognition of suitable host plants (Holmgren & Getz, 2000; Bernays, 2001).

Here we review some of our work on evolution of host-plant selection in insects using artificial neural networks as models for the plant recognition mechanism in insects (Holmgren & Getz, 2000; Norrström *et al.*, 2006). We present some new insights from the synthesis of our results. We explain why insects can become specialised, and how disruptive selection creates guilds of specialists as a result of evolution on recognition mechanisms. In addition we examine the evolutionary dynamics during coevolution of the exploiter species and their hosts. Finally we show that reproductive isolation can evolve in diploid and sexual populations, without being correlated to host-recognition genes. The results reveal a new and unexpected process that drives sympatric speciation. Although the model is inspired by insect-plant systems, it may be regarded as an example of a more general exploiter–victim system in which the evolution of the exploiters' recognition systems is critically influenced by the ability to assess resource quality of hosts/victims. We set out to present a framework for exploring the importance of specialisation for speciation, with the aim to stimulate further theoretical and empirical work.

11.2 Recognition system and ANN design

A niche-breadth model of exploiter evolution under neural constraints needs to be sufficiently detailed with regard to signals produced by victims and the ability of the exploiters to perceive and respond to these signals to adequately address the questions at hand. The plant victims, for example, produce signals of varying complexity dependent on a few key chemical compounds that occur in plant-specific ratios, but are collectively known as 'green odour' (Visser, 1986). To keep things simple, we let the plant signals in our model be represented by two odorants, the minimum needed for odour quality to depend on component ratios (Getz & Chapman, 1987). Insects, as exploiters for example, perceive these plant signals and compute an output signal coding for a behavioural action – in our models we take this action to be laying versus not laying eggs on a potential host plant.

For simplicity, we modelled the perceptual system as a perceptron (Haykin, 1994), rather than a dynamic neural network (Getz & Lutz, 1999), which captures the perceptual constraint feature seen in insect exploiters when selecting among victims with different phenotypes (c.f. Getz & Smith, 1990; Getz & Akers, 1997). The perceptron is a three-layered feedforward network with an ability to differentiate and categorise input signals once the perceptron has an appropriate set of weighting values for passing on information from one layer of nodes to the next. Our perceptrons had two inputs, implying that victim signals were points in a two dimensional odour space. The output layer has only one node which state corresponds to a preference/rejection response. In our simulations, these weightings evolve between generations by mutations and the relative fitness achieved from host-plant (i.e. victim) selection, where fitness is measured as the expected number of eggs that will successfully mature into new adults. In short, we have a mutation-selection algorithm on the synaptic weights of replicating perceptrons (see Appendix for more details).

In an initial study we assumed that the exploiters were represented by a unique perceptron reproduced as haploid clones (Holmgren & Getz, 2000): that is, from one generation to the next depending on the fitness of the represented individual, zero to two new perceptrons were created and then mutated with predetermined probabilities and size of weighting perturbations. Models with clonal reproduction exclude genetic exchange between exploiter lineages, and are thus insufficient to model how reproductive isolation and species arise in sexual populations. In a follow-up study, we developed a diploid genetic structure coding for the synaptic weights (Norrström *et al.*, unpublished).

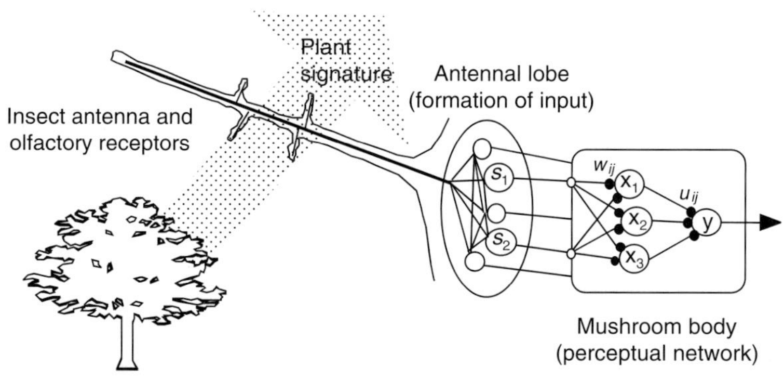

Figure 11.1. A diagram of the perceptual component of the model. Each plant type produces its own unique signature that stimulates the olfactory receptor cells located on the antenna of individual insects. The response of these cells is processed in the antennal lobes to produce an antennal lobe output S_i that we regard as input to a perceptual neural network located in the mushroom bodies of the protocerebrum. Our highly idealised model of this perceptual system is a three-layered feedforward neural network. For simplicity, we assume all compound specific signals are represented by the two inputs (S_1 and S_2), which are then propagated to a layer of hidden units (large labelled spheres). The strength of these input signals is modified by synaptic weights (small solid spheres) $w_{i,j}$, $i = 1, 2, j = 1, 2, 3$. The output x_j from each of the hidden units when stimulated, is the result of passing the input activity through a sigmoidal activation function. The activity impinging on the output unit is similarly modified by the synaptic weights u_j. The response y of the output neuron is characterised by the same activation function as in the hidden units. From Holmgren & Getz (2000).

11.3 Stimulus-response functions and specialisation

How well the recognition system of an exploiter is performing, in terms of identifying suitable victims, is ultimately determined by the number of eggs in each generation maturing into new adults. We constructed exploiter-fitness functions using an insect herbivore as our leitmotif (see Appendix for fitness function of exploiters). The fitness of each exploiter is a function of the response to the input signals from its victims. One may think of an insect herbivore as having the option of choosing among a number of different types of plants, where each type produces a characteristic odour. In the model, each insect samples the odours of all plants in the environment. The response or preference of insect g, $g = 1, \cdots, G$, for plant type h, $h = 1, \cdots, H$, is identified with the output $y_{g,h}$ of perceptron g to input signal h. One approach to constructing a fitness function is to assume that the decision to lay a clutch of eggs on a specific host plant depends on the strength of this response relative to alternative hosts. This may seem to be a reasonable approach at first. However, if the absolute signal strength is an excluded factor, genetic drift decreases the insects' sensitivity to plant odours because there is no selection on sensitivity (confirmed in unpublished simulations). Insects searching for host-plants respond behaviourally by flying upwind in plumes of host-volatiles only if these volatiles are presented over a threshold concentration (Olsson *et al.*, 2006). Thus, there is biological reason and a computational necessity to add a dependency on the signal strength to the relative response to each plant odour type. Thus we used the expression

$$e_{g,h} = \frac{y_{g,h}^2}{\sum_{h=1}^{H} y_{g,h}} \tag{1}$$

to calculate the relative clutch size, i.e. the number of eggs, e, laid by exploiter g on plant type h. The egg load affects the fitness function used to calculate the number of insect offspring (Eq. A1).

If response function (Eq. 1) is at all realistic, it has some significant consequences for insect diet breadth. The function selects for an all-or-none response of insects to their available hosts. For example, in an environment of two plants, an insect with the intermediate response $[y_{g,1} = 0.5, y_{g,2} = 0.5]$ will according to Eq. 1 lay $[e_{g,1} = 0.25, e_{g,2} = 0.25]$, i.e. in total only half of its egg complement. In contrast, an insect with a maximum response to one plant and a zero response to the other $[y_{g,1} = 1, y_{g,2} = 0]$ will lay $[e_{g,1} = 1, e_{g,2} = 0]$, i.e. its total egg complement on plant one. Thus, intermediate responses have the logical consequence that insects do not lay their full egg complement, and is as a strategy less fit than all-or-none response strategies.

Now consider an environment of various host plants that are neither perfect hosts nor completely noxious, but provide some intermediate resource in terms of the number of eggs that can successfully hatch and produce viable offspring. With the bimodal preference/non-preference response of insects, one generalist type cannot exploit this environment. A complete generalist will distribute their eggs evenly among available plants. Intra-phenotypic competition on the plant of the lowest resource value will limit population growth, thereby leaving the more valuable plants underutilised. Under these circumstances, selection favours guilds of insects with relative numbers selected to match the values and optimally exploit the plant resources in the environment. The same conclusion holds if competition on host plants affects the quality of offspring, e.g. the size or fecundity of mature offspring.

In all its simplicity, the hypothesis that insects behave in accordance with both their absolute and relative responses to plants of different types may explain the observation that many insects are more restrictive in their diet than needs be from a nutritional point of view (e.g. Wiklund, 1975; Ballabeni & Rahier, 2000). As discussed below, our work suggests that the herbivorous insects exploiting a particular ecosystem have evolved into a guild where the different ecological niches arising from host plant variation are occupied by a number of more or less specialised species.

11.4 Specialisation when resources are fixed

In order to study the evolutionary process of specialisation versus generalisation, a range of resources must be included in the model. In the initial study focusing on the evolution of niche breadth we used four fixed (i.e. non-evolving) resource types (Holmgren & Getz, 2000). In particular, we investigated the evolutionary process of insects in several different ecological backgrounds from both a resource-signalling and resource-value point of view. Some of these environments represented a more difficult resource-discrimination than others. The victims reproduced clonally and an insect–plant leitmotif was used to discuss and interpret the results.

11.4.1 The ideal free distribution

The spectrum of plant types used in our first analysis constituted an ecological resource space or, equivalently, a set of ecological niches (Holmgren & Getz, 2000). We identified the niches with the plants themselves, and assigned a niche value v_h to the h-th population of plants of type h, $h = 1, \ldots, 4$. Thus we identified the resource space using the set $[v1, v2, v3, v4]$ and normalised the analysis by setting $\sum_{h=1}^{4} v_h = 100$, which we interpreted as the carrying capacity of the environment (in our simulations this normalisation to 100 represented the actual number of exploiters that could survive to reproduce from one exploiter generation to the next, but could also be interpreted in terms of relative units).

Natural selection will favour exploiters occupying empty niches – i.e. the number of individuals produced by these exploiters will increase until the ecological niches are all fully occupied. In analogy to the ideal free distribution (Fretwell & Lucas, 1970; Holmgren, 1995), we expect the number of exploiters to match the resources. In several different environments with four non-evolving plant types, we showed that insect phenotypes evolved to equilibrium levels (numbers) that matched the plants' resource values (Holmgren & Getz, 2000). For example, in the case of the resource values of the

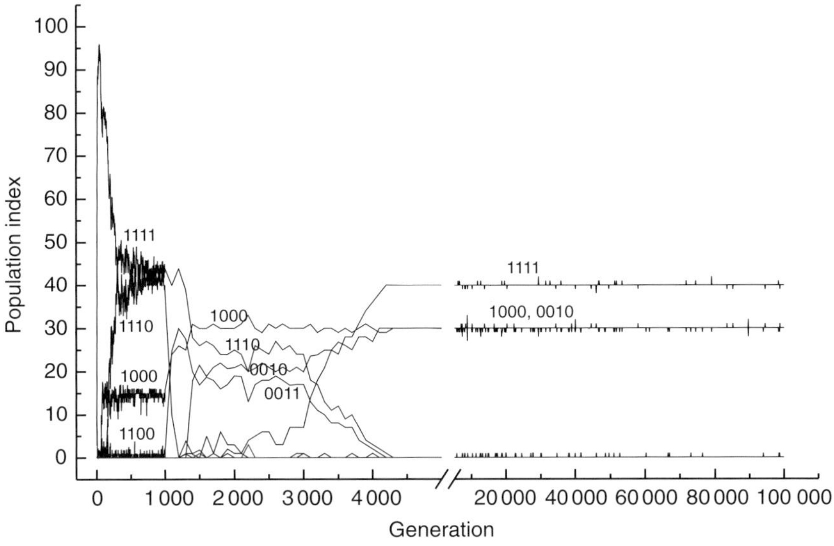

Figure 11.2. The values of the population indices of the phenotypes (as labelled on the graph) in the population are plotted for one of the simulations of the population evolving in the environment four plant types of the values as resources: 40, 10, 40, 10. The population index reflects the number phenotypes and their purity. Phenotype labels denote an array of preference to the four plants, in which 1 is preference and 0 is rejection. Because of the response function chosen for the insect phenotypes, they will tend to be all-or-none responses to each plant (see text for details). Values obtained every generation until 1000 generations, and thereafter every 100 generation are plotted. The scale of the abscissa is varied to portray both short- and long-term trajectories. From Holmgren & Getz (2000).

four plants being [40, 10, 40, 10], insect phenotypes evolved to match so that the numbers produced by each plant type was also [40, 10, 40, 10] (Figure 11.2).

Recalling that insect phenotypes evolve an all-or-none response (or close to it), we can conveniently represent each insect phenotype by an array of preference digits, one for each plant. For example, a phenotype denoted [1010] will lay eggs on plant types one and three, but reject plant types two and four. Simulations were initiated with naïve perceptrons, in which the synaptic weights were randomly set to small values. The simulations were then run for 100 000 generations (for more details see Holmgren & Getz, 2000). In the period of 1000–3000 generations (Figure 11.2) the first niche had 40 insects: 30 of the specialist phenotype [1000] and 10 of the generalist [1110], the second niche had 10 [1110] phenotypes, the third niche 40 comprising of 10 [1110] phenotypes, 20 [0010] specialist phenotypes and 10 [0011] phenotypes, and the last niche had 10 [0011] phenotypes. Simple arithmetic indicates that the phenotypes of this guild are numerically matching the value of the plant resources.

Interestingly, the above guild begins an evolutionary transformation after 3000 generations so that at 4000 generations the matching of plant phenotypes is accomplished by 40 [1111] generalist phenotypes in concert with the two specialists phenotypes, [1000] and [0010] of 30 individuals each (i.e. occupying niches 1 and 3, respectively). For reasons discussed below, this latter guild is more stable than the one above that arose first. Note that, resource matching is a predicted equilibrium of any exploiter-resource system unless the spectrum of exploiter types (i.e. guild) gets trapped in a non-optimal solution, from which there is no evolutionary escape if mutational perturbations are absent (as in simulated annealing optimisation processes: Haykin, 1994).

11.4.2 The evolution of guilds

Resource matching is a game-theoretic outcome in which no single exploiter phenotype can evolve independently of the frequencies of other phenotypes. From the simulations we conducted, we concluded that exploiters evolve in guilds of several exploiters utilising available resources in concert and in numbers matching the resource abundance and resource value. In Figure 11.2, the transitorily stable resource-matching guild of 30 [1000], 30 [1110], 20 [0010] and 20 [0011] phenotypes is ultimately replaced by a guild of 40 [1111], 30 [1000] and 30 [0010] phenotypes. The transition period is short in comparison with the phases during which the guilds prevail.

The geological record indicates that sudden turnovers of whole guilds of species seem to occur (Gould, 2002). Some rapid turnovers of extinct guilds may be the result of catastrophes, such as meteorite impacts on earth (Alvarez *et al.*, 1980). Species compositions in terrestrial and aquatic systems are also known to exhibit rapid turnovers when alien species are introduced (Crooks, 2002). In addition to catastrophes and major perturbations, guilds of specialists and generalists, unable to match their resources, are vulnerable to invasions of new species forming new guilds. In Figure 11.2, the guild prevailing from generation 1000 to 3000 is matching its plant environment less robustly or resiliently (Amemiya *et al.*, 2005) than the succeeding guild (after 4000 generations), because the former guild is more sensitive to mutations than the latter and more affected by inter-phenotypic competition. In reality, environments may change catastrophically or gradually, and a guild of species whose interactions are characterised by inter-specific competition may be unable to track those changes. As a consequence, a rapid turnover of species will follow, thereby establishing a new guild. Species in guilds have frequencies that are mutually dependent because of resource matching. As such, they are resistant to invasion of phenotypes that temporarily disrupts this matching. The more mutations and new exploiter phenotypes required to obtain a new matching guild, the greater the resilience of the existing guild to persist.

Appearance of a new species may not be sufficient to over-throw an existing guild. Sometimes two or more are required to invade in concert (Holmgren & Getz, 2000). When an invasion has started, existing phenotypes quickly lose fitness as their interdependence with other phenotypes weakens. Because a simultaneous increase in fitness of new phenotypes and fitness loss in old ones is required, turnover rates of guilds when they occur are relatively fast. This is not a group-selection argument: selection still acts at the individual level although the fitness of individuals is dependent on the composition and frequency of the different species in the guild of competitors.

11.4.3 Evolution of specialists versus generalists

Returning to the fact that the response of each clonally evolving insect phenotype in our system is close to all or none, in many situations resource matching can only be accomplished by a guild where some insect phenotypes are specialised on one or a few plants. In our simulations we found that the most stable guilds evolving to match their host-plant environment are those that minimise the degree to which niches overlap among the members of the guild. The reason is that guilds exhibiting considerable niche overlap are more vulnerable to changes in phenotype numbers due to inter-phenotypic competition within shared niches. If mutations of phenotypes lead to erroneous host choices and small deviations from the ideal free distribution, generalists are more likely than specialists to experience reduced fitness from over-crowded plants. Reduced fitness leads to decreased population size of the phenotype, which will lead to other plants in its diet being underutilised.

In reality, deviations from the ideal free distribution can be due to mutations on other phenotypes or changes in resource abundances. Guilds composed of specialists are less affected by inter-phenotypic competition and can more readily track a changing environment. In addition, specialists have a more simple discrimination task than generalists, given that simpler tasks involve fewer critical synapse settings in perception networks than more complex tasks. As such, the perceptual networks associated with simpler tasks are potentially less likely to be hampered by harmful mutations. An exploiter phenotype that utilises two types of plants with sufficiently similar chemical signatures to be able to lump the two plant types into one perceptual category, has no more complicated a task than a specialist exploiter that needs to identify a single plant type. The most extreme case of this would be the generalist that treats all existing plants as one category. For this reason we should expect guilds to be made up by specialists on single plant species (monophages), intermediate specialists utilising a few similar host plants (oligophages) and indiscriminant generalists (heterophages).

In summary, our simulations suggest that disruptive selection for specialisation in a heterogeneous resource environment could arise because of the following two mechanisms. First, selection for sensitivity to signals produces individuals that have an all-or-none response to the different host phenotypes so that only guilds of insects that specialise to some degree will be able to fill up available resource niches. Second, selection favours resource-matching guilds of exploiter phenotypes that perform relatively simple host-choice perceptual tasks robust to mutations.

11.5 Specialisation when resources evolve

In a follow-on study, we allowed the victims to evolve in terms of the signals used by the exploiters to detect these victims (Norrström *et al.*, 2006). As in the previous study, each exploiter was identified with a 3-layer perceptron with weights subject to mutations. Again, with focus on specialisation and disruptive selection, reproduction was assumed to be clonal for simplicity and to eliminate gene flows between genetic lineages. Specialisation and disruptive selection among sexual organisms is within the scope of future work. Also, victims were still represented by a point in a 2D signal (odour) space. Initially three groups of victims were introduced, differing in their relative palatability: high, intermediate and low. The fitness of each victim depended on the number of victims within the same group and on the attack rate represented by a weighted sum of all exploiters – the weighting being determined by the exploiter response functions to that particular victim.

11.5.1 Red queen evolution

The conflicting interests of exploiters and victims induce a continuously changing system that cycles over time. Exploiters are continuously selected to discriminate among victims of different palatability. Victims of high palatability are selected to become similar to victims of low palatability thereby reducing the intensity of attacks. Victims of low palatability are in turn selected in the signal space to escape from approaching highly palatable victims – that is, to move away from their high palatable mimics. This results in a directional movement of victims in the signal space, driven by the exploiters' continuous adaptation to discriminate among victim types. By measuring all exploiters' responses to many locations in the signal space and calculating a response average in these locations we create a response landscape. Lowlands in the response landscape mean low average response, hence little exploitation, and highlands mean high average response and high exploitation. In the exploiter response landscape (Figure 11.3), the victims move downhill to avoid attacks (e.g. Figure 11.3c), with the least palatable victims in the lead. Because there is variation in each victim cluster, as indicated by the width of the tubes in Figure 11.3, there can be differential selection on the victims within each palatability cluster. While the signals of victims evolve, the ability of exploiters to discriminate among these victims also evolves. For most of the time, the relative distances among the mobile victim clusters in the signal space are more or less constant, reflecting the presence of a red-queen evolutionary process (Van Valen, 1973). For a short time, though, this process is arrested when threshold and saturation constraints come into play (Figure 11.3d)

11.5.2 Mimicry evolution

Geometrically, the red-queen process arrests in 'corners' of the signal space (Figure 11.3d). In these corners signal cues are either saturated or absent. Selection now enables the most palatable victims to become perfect mimics of least palatable victims. Holmgren & Enquist (1999) suggested that an equivalent process can explain the evolution of Batesian mimicry, including the saturated colouration visual mimics and models often exhibit. In this phase of the process, the exploiters are unable to distinguish between these two victim types, and hence their individual response surfaces will relax and become flat over the whole signal space. This releases the palatable and unpalatable victims from differential selection due to lack of discrimination by the exploiters, thereby allowing the victims to drift apart in the signal space. In this way the mimetic resemblance is degraded.

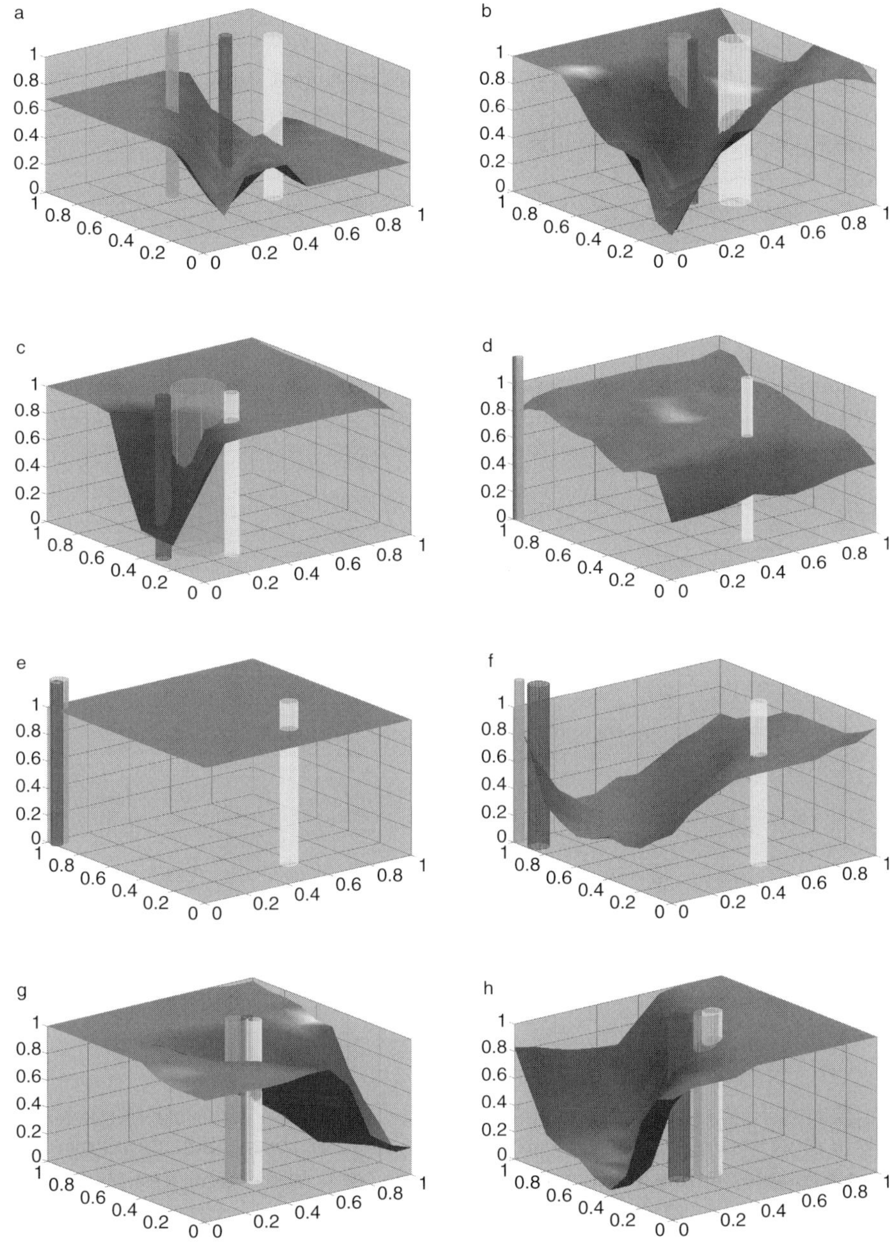

Figure 11.3. Victim-cue (e.g. host plants) phenotypes and exploiter-response phenotypes (e.g. phytophagous insects) are plotted above the 2D signal space. The two bottom axes represent signal strengths of cue phenotype. The surface shows the average response of all exploiters to a hypothetical signal at any point in the signal space. The vertical columns represent victim clusters: medium grey – undefended, light grey – intermediate, and dark grey – defended. The centre of each column is at the average of the victim-cuephenotypes in the cluster in question and the radius is the

11.5.3 Cyclicity of specialists and generalists

When host clusters are discriminable, the exploiters evolve to specialise on the palatable and intermediate host clusters. Once the unpalatable model and its mimics are driven to perfect mimicry in one of the corners of signal space, the exploiters become complete generalists. This process is cyclic (Figure 11.3) with the period length determined by the evolutionary response, i.e. changes from one generation to the next, according to Fisher's (1930) fundamental theorem of natural selection. The evolutionary response is a function of additive genetic variance, in our model determined by mutation rates, and the selection differential (Maynard Smith, 1998) given by the elevation differences in the response landscape of the perceptrons (Figure 11.3). The cyclicity is a consequence of the continuous changes in the host signal phenotype and the constraints on the strength on each of the two components of the signal. At signal saturation, first variation in signal traits degrades; second, differential selection on plants becomes vanishingly small. In the next phase, changes are determined by mutation rates alone. In this evolving plant environment, host races readily evolve. If exploiters reproduce sexually by mating among individuals sharing host plants, the reproductive isolation between host races will disappear when they come together on the same host. In this case, sympatric speciation would not be possible unless reproductive isolation is upheld by other mechanisms than assortative mating linked to host-plant preference.

11.6 Exploiter asexual versus sexual reproduction

The models described above are based on clonal reproduction in the exploiter population. No attention was paid to the homogenising effect of genetic recombination among lineages produced by sexual reproduction and recombination (Rice, 1984). In a recent study (Norrström *et al.*, unpublished), we extended our model by adding sexual reproduction and diploid genetic coding of the synaptic weightings in our perceptron representations of individual exploiters. An unlinked mating gene that could hold an allele for random or for assortative mating was added. In this case assortative matings were confined to individuals on the same

Caption for Figure 11.3 (cont.)
standard deviation. The images are captured after a simulation of exploiter–victim coevolution. (a) The simulation is initialised with the plants lined up on the diagonal. The insects have learnt to respond to the plants: two specialists, one on each edible plant has evolved (not seen in figure). (b) The most edible plant cluster has approached the noxious plant cluster. The runaway movement with the most edible plant following the least edible one, has started. (c) The chase moves to the border of the signal space. The intermediate plant is following behind. (d) The chase has been arrested in a corner, and the intermediate plant has been left behind. (e) The insects have stopped discriminating between plants. (f) The two plants in the corner have drifted apart because of the lack of selection on signals. Becoming separated, the insects are now starting to discriminate the plants again. The chase has been re-initiated. (g) The chase has been moving towards the centre and attracted all plants in the middle. (h) The chase is now continued toward a border of the signal space, and the procedure repeats from (c) and onwards. The figure is from Norrström *et al.* (2006).

host plant. We allowed for the equivalent of genetic cross-over to occur during meiosis by rearranging genes for perceptron weights among chromosomal-like structures (see Appendix for more details). We kept our four resources: two of equal value and two noxious. We can repeatedly confirm that also sexual populations evolve two specialists. In this case they are reproductively isolated homozygotes, i.e. sympatric speciation by definition.

We inhibited mutations of the mating gene during the first 20 000 generations. During this period, a polymorphism of two chromosomal haplotypes, here named A and B, arise. The two haplotypes are combined in three genotypes, two (AA and AB) expressed the same specialist phenotype on one resource and the third genotype (BB) express the specialist on the other resource. When mutations are allowed on the mating gene, the assortative mating allele invades and becomes fixed in the population. The protected polymorphism turns into two reproductively isolated, homozygote specialists (AA and BB). This suggests that stable polymorphisms can evolve as a result of disruptive selection, being a prior state to reproductive isolation and sympatric speciation. It also shows that genes for assortative mating can invade uncorrelated with ecological genes.

The initial evolution of a protected polymorphism evolves to match the available resources through mapping three genotypes on two phenotypes. This is possible with a multi-locus model and synergistically interacting genes. With genotypes expressing functional phenotypes, including heterozygotes, it is not obvious why assortative mating invades and eliminates the heterozygotes. The number of erroneous phenotypes in the population diminishes from 19.6% to 8.5% after invasion of assortative mating, thereby the mutation-selection balance for deleterious alleles settles at a lower level. Thus, there is selection for assortative mating either to get phenotypes more robust to mutations, or enhance the reductional selection of deleterious mutations. Experimental mutations on all prevailing phenotypes exhibit similar proportions of increased erroneous phenotypes and actually more during assortative mating (15.4%) than before (14.6%). This leaves us with the hypothesis that selection against deleterious mutations is more effective during assortative mating, which could be the case if deleterious mutations are silent in heterozygotes. When analysing the genotypes from our simulations, we see that the heterozygote actually has a number of silent alleles that are expressed in homozygote form. To understand how these alleles work, we have to recall the structure of the neural net and how the alleles affect its performance. Our nets have three sensory neurons, each capable of having a threshold response at a given ratio of the two input compounds. This threshold response is a linear discrimination controlled by four alleles, which in our case is an allele complex constituting a functional unit. The output of the three sensory neurons are thus governed by three allele complexes, and combined in the output neuron which can perform nonlinear responses. Simple discriminations require only one critical allele complex, whereas a complicated discrimination requires critical settings of many allele complexes. In our simulations, critical allele complexes exhibit much less variation than noncritical ones. We can thus understand the underlying genetics as three super-loci occupied with super-alleles (but let us call them loci and alleles from now on for simplicity). The output neuron later combines the output from the alleles but the combination is elemental and left out of the description for clarity. The loci can have three discrimination alleles: *d1* discriminates between resource 1 and 2 in homozygote form, *d2* does the same but between resource 2 and 3 and *d3* discriminates between resource 3 and 4. We also find a modifier allele, *m*, which modifies the expression of *d2* to become like *d1*. The allele m does not perform any discrimination in homozygous form. There is also an inhibitor allele, *i*, that inhibits the expression of *d*-alleles so that the sensory neuron does not perform any discrimination. In homozygous form the *i*-allele also does not perform any discrimination. The allele arrangements of the genotypes in the polymorphism are:

	Genotypes		
Locus	AA	BB	AB
1	*m m*	*d2 d2*	*m d2*
2	*d1 d1*	*i i*	*d1 i*
3	*i i*	*d3 d3*	*i d3*

We can see that AA expresses *d1* to become specialist on resource 1, whereas the other loci are silent. Genotype BB expresses *d2* and *d3*, which is required to become specialist on resource 2. The heterozygote AB expresses *d1*, because the expression of *d2* is modified by *m* in locus 1. Loci 2 and 3 are silent because of the inhibitor alleles. Hence, deleterious mutations on *d1* and *d3* are silent in the heterozygote, and thus not subject to selection. Since homozygote lineages, mating assortatively, need no modifier or inhibitor alleles the selection against deleterious mutations is stronger in homozygote specialists than in the polymorphism. Thus, the model demonstrates that species can arise sympatrically because deleterious mutations can more effectively be removed from the population. As the pre-zygotic barrier (of assortative mating) has evolved, heterozygotes are rare and the selection on *m* as a modifier is relaxed. As *m* loses its modifying property, the homozygotes in addition become post-zygotically isolated (Norrström *et al.*, unpublished). This is the evolutionary end result we can expect to observe in nature.

F_1 hybrids between host races of phytophagous flies have a markedly reduced behavioural response to parent host odours and mixtures of them (Linn *et al.*, 2004). This is convincing evidence for a significant genetic component of the host preference trait. The lack of intermediate and parental-like responses indicates a synergistic interaction between 'preference alleles' in heterozygote form. Electrophysiological measurements on olfactory receptor neurons suggests that it is genes for olfactory receptor proteins in dendritic membranes that interact synergistically and elicit hybrid-specific responses to host volatiles (Olsson *et al.*, 2006). The host races of these insects exhibit already pre-zygotic (by mating only on host plants) and post-zygotic isolation (by the disability of hybrid offspring to find host plants), which is analogous to the end state of our simulations. The fact that host-related, reproductively isolated lineages in some insects are morphologically indistinguishable and lack other allozyme or genetic differences (see Olsson *et al.*, 2006 and references therein), suggest that the speciation process is driven by disruptive selection on the host-recognition trait. These may be incipient species in early stages of speciation, or recently arisen species according to the species definition of Orr & Smith (1998).

11.7 Conclusions

It has been shown that disruptive selection on morphological traits with innate limitation effects on niche-breadth can give rise to sympatric speciation if reproductive isolation evolves as a correlated character (Dieckmann & Doebeli, 1999). Such traits are typically related to the feeding apparatus, and its size sets the boundaries of suitable resources. Trait examples from cases of sympatric diversification are shell size in marine snails (Panova *et al.*, 2006) or jaw length in fish (Rundle & Schluter, 2004; Barluenga *et al.*, 2006). In many phytophagous insects, the recognition mechanism itself seems to be under disruptive selection (Linn *et al.*, 2004). There are no innate niche-limitations to this trait, so we need to understand why specialisations evolve before we can understand speciation based on the resource recognition trait. We have shown that generalists are more susceptible to mutations and their larger niches over-lap link perturbations (due to mutations) between genetic lineages (Holmgren & Getz, 2000; Norrström *et al.*, 2006). In either case, disruptive selection depends on resource competition between individuals.

The underlying genetics of the trait under disruptive selection is of importance for the evolutionary response. Traits governed by additive genetics typically produce hybrids between incipient branches that are intermediates, which prevents the population from branching into two lineages (Geritz & Kisdi, 2000). Even if there is selection against intermediates because they experience the highest resource competition, it is not strong enough to counteract the homogenising effect on an incipient branching population. For traits expressed by synergistic interaction between genes, however, hybrids may express any phenotype. In this case disruptive selection results in a polymorphism (Norrström *et al.*, unpublished). The population has branched into two phenotypes, expressed by two haplotypes in Hardy–Weinberg equilibrium.

To meet the criteria of species, branched lineages in a population have to become reproductively isolated (e.g. Orr & Smith, 1998). The two scenarios with different genetics also have different causes for the selection of assortative mating. When ecological traits have additive genetics, assortative mating arises to avoid the less fit intermediate hybrids. Hence, the population branches in concert with the establishment of assortative mating. Therefore, ecological and mating genes need to be correlated (e.g. Felsenstein, 1981; Dieckmann & Doebeli, 1999). When ecological traits are expressed by synergistically interacting genes, the population is likely to be branched when an assortative mating mutant arises in the population. Hybrids or heterozygotes are as fit as other genotypes, but they are likely to have genes that are silent in heterozygote form. We see that regulatory genes evolve in order to make a haplotype compatible both to itself and its match in the stable polymorphism (as described above). Silent genes are not subject to selection and accumulate harmful mutations that are expressed in, and deleterious to

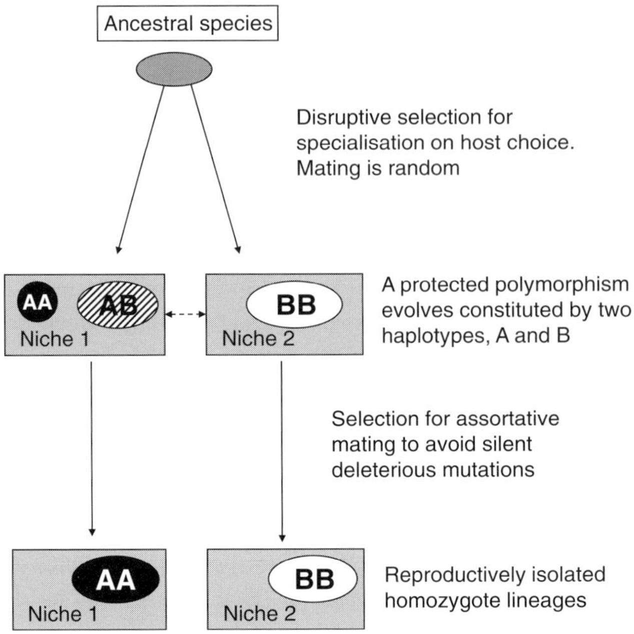

Figure 11.4. Selection on the ecological niche recognition mechanism can result in sympatric speciation. First, there is disruptive selection for specialisation on two niches represented by squares. Specialists evolve in a protected polymorphism with two haplotypes A and B. The genotypes mate randomly and consequently are in Hardy–Weinberg equilibrium. Heterozygotes have silent alleles that carry harmful mutations. These are expressed in homozogyte offspring. Assortative mating homozygotes invade the population and thereby avoid harmful mutations in their offspring. Two species reproductively isolated by pre- and post-zygotic have evolved sympatrically.

homozygote offspring. So in this case, mutant assortative mating homozygotes invade the population because they avoid silent deleterious mutations carried by heterozygotes that will be expressed in their offspring (Figure 11.4).

If sympatric speciation results from specialisation on biological resources, the resource may coevolve with its exploiters. When exploiters are entrained in a cyclic coevolutionary processes with their victims, selection for generalists and specialists may shift back and forth (Norrström *et al.*, 2006). Recent investigations reveal new dynamic properties of the specialisation process (Janz *et al.*, 2001; Nosil, 2002). They question the view of the specialisation process as always going from generalisation towards specialisation; hence suggesting that specialisation is not an evolutionary end-point. Janz *et al.* (2001) investigated the phylogeny of the nymphali butterfly tribe Nymphalini. They concluded that there is no directed evolution towards specialisation and that the changes in host range show a very dynamic pattern. Nosil (2002) used phylogenies from 15 groups of phytophagous insects to investigate the rates of evolution towards specialisation and generalisation. They found that the rate of the evolution towards specialisation is significantly higher than the rate toward generalisation. In some cases, however, the rate of generalisation was higher, or equal to the rate of specialisation. These observations that niche breadth is a variable trait in some taxa, and it can both be widen and narrowed by evolution.

11.8 Appendix: Model description

The model has a structure with a population of exploiters (e.g. insects) and an environment of victims (e.g. plants). Both exploiters and victims are equipped with traits essential to the model and will be explained below. Traits are constants or subject to mutations. Each individual exploiter and victim is evaluated by a fitness function. New generations are created by letting the most fit individuals reproduce. The offspring inherit the parents' traits but mutational changes and crossover (in the sexual model only) may occur. Generation times in evolving plants are longer than in insects. The model is a general exploiter-victim model, but was originally inspired by phytophagous insects and their host plants. Below, the models are described, but for all details we refer the reader to our original publications.

11.8.1 The exploiters

The exploiter population is modelled individually, each being equipped with an artificial neural net. We use a three-layer, feedforward perceptron as a model for the perceptual system of the exploiters. The perceptron has two input nodes and one output node. The number of hidden nodes in the intermediate layer varies between 3–6. The net is fully connected with synapses in a feedforward fashion. So called 'bias signals' were applied to the nodes. Synapse values were limited to ± 10. Each node (except the input nodes) is activated by a standard sigmoid threshold function (see Holmgren & Getz, 2000). The output of the ANN is bounded between 0–1, and represents the exploiter's preference to the victim whose two signal cues are applied to the input nodes.

The fitness W_g^E of exploiter g depends on the resource value of the victims h it attacks (v_h), the intensity of the attack (e.g. the number of eggs the exploiter lays) on the victims ($e_{g,h}$; Eq 1), and the probability of attack success (e.g. the eggs hatching), and is given by the function

$$W_g^E = \sum_h e_{g,h} \frac{1}{1 + \left(\frac{e_h}{v_h \varepsilon_{1/2}}\right)^a} \qquad (2)$$

Here a is a parameter that determines the abruptness of the effects of density dependence. Parameter $\varepsilon_{1/2}$ sets at which value of the total attack of plant (e_h) the sigmoid fitness function returns half its maximum value. Hence, the number of offspring in the next generation is density dependent on each host. When individuals of the generation are created the nodes are subject to point mutations with a given probability. A mutation incurs a change to the weight drawn from a rectangular distribution with given limits. The probability of mutations and the limits of the rectangular distribution varied as a part of a sensitivity analysis. The fitness determines the number of offspring in the next generation.

In the sexual model with diploid, sexually reproducing insects, each insect has two vectors representing the parts of the insect's genome that codes for the structures associated with host plant selection in the insect's nervous system. Stored in the vectors are values representing the genetic expression of each gene in the above-mentioned parts of the genome. Each position in the vectors corresponds to a specific synaptic weight in the insect's ANN. The value of a specific synaptic weight is the intermediate value of the two corresponding values in the vectors. During reproduction each parent produces a gamete that will become one of the offspring's vectors. The gamete is created by copying values from one of the vectors into the gamete and proceeding down the vector. With a given probability, crossover occurs, and the values of the gamete are instead read from the other parent continuing at the position where the crossover occurred. There are no restrictions on the number of crossovers that can occur during the creation of a gamete. The direction of the copying is always the same; hence the positions of the new values in the vectors correspond to the same synaptic weights as in the parents. A value in the gamete may mutate with a given probability. During mutation the value is modified with a random value within a fixed range.

11.8.2 The victims

In simulations where victims, e.g. plants, do not evolve, they are represented as homogeneous populations with different traits. When plants evolve, plant populations are modelled as a group of individuals that share their value as a resource to the exploiters, but exhibit variation in their signals. The signals consist of two cues that are subject to mutational changes with a fixed probability and drawn from a rectangular distribution of a given range. Range and probability have been varied as a part of a sensitivity analysis.

The fitness W_h^E of victim h is a sigmoid function of the size of the population of which it is a member (P_h), its egg-load (e_h) and its resource value to exploiters (v_h):

$$W_h^V = \frac{\beta}{1 + e^{\delta(dP_h + c_1 e_h + c_2(v_{MAX} - v_h) - \gamma)}} \tag{3}$$

Parameter β determines maximum fitness, δ is a slope parameter and d is the intensity of density dependence. Parameter γ is a population growth rate parameter. Parameter v_{MAX} sets the maximum resource value a victim can have. The cost parameter for an egg-load is c_1, and for the defence against exploiters c_2. We require $c_1 > c_2$ since the victims trade off predation costs by expending fitness currency on defence (Norrström *et al.*, 2006). The number of offspring of each plant in the next generation is given by rounding off the fitness upwards or downwards to the nearest integer, which is determined by distance-weighted probability.

Acknowledgement

This work was funded by a James S. McDonnell Foundation 21st Century Science Initiative Award to WMG.

References

Abrahamson, W. G., Eubanks, M. D., Blair, C. P. & Whipple, A. V. 2001. Gall flies, inquilines, and goldenrods: A model for host-race formation and sympatric speciation. *Am Zool* **41**, 928–938.

Alvarez, L. W., Alvarez, W., Asaro, F., & Michel, H. V. 1980. Extraterrestrial cause for the Cretaceous-Tertiary extinction. *Science* **208**, 1095–1108.

Amemiya, T., Enomoto, T., Rossberg, A. G., Talamura, N. & Itoh, K. 2005. Lake restoration in terms of ecological resilience: a numerical study of biomanipulations under bistable conditions. *Ecol Soc* **10**, 3.

Ballabeni, P. & Rahier, M. 2000. Performance leaf beetle larvae on sympatric host and non-host plants. *Entomol ExperApplicata* **97**, 175–181.

Barluenga, M., Stöltig, K. N., Salzburger, W., Muschick, M. & Meyer, A. 2006. Sympatric speciation in Nicaraguan crater lake cichlid fish. *Nature* **439**, 719–723.

Bernays, E. A. 2001. Neural limitations in phytophagous insects: implications for diet breadth and evolution of host affiliation. *Ann Rev Entomol* **46**, 703–727.

Crooks, J. A. 2002. Characterizing ecosystem-level consequences of biological invasions: the role of ecosystem engineers. *Oikos* **97**, 153–166.

Dethier, V. G. 1947. *Chemical Insect Attractants and Repellents.* Blakiston Co.

Dieckmann, U. & Doebeli, M. 1999. On the origin of species by sympatric speciation. *Nature* **400**, 354–357.

Dieckmann, U., Doebeli, M., Metz, J. A. J. & Tautz, D. 2004. *Adaptive Speciation.* Cambridge University Press.

Doebeli, M., Dieckmann, U., Metz, J. A. J. & Tautz, D. 2005. What we have also learned: adaptive speciation is theoretically plausible. *Evolution* **59**, 691–695.

Felsenstein, J. 1981. Skepticism towards Santa Rosalia, or why are there so few kinds of animals. *Evolution* **35**, 124–138.

Fisher, R. A. 1930. *The Genetical Theory of Natural Selection.* Clarendon Press.

Fretwell, S. D. & Lucas, H. L. 1970. On territorial behaviour and other factors influencing habitat distribution in birds. *Acta Biotheoretica* **19**, 16–36.

Futuyma, D. J. 1991. Evolution of host specificity in herbivorous insects: genetic, ecological, and phylogenetic aspects. In *Plant–Animal Interactions: Evolutionary Ecology in Tropical and Temperate Regions* (eds P. W. Price *et al.*), pp. 431–454. John Wiley & Sons, Inc.

Futuyma, D. J. & Mayer, G. C. 1980. Non-allopatric speciation in animals. *SystZool* **29**, 254–271.

Futuyma, D. J. & Moreno, G. 1988. The evolution of ecological specialization. *Ann Rev Ecol Syst* **19**, 207–233.

Geritz, S. A. H. & Kisdi, E. 2000. Adaptive dynamics in diploid, sexual populations and the evolution of reproductive isolation. *Proc R Soc B* **267**, 1671–1678.

Getz, W. M. & Akers, R. P. 1997. Response of American cockroach (Periplaneta americana) olfactory receptors to selected alcohol odorants and their binary combinations. *J Comp Physiol* A **180**, 701–709.

Getz, W. M. & Chapman, R. F. 1987. An odor perception model with application to kin discrimination in social insects. *Int J Neuriosci* **32**, 963–978.

Getz, W. M. & Lutz, A. 1999. A neural network model of general olfactory coding in the insect antennal lobe. *Chem Senses* **24**, 351–372.

Getz, W. M. & Smith, K. B. 1990. Odorant moiety and odor mixture perception in free flying honey bees (*Apis mellifera*). *Chem Senses* **15**, 111–128.

Gould, S. J. 2002. *The Structure of Evolutionary Theory*. Harvard University Press.
Haykin, S. 1994. *Neural Networks. A Comprehensive Foundation*. MacMillan College Publishing Company.
Holmgren, N. 1995. The ideal free distribution of unequal competitors: predictions from a behaviour-based functional response. *J Anim Ecol* **64**, 197–212.
Holmgren, N. M. A. & Enquist, M. 1999. Dynamics of mimicry evolution. *Biol J Linn Soc* **66**, 145–158.
Holmgren, N. M. A. & Getz, W. M. 2000. Evolution of host plant selection in insects under perceptual constraints: a simulation study. *Evol Ecol Res* **2**, 81–106.
Jaenike, J. 1990. Host specialization in phytophagous insects. *Ann Rev Ecol Syst* **21**, 243–273.
Janz, N., Nyblom, K. & Nylin, S. 2001. Evolutionary dynamics of host-plant specialization: a case study of the tribe Nymphalini. *Evolution* **55**, 783–796.
Jermy, T. 1984. Evolution of insect/host plant relationships. *Am Nat* **124**, 609–630.
Linn, Jr. C. E., Dambroski, H. R., Feder, J. L. *et al.* 2004. Postzygotic isolating factor in sympatric speciation in Rhagoletis flies: reduced response of hybrids to parental host-fruit odors. *Proc Natl Acad Sci USA* **101**, 17753–17758.
Maynard Smith, J. 1998. *Evolutionary Genetics*. 2nd Edn. Oxford University Press.
Norrström, N., Getz, W. M. & Holmgren, N. M. A. 2006. Coevolution of exploiter specialization and victim mimicry can be cyclic and saltational. *Evol Bioinform Online* **2**, 1–9.
Nosil, P. 2002. Transition rates between specialization and generalization in phytophagous insects. *Evolution* **56**, 1701–1706.
Olsson, S. B., Linn, Jr. C. E., Michel, A. *et al.* 2006. Receptor expression and sympatric speciation: unique olfactory receptor neuron responses in F1 hybrid *Rhagoletis* populations. *J Exp Biol* **209**, 3729–3741.
Orr, M. R. & Smith, T. B. 1998. Ecology and speciation. *Trends Ecol Evol* **13**, 502–506.
Panova, M., Hollander, J. & Johannesson, K. 2006. Site-specific genetic divergence in parallel hybrid zones suggests nonallopatric evolution of reproductive barriers. *Mol Ecol* **15**, 4021–4031.
Rice, W. R. 1984. Disruptive selection on habitat preference and the evoultion of reproductive isolation: a simulation study. *Evolution* **38**, 1251–1260.
Rice, W. R. & Salt, G. W. 1988. Speciation via disruptive selection of habitat preference: experimental evidence. *Am Nat* **131**, 911–917.
Rice, W. R. & Salt, G. W. 1990. The evolution of reproductive isolation as a correlated character under sympatric conditions: experimental evidence. *Evolution* **44**, 1140–1152.
Rundle, H. D. & Nosil, P. 2005. Ecological speciation. *Ecol Lett* **8**, 336–352.
Rundle, H. D. & Schluter, D. 2004. Natural selection and ecological speciation in sticklebacks. In *Adaptive Speciation* (eds U. Dieckmann *et al.*), pp. 192–209. Cambridge University Press.
Savolainen, V., Anstett, M.-C., Lexer, C. *et al.* 2006. Sympatric speciation in palms on an oceanic island. *Nature* **441**, 210–213.
Van Valen, L. 1973. A new evolutionary law. *Evol Theor* **1**, 1–30.
Visser, J. H. 1986. Host odor perception in phytophagous insects. *Ann Rev Ent* **31**, 121–144.
Walsh, B. D. 1864. On phytophagic varieties and phytophagous species. *Proc Ent Soc Phila* **3**, 403–430.
Wiklund, C. 1975. The evolutionary relationship between adult oviposition preferences and larval host plant range in *Papilio machaon L. Oecologia* **18**, 185–197.

12

Probabilistic design principles for robust multi-modal communication networks

David C. Krakauer, Jessica Flack and Nihat Ay

12.1 Stochastic multi-modal communication

Biological systems are inherently noisy and typically comprised of distributed, partially autonomous components. These features require that we understand evolutionary traits in terms of probabilistic design principles, rather than traditional deterministic, engineering frameworks. This characterisation is particularly relevant for signalling systems. Signals, whether between cells or individuals, provide essential integrative mechanisms for building complex, collective, structures. These signalling mechanisms need to integrate, or average, information from distributed sources in order to generate reliable responses. Thus there are two primary pressures operating on signals: the need to process information from multiple sources, and the need to ensure that this information is not corrupted or effaced. In this chapter we provide an information-theoretic framework for thinking about the probabilistic logic of animal communication in relation to robust, multi-modal, signals.

There are many types of signals that have evolved to allow for animal communication. These signals can be classified according to five features: modality (the number of sensory systems involved in signal production), channels (the number of channels involved in each modality), components (the number of communicative units within modalities and channels), context (variation in signal meaning due to social or environmental factors) and combinatoriality (whether modalities, channels, components and/or contextual usage can be rearranged to create different meaning). In this paper we focus on multi-channel and multi-modal signals, exploring how the capacity for multi-modality could have arisen and whether it is likely to have been dependent on selection for increased information flow or on selection for signalling system robustness. The robustness hypothesis argues that multiple modalities ensure message delivery (backup signals) (Johnstone, 1996) when one modality is occluded by noise in the environment (Hauser, 1997) or noise in the perceptual system of the receiver (Rowe, 1999).

Some multi-modal signals will have nonredundant features in that each modality elicits a different response. A compound stimulus can either elicit responses to both components (OR function), only one of the two original components (XOR), a modulated version of

Modelling Perception with Artificial Neural Networks, ed. C. R. Tosh and G. D. Ruxton. Published by Cambridge University Press.

the response to one of the two original components, or the emergence of an entirely new response (Partan & Marler, 1999). For example, male jumping spiders (*Habronattus dossenus*) appear to communicate quality through the coordination of seismic and visual displays (Elias *et al.*, 2003). In contrast to uni-modal, multi-channel signals, multi-modal signals, like those used by the jumping spider, are not typically perceived as a single stimulus (Hillis *et al.*, 2002). Receiver discrimination makes sense when the information content in each modality is not perfectly correlated (multiple messages) (Johnstone, 1996).

We consider a simple model in which a signaller transmits a message to a receiver. Signaller and receiver are assumed to have matching interests and there is no advantage to deception. The signaller transmits the message through an arbitrary number of channels using, for example, multiple frequencies (e.g. the fundamental frequency, second harmonic, etc.) in a vocalisation. The receiver is free to attend to as few or as many channels as it wishes. The signalling strategy is to generate correlations among the channels in such a way as to allow the receiver to decode the intended meaning. We ask how many channels and what correlational structure among the channels the signaller should employ in order to allow the receiver to decode a message assuming random subsets of channels become occluded.

Accurate receiver decoding, or robust encoding, depends on two factors: the causal contribution of each signalling channel to the message meaning and the sensitivity, or exclusion dependence, of the message meaning to the elimination of a channel. The causal contribution refers to the unique information provided by each channel. As channels are duplicated, any one channel necessarily makes a smaller contribution to message meaning. Exclusion dependence refers to the consequences for accurate decoding of occluding a single channel or set of channels. The exclusion dependence can be minimised in two ways. One way is to reduce the causal contribution of a channel by having many duplicates. This we call duplication-based buffering or redundancy. The second way is to build into the signalling system an error-correction mechanism that prevents the perturbation (deletion of the channel) from impacting receiver decoding. Error-correcting signalling mechanisms, which likely require repeated signalling interactions, are beyond the scope of this paper.

To understand how the causal contribution / exclusion dependence distinction in the context of redundancy maps onto communication in the natural world, consider the following vocalisation example. Experimental studies of the combination long calls of golden lion tamarins (Sanguinus oedipus) indicate that tamarins treat unmanipulated long calls as perceptually equivalent to long calls with deleted fundamental frequencies or second harmonics (Weiss & Hauser, 2002). Assuming there is no error-correction mechanism operating, the experiment suggests that the causal contribution of the second harmonic is low in so far as its absence does not affect how well, or whether, a receiver can decode a call. This is likely to be the case because the information in this channel is duplicated at other frequencies. Furthermore, tamarins do distinguish between unmanipulated calls and synthetic calls in which all of the harmonics above the fundamental have been deleted (Weiss & Hauser, 2002). This illustrates the principle of exclusion dependence: removing sets of channels – channels that together make a relatively large contribution to meaning but independently make almost none – can jeopardise accurate receiver decoding.

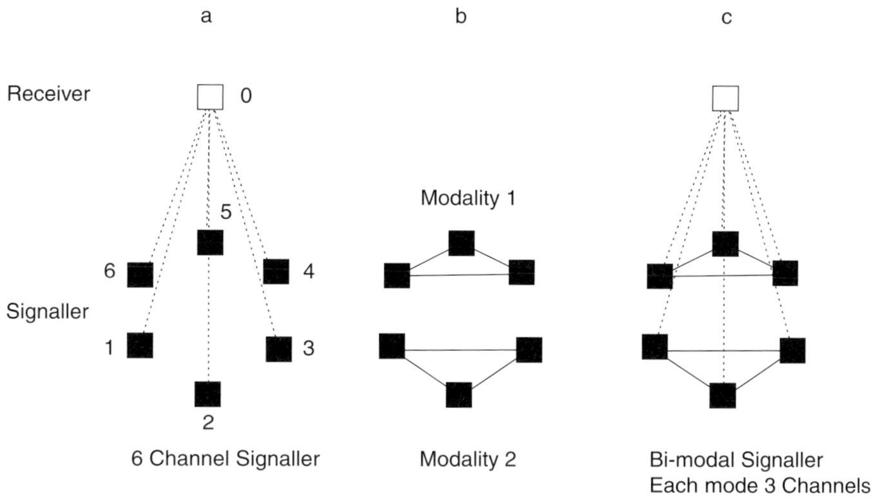

Figure 12.1. In (a) we show the basic perceptron architecture illustrated with six nodes of a signaller (numbered black squares) and six corresponding channels used to convey a message generated by node activity to a receiver (white square numbered 0). In (b) we show through connections among signaller nodes, patterns of correlated activity. Two sets of three nodes show highly correlated activity: Nodes 1, 2 and 3 are highly correlated, and nodes 4, 5 and 6 are highly correlated. Each correlated cluster we refer to as a modality. In (c) we represent the receiver integrating inputs from the six channels constituting two signalling modalities.

The question raised by this example is how do correlated clusters of channels arise? To address this issue, we represent the multi-channel and multi-modal property of signals through two different types of connectivity in a modified perceptron network. Channels connect the receiver to the nodes of a signaller (Figure 12.1a), whereas clusters of highly correlated nodes of the signaller define modalities (Figure 12.1b). In this paper we show mathematically that the clusters of correlated activity constituting modalities emerge as robust solutions to a channel occlusion problem. We also find that our robustness measure acts as a lower bound on a well-known network complexity measure arising from maximising information flow (Tononi *et al.*, 1994). We end by discussing the potential implications for the evolution of combinatorial signals.

12.2 Receiving as a feedforward pathway

To describe the response of a receiver to incoming signals we consider a simple neural network structure in terms of a map T, which describes how a node labelled as 0 generates an output y based on information from an input vector $x_1,\dots,x_N$. (Figure 12.1a). Write Λ for the set $\{1,\dots, N\}$ of input units and Λ_0 for the set $\Lambda \cup \{0\}$ of all units. The states of a unit $i \in \Lambda_0$ are denoted by $\boldsymbol{X}_i$. The formal description of the transformation T is given by a Markov transition matrix

$$T : \boldsymbol{X}_\Lambda \times \boldsymbol{X}_0 \to [0, 1], (x, y) \to T(y|x),$$

where T is the function performed by the network, and the input set is given by $\boldsymbol{X}_\Lambda := \boldsymbol{X}_1 \times \cdots \times \boldsymbol{X}_N$. The value $T(y|x)$ is the conditional probability of generating the output y given the input $x = (x_1, \ldots, x_N)$. This implies that for every $x \in \boldsymbol{X}_\Lambda$,

$$\sum_{y \in \boldsymbol{X}_0} T(y|x) = 1. \tag{1}$$

Example: Nodes have two states '0 = not active' and '1 = active', corresponding to the presence or absence of an active input. The system parameters are the *edge/channel weights* w_i, $i \in \Lambda$, which describe the strength of interaction among the individual input nodes i and the output node 0, and a *threshold value* θ for the output node which controls its sensitivity to the input. In neuroscience the weight describes the product of the density of post-synaptic receptors and neurostransmitter, and the threshold controls the sensitivity. We assume that given an input vector $x = (x_i)_{i\Lambda} \in \{0,1\}^N$ in a first step, the output node assumes a value given by the function

$$h(x) = \sum_{i \in \Lambda} w_i x_i - \theta,$$

and then, in a second step, it generates the output 1 with the probability

$$T^\beta(1|x) = \frac{1}{1 + e^{-\beta h(x)}}.$$

The normalisation property (1) then implies that the output 0 is generated with probability $1 - T(1|x)$. Here, the *inverse temperature* β controls the stochasticity of the map T^β. This is the familiar perceptron neural network (McCulloch & Pitts, 1943).

12.2.1 Information measures

In this section we introduce a number of measures of information that form the basis of a formal definition of robustness in networks. The argument proceeds by relating the perceptron architecture, interpreted as a simple stochastic map, to Shannon information or Shannon entropy. The connection of entropy to information derives from their common roots in deriving an extensive measure of the degree of ignorance we possess about the state of a coarse-grained system. The greater our ignorance the greater the information value of a signal. Stated differently, signals emitted in a low entropy or highly regular system provide very little information. Given an arbitrary subset $S \subseteq \Lambda_0$, we write $\boldsymbol{X}_S$ rather the more cumbersome notation $\prod_{i \in S} X_i$ and we consider the projection

$$X_S : \boldsymbol{X}_{\Lambda 0} \to \boldsymbol{X}_S, \quad x = (x_i)_{i \in \Lambda} \mapsto x_S = (x_i)_{i \in S}.$$

With an input distribution p on $\boldsymbol{X}_\Lambda$ and a stochastic map T from $\boldsymbol{X}_\Lambda$ to $\boldsymbol{X}_0$ we have the joint probability vector

$$P(x, y) = p(x)T(y|x), \quad x \in \boldsymbol{X}_\Lambda, \; y \in \boldsymbol{X}_0. \tag{2}$$

The projection X_S becomes a random variable with respect to P. Now consider three subsets $A, B, C \subseteq \Lambda_0$. The *entropy* of X_C or Shannon information is then defined as

$$H_P(X_C) = -\sum_{z \in \mathbf{X}_c} P\{X_C = z\} \ln (P\{X_C = z\}).$$

This quantity is a measure of the uncertainty that one has about the outcome of X_C (Cover & Thomas, 2001). Once we know the outcome, this uncertainty is then reduced to zero. This justifies the interpretation of $H_P(X_C)$ as the *information gain* after knowing the outcome of X_C. Now, having information about the outcome of the second variable X_B reduces the uncertainty about X_C. More precisely, the *conditional entropy* of X_C given X_B is defined as

$$H_P(X_C|X_B) = -\sum_{y \in \mathbf{X}_B, z \in \mathbf{X}_C} P\{X_B = y,\ X_C = z\} \ln (P\{X_C = z|X_B = y\}),$$

and we have $H_P(X_C) \geq H_P(X_C|X_B)$. Using these entropy terms, the *mutual information* of X_C and X_B is then defined as the uncertainty of X_C minus the uncertainty of X_C given X_B:

$$I_P(X_C : X_B) = H_P(X_C) - H_P(X_C|X_B).$$

The *conditional mutual information* of X_C and X_B given X_A is defined in a similar way:

$$I_P(X_C : X_B|X_A) = H_P(X_C|X_A) - H_P(X_C|X_A, X_B).$$

We simplify the notation by writing these quantities without explicitly mentioning P.

Thus we have a measure of the information that is output by the network as a function of the information present at the input units.

12.3 Network complexity measures and multi-modal modules

Now that we have defined a simple signalling network and appropriate information measures, we discuss a measure of network complexity. This measure will refer to the structure of correlations among the nodes of the signaller (Figure 12.1b) and lead to a statistical definition of a signalling modality. In a series of papers, Tononi *et al.* (1994) (TSE from here) consider information-theoretic measures of complexity in neural networks. The primary goal of this research is to determine which anatomical properties we should expect to observe in networks, such as a nervous sytem, where communication among cells plays a crucial role in promoting functional states of the system. TSE relate the functional connectivity of the network to statistical dependencies among neurons that arise through patterns of connectivity. The dependencies are measured using information theoretic expressions outlined in the previous section. TSE identify two principles of functional organisation. The first principle derives from the observation that groups of neurons are functionally segregated from one another; into modules, areas or columns. The second principle maintains that to achieve global coherence, segregated components need to become integrated. Segregation and integration combine to produce systems capable of both discrimination and generalisation. In an animal signalling context, segregation can be related to clusters of cells dedicated to generating different messages; in other words, different modalities of expression. Integration binds these signals into a compound meaning or function. According to TSE integration is a measure of the difference between the entropy expected on the basis of network connectivity and the observed entropy:

$$I(X_A) = \sum_{v \in A} H(X_v) - H(X_A)$$

The *TSE-Complexity* is then defined as the deviation of the normalised network integration from the integration measured over all possible bi-partitions of the network,

$$C(X_A) := \sum_{k=1}^{N} \left[\frac{k}{N} I(X_A) - \frac{1}{\binom{N}{k}} \sum_{\substack{A \subseteq A \\ |A|=k}} I(X_A) \right] \tag{3}$$

It has been shown that this complexity measure is low for systems whose components are characterised either by total independence or total dependence. It is high for systems whose components show simultaneous evidence of independence in small subsets and increasing dependence in subsets of increasing size. The TSE-Complexity can be written in terms of mutual information:

$$C(X_A) := \sum_{A \subseteq A} \frac{|A|}{N} \frac{1}{\binom{N}{k}} I(X_A : X_{A/A})$$

Now that we have defined TSE-Complexity in closed form in terms of information measures we can relate this back to information flows through the perceptron architecture.

12.4 Robustness as a complexity catalyst

12.4.1 A definition of robustness

In this section we relate the TSE-Complexity to a robustness measure. In order to capture the main idea behind this approach we consider two random input variables X and Y with distribution $p(x,y)$, $x \in \boldsymbol{X}$, $y \in \boldsymbol{Y}$, and one output variable Z which is generated by a stochastic map

$$T : (\boldsymbol{X} \times \boldsymbol{Y}) \times Z \to [0, 1], \quad (x, y, z) \mapsto T(z|x, y)$$

Now we assume that Y is knocked out, and we want to have a measure for the robustness of the function T against this knockout. First, robustness should include some notion of invariance with respect to this knockout pertubation. When the invariance property is satisfied, we say that the *exclusion dependence* is low. On the other hand, trivially vanishing exclusion dependence can be achieved if T does not depend on Y. In order to have a robustness measure that captures non-trivial invariance properties, we have to take the *contribution* of Y to the function T into account (Pearl, 2000). Our robustness measure is then defined as follows:

Informal Definition of Robustness: *We define the robustness of T against knockout of Y as the causal contribution of Y to the function T minus the exclusion dependence of T with respect to the knockout of Y.*

We consider the case where channels are occluded rather than simply noisy, as a limiting case that maps more naturally onto the biological problem that we are considering. Namely, under conditions where channels are not available for inspection, how might alternative channels be used to extract adaptive information? We have formalised this probablistic notion of robustness in terms of informational metrics (Amari, 1985; Ay & Krakauer, 2007) and derive the following formula:

$$R(Y,p,T) := \sum_x \sum_z \sum_y p(x,y)\ \ T(z|x,y)\ \ln \frac{\sum_{y'} p(x,y')T(z|x,y')}{p(x)\sum_{y'} p(y')T(z|x,y')}$$

where we sum over all values of x, y, z. This measures the amount of statistical dependence between X and Y that is used for computing Z in order to compensate for the exclusion of y. The robustness vanishes if for all x and all z

$$\sum_y p(x,y)T(z|x,y) = p(x)\sum_y p(y)T(z|x,y),$$

or equivalently

$$\sum_y T(z|x,y)(p(x,y) - p(x)p(y)) = 0 \tag{4}$$

There are two extreme cases where this equality holds. The first case is when there is no statistical dependence between x and y that can be used for compensation. Then $p(x,y) = p(x)p(y)$, and the equality (5) holds. The other extreme case is when there is statistical dependence, but this dependence is not used by T. In this case $T(z|x, y) = T(z|x, y')$ for all y, y'.

The ability of the perceptron to make use of redundant information by integrating over input channels is functionally analogous to von Neumann's theory for achieving reliable computation with probabilistic (stochastically failing) logic elements (von Neumann, 1956).

12.4.2 An example: duplication and robustness

In this let T: $\boldsymbol{X} \times \boldsymbol{Y} \rightarrow [0,1]$ be a stochastic map, and let p be a probability distribution on $\boldsymbol{X}$. In this example we seek to measure robustness as we duplicate T. In order to have several copies of this map, we consider the N-fold cartesian product $\boldsymbol{X}^N$, and we define the input probability distribution

$$\widetilde{p} := \sum_{x \in \boldsymbol{X}} p(x)\, \delta_{(x,\ldots,x)}.$$

We define the extension of the map T to the set of N identical inputs by choosing one input node with probability $\frac{1}{N}$ and then applying the map T to that node. This leads to

$$\tilde{T}(y|x_1,\ldots,x_N) := \frac{1}{N}(T(y|x_1) + T(y|x_2) + \cdots + T(y|x_N)).$$

With a the probability $1 - a$ for the exclusion of an input node $\upsilon \in \{1,\ldots, N\}$ we define the probability for a subset $A \subseteq \{1,\ldots, N\}$ to remain as input node set after knockout as

$$r(A) := a^{|A|}(1-a)^{N-|A|}.$$

We find the mean robustness of $\tilde{T}$ with respect to this knockout distribution:

$$R(r,p,\tilde{T}) := \sum_{A \subseteq \{1,\ldots,N\}} r(A)\left[\textit{Robustness of } \tilde{T} \textit{ against knockout of the complement of } A\right]$$

Now we want to show the robustness properties with respect to the number N of channels and the probability a by specifying T in terms of the identity map on the set $\{\pm 1\}$ with uniform distribution $p(-1) = p(+1) = \frac{1}{2}$. The output node just copies the input: $x \mapsto x$. Following our concept of robustness, we can show that the robustness is given by

$$-\sum_{k=0}^{N} \binom{N}{k} a^k (1-a)^{N-k} \ln\left(\frac{N+k}{2N}\right) - (1-a)^N \ln(2).$$

Figure 12.2 shows that channel duplication first leads to an increase of robustness but then declines as the number N increases. Note that the duplication of identical inputs is not optimal for robustness as each input transmits identical information, and thereby lessens its contribution to the signal. Robustness would increase if each input overlapped but included some uncorrelated information.

12.4.3 Extension of robustness measure to recurrent networks

In order to formally connect robustness with complexity we extend the robustness measure to the network setting. The set of network nodes is denoted by V, and the set of edges is denoted by $E \subseteq V \times V$. Given a unit $\upsilon \in V$, $\text{pa}(\upsilon) := \{u \in V : (u, \upsilon) \in E\}$ is the set of units that provide direct information to v. With the state sets X_υ, $\upsilon \in V$, we consider a family of stochastic maps denoted as

$$T^\upsilon : \boldsymbol{X}_{pa(\upsilon)} \times \boldsymbol{X}_\upsilon \to [0,1], \quad P(x,y) \mapsto T^\upsilon(y|x).$$

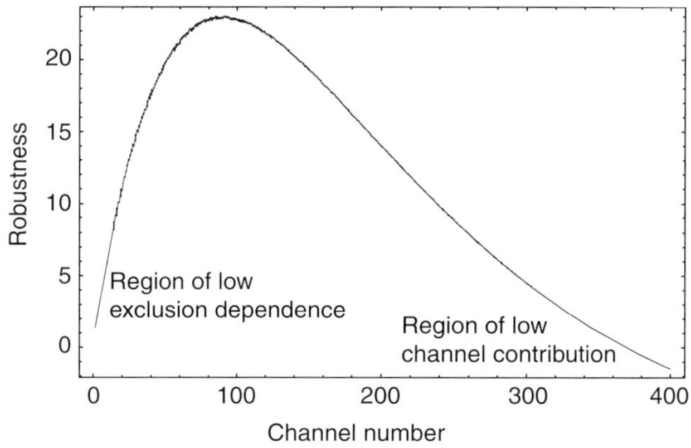

Figure 12.2. The robustness value as a function of channel number. At low levels of duplication individual channels increase the robustness by lowering the exclusion dependence. At high levels of duplication individual channels make very low contribution to function, and thereby lower the robustness value. The figure illustrates that systems of large non-integrated elements should not be deemed robust as channel removal does not influence behaviour. For increasing numbers of channels to increase robustness we need more than duplication, we require the emergence of correlated modules or statistical modalities. Parameter $a = 10^{-3}$.

The global dynamics $T\colon \boldsymbol{X}_V \times \boldsymbol{X}_V \to [0, 1]$ is then defined by

$$T\big((y_\upsilon)_{\upsilon\in V}\big|(x_\upsilon)_{\upsilon\in V}\big) := \prod_{\upsilon\in V} T^\upsilon(y_\upsilon|x_\upsilon).$$

Now we consider exclusions of subsets of the set V and associate a robustness measure for the network with regard to these exclusions. After knockout we have a remaining set A, and for a node $\upsilon \in A$ we consider the part of pa(υ) that is contained in A and the part pa(υ) $\backslash A$ that has been knocked out. Then we have the following robustness of T^υ against this exclusion:

$$\textit{Robustness of } T^\upsilon \textit{ against exclusion of } \mathrm{pa}(\upsilon)\backslash A$$
$$= \sum_{x\in \boldsymbol{X}_{\mathrm{pa}(\upsilon)\cap A},\, y\in \boldsymbol{X}_{\mathrm{pa}(\upsilon)\backslash A},\, z\in \boldsymbol{X}_\upsilon} p(x,y)T^\upsilon(z|x,y)\,\ln\left(\frac{\sum_{y'\in X\mathrm{pa}(\upsilon)\backslash A} p(x,y')T^\upsilon(z|x,y')}{p(x)\sum_{y'\in X_{\mathrm{pa}(\upsilon)\backslash A}} p(y')T^\upsilon(z|x,y)}\right).$$

With a probability distribution r, we define the following total robustness of the network:

$$R(r,p,T) = \sum_{A\subseteq V} r(A)\left\{\frac{1}{|A|}\sum_{\upsilon\in A}\left[\textit{Robustness of } \upsilon \textit{ against exclusion of } \mathrm{pa}(\upsilon)\backslash A\right]\right\}$$

In this formula we assume that p is a stationary distribution of T. Note that our robustness measure is a temporal quantity. It is surprising that one can relate this quantity to a purely spatial quantity, which depends only on the stationary distribution p. More precisely, we have the following upper bound for the robustness:

$$R(r,p,T) \leq \sum_{A \subseteq V} r(A) I(X_A : X_{V \setminus A}). \tag{6}$$

Let's compare this upper bound with the TSE-Complexity (3). For appropriate coefficients

$$r(A) = \frac{2|A|}{N(N+1)\binom{N}{|A|}} \tag{7}$$

we have the following connection:

$$\sum_{A \subseteq \Lambda} r(A) I\big(X_A : X_{\Lambda \setminus A}\big) = \frac{2}{N+1} C(X)$$

Here the coefficients are normalised, that is $\sum_{\mathrm{A}} r\,(A) = 1$. With (6) this directly implies

$$R(r,p,T) \leq \frac{2}{N+1} C(X) \tag{8}$$

We see that systems with high robustness display a high value of TSE-Complexity. One might suppose that this relation holds only for a special distribution r. To make the more general statement we generalise TSE-Complexity to cases where we can arbitrarily select r. We do this

$$C(r,X) := \sum_{A \subseteq V} r(A) \left(I(X) - \frac{N}{|A|} I(X_A) \right) = \sum_{A \subseteq V} r(A) I\big(X_A : X_{V \setminus A}\big)$$

In order to see how this extends the definition (3) we consider distributions r with $r(A) = r(B)$ whenever $|A| = |B|$. Such a distribution is uniquely defined by a map $r : \{0, 1, \ldots, N\} \rightarrow [0, 1]$ with $\sum_{k=0}^{N} \binom{N}{k} r\,(k) = 1$. This implies

$$C(r,X) = \sum_{k=0}^{N} \frac{r(k)N}{k} \binom{N}{k} \left[\frac{k}{N} I(X_\Lambda) - \frac{1}{\binom{N}{k}} \sum_{\substack{A \subseteq \Lambda \\ |A| = k}} I(X_A) \right]$$

We find that this is closely related to the complexity definition (3). The only difference is that in the generalised definition the sum is weighted appropriately. In any case, we have an extension of the inequality (8)

$$R(r,p,T) \leq C(r,X) \tag{9}$$

This inequality makes explicit that network robustness is a lower bound on network complexity. Hence pressure towards an increase in robustness, through for example multi-modality, will always lead to an increase in the complexity of the message by promoting structured correlations in the channels. TSE have found through extensive numerical simulation that the complexity measure $C(r, p, T)$ is maximised by network structures in which cells form densely connected clusters which are themselves sparsely connected. In other words, partially overlapping functional modules. In the context of signalling networks, segregated modules can be interpreted as different signalling modalities, where each modality emphasises a different feature of the system. Of interest to us in this chapter is the way in which signal complexity arises naturally out of pressures favouring robust signal detection.

12.5 Multi-modality and combinatoriality

We have derived a measure to quantify the impact of signalling channel perturbations on the ability of a receiver to process a signal. The measure quantifies a functional distance between a 'perfect information' condition between signaller and receiver, and a reconfigured condition, in which a subset of input channels have been occluded. Using multiple channels through which correlated activity can be transmitted by a signaller allows a receiver to choose from a different channel once one has been lost. However, if all channels are identical, then it makes little sense to refer to a signalling system as robust following perturbation. This is because as channels become more numerous each channel makes a diminishing contribution to decoding the message (Pearl, 2000). In the large channel limit each channel becomes effectively independent from the message (Figure 12.3a) as its unitary contribution is just $1/N$. From the selection perspective irrelevant channels are expected to be lost (Krakauer & Nowak, 1999). By reducing correlations among channels we increase their individual causal contribution to the message. However, if they become completely decorrelated (Figure 12.3b), the removal of any one has a significant impact on the receiver's ability to decode the message. The optimal solution is to generate clusters within which channel activity is highly correlated and between which activity is weakly correlated (Figure 12.3c). Hence an optimal signaller distributes a message among weakly correlated modalities, within which multi-channel redundancy remains high. Each cluster we might think of as a primordial modality promoting specialised interpretive means by the receiver.

Somewhat surprisingly the modular structure of a robust signal is what makes for an effective information-processing neural network (Tononi *et al.*, 1994). Selection for high information flow and thereby reduced sensitivity to channel occlusion leads naturally to signalling networks with high levels of integration and segregation (clustering). High levels of segregation and integration in turn promote the development of an effective computational system (in the signaller) which trades off specificity with generalisation. This property has been referred to as network complexity (Tononi *et al.*, 1994).

In terms of natural communication systems, our results suggest that multi-modality might arise in the following way: senders emit multi-channel signals (like the tamarin combination long call or chimpanzee pant-grunts, each of which is characterised by the presence of multiple frequencies). Over time, selection for robustness generates

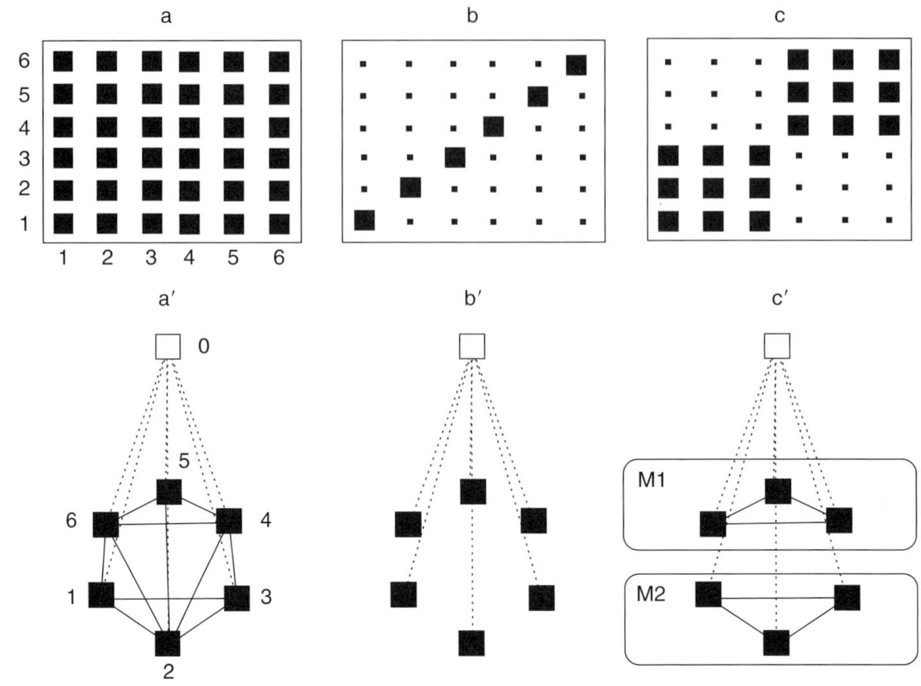

Figure 12.3. (a) Correlation matrix indicating by the size of black squares the magnitude of correlation or mutual information between pairs of nodes of the signaller. Here all nodes are perfectly correlated in their acitivity. (a′) The perceptron connectivity corresponding to the correlation matrix (a). (b) Signaller nodes are only weakly correlated with each other and constitute approximately independent channels (b′) for the receiver to integrate. (c) Channels form correlated clusters of activity, with weak correlations among clusters. This corresponds to a two-modality signalling system (sets of channels M1 and M2). The two-modality case is both more robust and more complex. Robustness derives from insensitivity to channel occlusion through channel redundancy coupled to high information flow through weak modality decoupling.

correlations among channels and, consequently, channel clustering. From the signal production/encoding perspective, this means that neural and behavioural substrates underlying production are becoming modularised, setting the stage for the evolution of alternative sensory modalities. Pant grunts, a subordination signal, are not produced alone, but are emitted in conjunction with gestural or behavioural displays, like bobbing and bowing (de Waal, 1982). From the decoding perspective, multi-modal signals that arise in the way we have outlined in this chapter, maximise encoding robustness and by doing so also minimise the cognitive cost of processing signals with multiple channels. The reason for this is that the overlap between modalities (clusters of channels) must be large enough to ensure sufficient redundancy should one mode be knocked out or occluded, but small enough to ensure that each modality contributes cumulatively to signal meaning. This means that the decoding algorithm used by a receiver to decode one part of a signal can be useful for decoding another part of the same signal.

12.6 Robustness and measures of complexity

It is worth saying a few more words about the relationship between complexity and robustness. In the complexity measure random bi-partitions of a network (Figure 12.1b) are used to assess the extent of communication among network regions of a signaller; whereas in the robustness measure, bi-partitions reflect the excision of large sets of communicating channels. In both cases information flow is required. In the complexity case, information flow is assumed to reflect increased associative power among modules, whereas in the robustness case information flow is required in order that non-occluded channels can be used as alternative information sources. It is not obvious why selection should favour information flow among channels of a signaller, but it is obvious that this information flow can be used by the receiver in case of occlusion. If we were considering a signal internal to the sender, the situation would be different, as in this case information flow could come under selection for more effective integration for cognitive function.

The net result is that signallers seeking to transmit complex messages benefit from a multi-modal strategy as it both increases the diversity of information flowing to a receiver and increases the robustness of the signal. When selection for robustness mechanisms leads to increased complexity (according to some formal metric) this is an instance of the *principle of robust overdesign* (Krakauer & Plotkin, 2004).

12.7 Signalling and evolutionary innovation

A secondary advantage of selection for increased information flow is that it provides the basis for novel combinatorial signalling. Signals can be built up through composition from segregated functional units, with each unit producing a different signal component. Integration ensures that at first these components have overlapping meanings (through correlated activity) and are thereby likely to be understood or learnable. This applies to multi-component signals (different communicative features in the same modality) as well as to multi-modal ones. The learnability problem (Flack & de Waal, 2007), which is particularly problematic when receivers are confronted with new signals that are spatially or temporally divorced from their objects and thereby hard to associate, can be mitigated by 'pointing' to a new signal object using a 'compound stimulus'. A compound stimulus signal is one comprised of two or more modes or components, one or more of which has an established meaning. Emitting these together allows the receiver to infer from overlap a new meaning. The capacity for combinatoriality also allows many meanings to be created from a small set of components, reducing the need for cognitively burdensome and error-prone storage of many one to one mappings (Nowak & Krakauer, 1999). For example, it is easier to generalise unknown word meaning from contextual usage (Grice, 1969), than it is to store every word (and their associated meanings) one is likely to encounter. Thus with the evolution of signal robustness, there arises the possibility of establishing novel functions through the combinatorial expansion of increased message encoding. Evolutionary innovation in communication is hypothesised to proceed through an early fault-tolerant stage, which provides a multi-modal substrate for subsequent semantic elaboration and signal diversification. Selection for perfect combinatoriality (independent channels or modalities), would result in overlap in meaning among signal components, thereby decreasing robustness in the long run. An important empirical and theoretical question is what level of component divergence is optimal given the dual selection pressures for communicative diversity and and informational redundancy.

References

Amari, S. 1985. *Differential-Geometric Methods in Statistics. Lecture Notes in Statistics 28*. Springer-Verlag.

Ay, N. & Krakauer, D. C. 2007. Geometric robustness theory and biological networks. *Theory in Biosciences* **125**(2), 93–121.

Cover, T. M. & Thomas, J. A. 2001. *Elements of Information Theory*. John Wiley and Sons.

de Waal, F. B. M. 1982. *Chimpanzee Politics: Power and Sex Among the Apes*. Johns Hopkins Press.

Elias, D. O., Mason, A. C., Maddison, W. P. & Hoy, R. R. 2003. Seismic signals in a courting male jumping spider (Araneae: Salticidae). *J Exp Biol* **206**, 4029–4039.

Flack, J. C. & de Waal, F. B. M. 2007. Context modulates signal meaning in primate communication. *Proc Natl Acad Sci USA* **104**, 1581–1586.

Grice, H. P. 1969. Utterer's meaning and intention. *Phil Rev* **68**, 147–177.

Hauser, M. D. 1997. *The Evolution of Communication*. MIT Press.

Hillis, J. M., Ernst, M. O., Banks, M. S. & Landy, M. S. 2002. Combining sensory information: mandatory fusion within, but not between senses. *Science* **298**, 1627–1630.

Johnstone, R. A. 1996. Multiple displays in animal communication: 'backup signals' and 'multiple messages'. *Phil Trans R Soc B* **351**, 329–338.

Krakauer, D. C. & Nowak, M. A. 1999. Evolutionary preservation of redundant duplicated genes. *Sem Cell Dev Biol* **10**, 555–559.

Krakauer, D. C. & Plotkin, J. B. 2004. Principles and parameters of molecular robustness. In *Robust Design: a Repertoire for Biology, Ecology and Engineering* (ed. E. Jen). Oxford University Press.

McCulloch, W. S. & Pitts, W. A. 1943. Logical calculus of the ideas immanent in nervous activity. *Bull Math Biophyss* **5**, 115–133.

Nowak, M. A. & Krakauer, D. C. 1999. The evolution of language. *Proc Natl Acad Sci USA* **96**, 8028–8033.

Partan, S. & Marler, P. 1999. Communication goes multimodal. *Science* **283**, 1272–1273.

Pearl, J. 2000. *Causality*. Cambridge University Press.

Rowe, C. 1999. Receiver psychology and the evolution of multicomponent signals. *Anim Behav* **58**, 921–931.

Tononi, G., Sporns, O. & Edelman, G. M. 1994. A measure for brain complexity: relating functional segregation and integration in the nervous system. *Proc Natl Acad Sci USA* **91**, 5033–5037.

von Neumann, J. 1956. Probabilistic logics and the synthesis of reliable organisms from unreliable components. In *Automata Studies* (ed. C. E. Shannon & J. McCarthy). Princeton University Press.

Weiss, D. J. & Hauser, M. D. 2002. Perception of harmonics in the combination long call of cottontop tamarins (*Saguinus oedipus*). *Anim Behav* **64**, 415–426.

13

Movement-based signalling and the physical world: modelling the changing perceptual task for receivers

Richard A. Peters

13.1 Introduction

Consideration of the design and use of animal signals is of fundamental importance for our understanding of the social organisation and the perceptual and cognitive abilities of animals (e.g. Endler & Basolo, 1998). Movement-based visual signals have proven particularly difficult to understand because (in contrast to colour and auditory signals) perception, environmental conditions at the time of signalling and information content of motion signals cannot be easily modelled. Image motion has to be computed by the brain from the temporal and spatial correlations of photoreceptor signals. Although the computational structure of motion perception is well understood, in most situations it is still practically impossible to accurately quantify image motion signals under natural conditions from the animal's perspective. This undermines our ability to understand the perceptual constraints on movement-based signal design.

Extrapolating from other signalling systems, the diversity of movement-based signals between species is likely to be a function of the characteristics of competing, irrelevant sensory stimulation, or 'noise', and sensory system capabilities. The extent to which the spatiotemporal properties of signal and noise overlap remains unclear, however, and indeed, the motion characteristics that reliably lead to segmentation of the signal from noise are largely unresolved. It is therefore difficult to know the circumstances in which signal detection is compromised. In this chapter, I begin to generate the kind of data that will help explain movement-based signal evolution by modelling the changing perceptual task facing the Australian lizard *Amphibolurus muricatus* in detecting conspecific communicative displays. These territorial agamid lizards use a stereotyped sequence of motor patterns during agonistic encounters (Peters & Ord, 2003), which convey information to receivers about the quality of the signaller. The distribution of these lizards includes large populations in coastal reserves (Cogger, 1996) that are densely vegetated with shrubs (Costermans, 2005). Communicative signals are therefore performed in environments in

Modelling Perception with Artificial Neural Networks, ed. C. R. Tosh and G. D. Ruxton. Published by Cambridge University Press.

which variable wind conditions generate irregular patterns of plant motion. Variation in site topography (Hannah *et al.*, 1995), plant shape and their mechanical properties (Wood, 1995), as well as the depth structure of the environment (Fleishman, 1986), ensure that image motion due to wind-blown plants varies considerably in space and time (Peters *et al.*, 2008).

The masking effect of environmental motion noise is becoming clearer. Plant image motion, under certain conditions, will lengthen the time taken to detect conspecific signals (Peters, 2008) and promote changes in signalling strategy by signallers to compensate for a likely reduction in signal efficacy (Peters *et al.*, 2007). Complementary studies of other lizards (genus: *Anolis*) also indicated that environmental conditions are potentially detrimental to signal detection as they too adjust signalling strategies to compensate (Ord *et al.*, 2007). Interestingly, the strategies used to compensate for adverse signalling conditions varied between the lizard species: *A. muricatus* lengthened the duration of introductory tail flicking (Peters *et al.* 2007), while *Anolis* lizards favoured increased speeds (Ord *et al.* 2007) or the addition of a facultative eye-catching motor pattern (Ord & Stamps, 2008).

To understand the nature of this diversity it would be useful to characterise in more detail the detrimental effect of plant motion noise. In earlier work, I trained various neural network models to distinguish between the image motion generated by displays of *A. muricatus* from that of wind-blown plants (Peters & Davis, 2006). The image motion of isolated displays and plant movement were calculated separately and summary values, collapsed across space, were then presented to neural network models. Although simple feedforward and dynamic networks could be trained to distinguish the two classes of image motion, the utility of the modelling approach was limited because the input classes were quite distinct to begin with. The choice of plant sequences, which were filmed under light to moderate wind conditions that did not truly sample the range of conditions experienced in the wild, as well as excluding spatial detail and keeping the two types of image sequences separate contributed to making the task oversimplistic.

A less rudimentary approach is required if modelling is to progress our understanding of movement-based signal evolution. A particular challenge will be the need to retain spatial detail and allow for signals to be embedded within noise. This dramatically changes the nature of the task and requires an alternative type of model. For the present purposes, a descriptive model that provided testable hypotheses about the perceptual constraints influencing movement-based signal evolution was sought, without necessarily seeking to meet the demands of biological realism. To this end, I have used a published model of visual attention that identifies salient regions of an image using a winner-take-all neural network (Itti *et al.*, 1998). The model assumes, with some foundations in biology, that certain regions of a scene are processed preferentially by the visual system as a function of their relative salience (Koch & Ullman, 1985). Computationally, this is achieved by identifying locations where local attributes significantly differ from their surroundings along one or more dimensions (Itti *et al.*, 1998). The typical implementation of the model has been for target detection among an array of distractors, or scanning natural scenes to predict saccadic eye movements. Here, I have used this model to identify salient regions within data matrices that represent angular speeds generated by motion detectors algorithms. Motion analysis was therefore done separately. Saliency analysis can be extended to consider movement as a feature (e.g. Walther, 2006), but I will consider such capabilities in follow-up work.

Three key questions have guided the modelling work presented in this chapter. The underlying theme concerns how detection of a given signal is influenced by the physical world within which it is performed:

(1) *How do changing environmental conditions affect signal efficacy?* The masking effect of plant motion can be demonstrated by digitally combining a given animal signal onto sequences of plant motion filmed under varying wind conditions. As illustrated in Peters *et al.* (2008), a given signal is expected to be highly conspicuous in wind-still conditions, but notably harder to segment when surrounded by greater motion noise. Assuming this is reflected in the saliency analysis, I was interested to see how efficacy declined as a function of environmental conditions, and how it varied between microhabitats. Also apparent in the simple illustration (figure 11 in Peters *et al.*, 2008) was the surprising suggestion that the efficacy of the motor pattern used by *A. muricatus* for attracting the attention of receivers (tail flicking) may be more affected by variation in plant motion than subsequent motor patterns. To consider further this intriguing contradiction, I isolated these two signal components and analysed them separately.

(2) *What influence does the geometry of the plant habitat have on signal efficacy?* As perceived angular speed is dependent on viewing distance, a given signal may appear more or less salient depending on its proximity to surrounding plants. I anticipated that signals performed at the same depth plane as plants would be more difficult to segment, under certain conditions, than those at greater relative signaller–plant distances.

(3) *Can signalling faster offset the masking effects of plant motion?* Such a strategy was suggested by the analyses of signalling *Anolis* lizards (Ord *et al.*, 2007), but contrasts with the strategy adopted by *A. muricatus*, which favour longer durations over changing signalling speed. Peters *et al.* (2008) speculated that it might not make sense to signal faster if you are signalling in the same depth plane as plants. Generating the necessary speeds in this context may in fact be beyond the physical capabilities of the animal. Following this, I expected that the efficacy benefit of faster signalling would only be realised at greater signaller–plant distances.

With these three questions in mind, I simulated a range of signalling contexts that varied in plant species (i.e. microhabitat), wind conditions and the signaller's distance from the nearest plant. Using techniques developed for modelling attention to objects in natural scenes (Walther & Koch, 2006), salient regions of the scene were identified and a winner-take-all neural network determined the most salient location; this was done for each frame of the sequence. By comparing the region of space that was identified as being the most salient (the 'winning location') from the composite sequences (signal + noise), with a control sequence featuring the same signal and no plant movement (signal), I determined the relative effectiveness of that signal under the environmental and geometric conditions imposed. Systematically varying the signalling conditions has provided a picture of relative signal efficacy that is predictive and ultimately testable.

13.2 Methods

13.2.1 Defining the input

To consider the relative conspicuousness of movement-based signals under varying environmental conditions, I selected digital video sequences of territorial displays by *A. muricatus* and wind-blown plants recorded for studies reported elsewhere (Ord *et al.*, 2002; Peters *et al.,* 2008 respectively). As the lizard displays were recorded against a static heterogeneous background I was able to isolate the animal from the background and overlay this onto the plant footage using custom written functions in Matlab (MathWorks Inc.). Image motion analysis was then used to calculate angular speeds of the composite sequences, which were used as input to the saliency model. The same signal could therefore be considered under a variety of environmental conditions.

13.2.1.1 Territorial displays by A. muricatus

Three representative display sequences from different lizards were selected from a library of archival footage that featured a displaying lizard on a wooden perch against a uniform light blue background (see Appendix). The lizards varied in size (90–108 mm snout–vent length) and weight (24.5–45 g), and to the naked eye, performed signals that varied in speed and amplitude. The nature of these differences is not relevant to the present work, but selecting signals that might vary in such detail does permit me to consider whether some displays may be more effective than others.

Each display was split into two and analysed separately. The opening part of the display features tail flicking, which I believe serves an alerting role and is the component of their display that lizards adjust in response to changing plant motion (Peters *et al.*, 2007). Tail flicking varies within and between individuals in terms of duration and the number of pauses between flicks. I selected windowed clips of each of the three tail-flick sequences that featured continuous tail movement and that did not disappear behind the perch, lasting 1–2 s in duration (23, 46 and 48 PAL frames). The duration of tail-flicking sequences was selected to match roughly the duration of the rest of the display, which featured four different motor patterns centred on a push-up (see Peters & Ord, 2003 for a full description). I selected a single push-up sequence for each lizard that also lasted 1–2 s (38, 29 and 29 frames); henceforth I will refer to this latter part of the signal as the 'display'. A link to video clips is available in the Appendix.

13.2.1.2 Plant motion backgrounds

Sequences of plant motion were chosen to sample a range of environmental conditions at four different microhabitats. These locations were identified in a previous study to be known perch sites, and represent the types of plants against which lizard signals must be detected (Peters *et al.*, 2008). The plant species at these sites were *Banksia integrifolia* (coastal banksia), *Leucopogon parviflorus* (coast bearded heath), *Acacia longifolia* (coast wattle) and *Lomandra longifolia* (mat rush). For each site separately, I first calculated the image motion characteristics of several minutes of footage, captured on separate days to sample varying wind conditions (angular speeds increase as the prevailing wind gets stronger; see Peters *et al.*, 2008 for details). A single summary value of calculated angular speeds was obtained by averaging across space for non-overlapping 2 s time windows. I sorted the distribution of summary speed values from lowest to highest and selected 26 equally spaced values within the distribution. The respective 2 s sequences of plant footage that generated each of these 26 values were used in the present modelling work as the background for the lizard displays. These steps were performed for each microhabitat separately. However, the distribution of measured values differed between sites and hence the 26 sequences for each site had slightly different profiles (see Figure 13.3).

13.2.1.3 Composite sequences of signal and background

Lizard displays and plant footage were combined in Matlab. The lizard was removed from each frame, scaled in size and 'pasted' onto the respective frame of the plant background (see Appendix). By scaling the lizard before overlaying onto the plant background, I varied the relative distance between the signalling animal and the nearest plant. As filming details varied between sites in the original study (camera zoom, filming distance), scaling values differed between the backgrounds. Because the filming geometry was known, however, I was able to determine the scaling value required to simulate a displaying animal in the same depth plane as the plants at each site. This value represented the first signaller–plant distance. The fourth simulated distance was double the original value, while the remaining two were equally spaced between these two extremes. The factorial combination of lizard (3), display type (tail flick, display), scale (4 levels), microhabitat (4 sites) and environmental conditions (26 levels) resulted in 2496 sequences. Although absolute sizes varied at a given scale across the four sites, the scaling approach that I used ensured that the position of lizards relative to plants were equivalent across the four sites. An additional 96 sequences were generated that combined each lizard, display type and the four levels of scale against a static background. These static background sequences were used to assess performance, as described below.

13.2.1.4 Image motion analysis

Image motion in the composite sequences was calculated using another custom-written Matlab program that has been described in detail elsewhere (Peters *et al.*, 2002; Peters & Evans, 2003). Briefly, the program is based on a gradient detector model and calculates the velocity field in image sequences based on temporal and spatial derivatives of filtered versions of image intensity (see Peters *et al.*, 2002 and references therein for details). Each image was down-sampled to 144×180 pixels to reduce computational load. The calculated velocity field comprises the angular speed and direction of movement at each location in the frame. The angular speed data matrix for each frame was used as the input to saliency analysis (see Appendix).

13.2.2 Saliency analysis: identifying the focus of attention

Itti *et al.* (1998) described a computational model for saliency-driven bottom-up selective attention to objects, which involved scanning a saliency map computed from local feature contrasts. They suggested, however, that their approach could be tailored to model attention to arbitrary tasks through the implementation of dedicated feature maps. Here, I follow their suggestion to model attention to salient image motion based on angular speed distributions within image sequences. To this end, I make use of a Matlab implementation of the aforementioned model, the Saliency Toolbox v2.1 (Walther & Koch, 2006). I introduce the model's architecture and my particular implementation below. I do not, however, reproduce the underlying mathematics and refer readers interested in such computational detail to the following papers: Koch & Ullman (1985), who first presented the saliency map-based model of bottom-up attention; Itti *et al.* (1998) who developed the model; and to Walther & Koch (2006) who extended further this model, and are the authors of the Saliency Toolbox that they provide as a free download from the World Wide Web.

The version of the model that I have used in the present work is illustrated in Figure 13.1. Input images are first separated into *feature maps* at multiple scales before being combined into a single *conspicuity map* for each feature. The conspicuity maps are subsequently combined to form a single *saliency map*. A winner-take-all neural network of leaky integrate-and-fire neurons scans the saliency map to identify the winning location. I summarise below each of these steps in the model, drawing particularly from the descriptions of Itti *et al.* (1998) and Walther & Koch (2006).

13.2.2.1 Feature and conspicuity maps

Two feature types are defined from the input image: *intensity* and *orientation.* In my implementation, intensity represents the angular speed values, while orientation denotes how discontinuities in angular speeds are aligned in space. Colour, an additional feature represented in the original work, was not relevant here and therefore not discussed below or presented in Figure 13.1.

Prior to extracting each of these features, a Gaussian pyramid is generated for each input image. This involves progressively low-pass filtering and subsampling the input to produce a sequence of reduced resolution images, each decimated by a factor of 2 (Adelson *et al.*, 1984). Given the size of the input [144,196], the pyramids generated here were in the range [0,7], where 0 represents the original image and 7 has a resolution of $1/2^7$ relative to the original; clearly, images of scale 7 (size of [1,1]) are uninformative, and were not utilised in subsequent processing.

Each feature is computed from the Gaussian pyramid by a number of centre-surround operations, which have proven sensitivity to local spatial discontinuities (Itti *et al.*, 1998), and implemented in the model as the difference between high and low resolutions. The model calculates feature maps as across-scale difference maps by interpolating to the

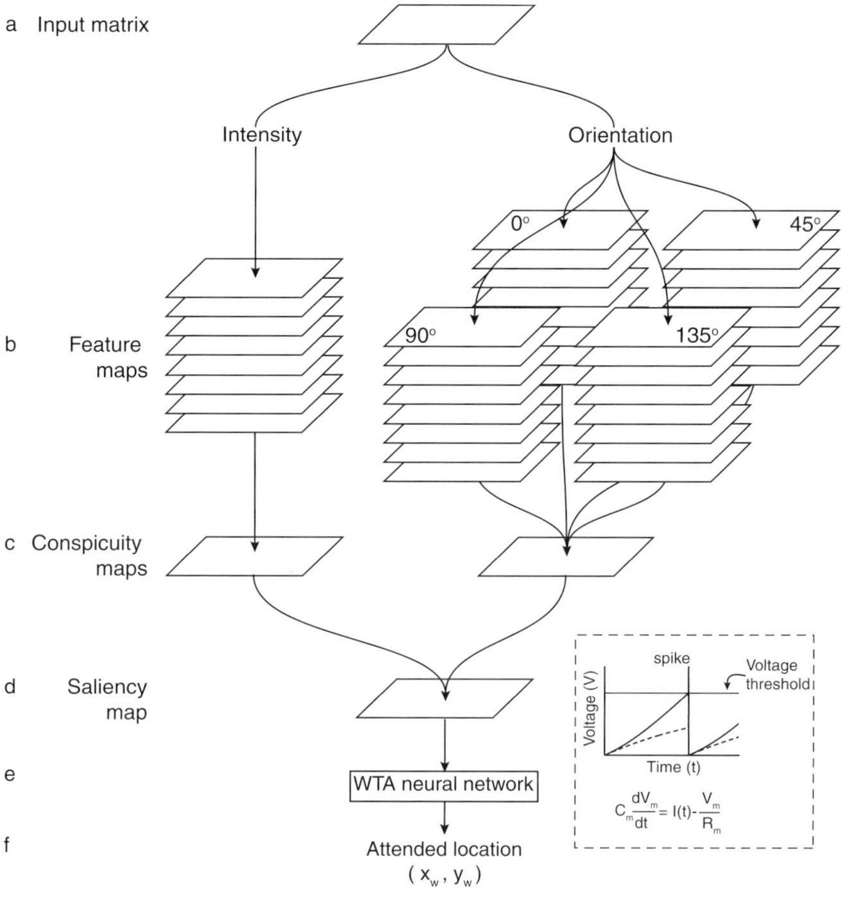

Figure 13.1. Schematic illustration of the saliency model adapted from Itti *et al.* (1998) that was used to consider the relative efficacy of lizard displays in plant motion noise. (a) Input matrices of angular speed values were subject to low-pass filtering and subsampling to create a Gaussian pyramid at eight spatial scales (not shown). (b) Two features were extracted at each of the eight spatial scales. Intensity features were extracted by centre-surround operations on raw speed values and then normalised to the range [0,1], while orientation information was first computed by convolving each level of the pyramid with oriented Gabor filters, followed by the same centre-surround operations and normalisation. (c) Across scale addition and subsequent normalisation of the feature maps led to conspicuity maps for each feature. These were combined by averaging at scale 3 to obtain a single saliency map (d), implemented as a 2D sheet of leaky integrate-and-fire neurons. A winner-take-all neural network also comprising integrate-and-fire neurons (e) was then used to detect the most salient location (f). Inset: The subthreshold time-course of leaky integrate-and-fire neurons. The solid line represents a location in which the voltage reached the threshold and generated a spike before being reset to zero; the dashed line reflecting another location was also reset to zero after the spike was generated elsewhere. Neuron voltage V_m is a function of the input current I, which is reduced over time according to a leak term characterised by the resistance flowing out of the cell R_m. C_m is a constant that defines the capacity of the model cell.

resolution of the finer scale image and taking the absolute value after point-by-point subtraction. The centre is a pixel at scale $c \in \{3,4,5\}$ and the surround is the corresponding pixel at scale $s = c + \delta$ with $\delta \in \{2,3,4\}$. I computed eight feature maps for intensity (Figure 13.1b left) with this combination of centre-surround scales and excluding values for s that exceed the number of levels on the Gaussian pyramid. Each feature map is then subject to normalisation that is fully described elsewhere (see Itti & Koch, 2000). Briefly, each feature map is normalised to a fixed range [0,1] and then iteratively convolved by a 2D difference-of-Gaussian filter yielding strong local excitation counterbalanced by inhibition at neighbouring locations. Between iterations, the original image is added to the new one and negative values are set to 0. The purpose of this normalisation process is to suppress any feature map that has numerous peaks of similar amplitude, while enhancing those that have only a few strong peaks. A single conspicuity map is then obtained for intensity by across-scale addition of the feature maps and normalisation (Figure 13.1c left).

Local orientation information was obtained by convolving the levels of the Gaussian pyramid with oriented Gabor filters (0°, 45°, 90°, 135°). Gabor filters are the product of a cosine grating and a 2D Gaussian envelope to encode local orientation contrast between centre and surround scales (see Walther & Koch, 2006 for formulae). Across-scale subtraction, as described above, is performed within each orientation and the outcome normalised. Eight feature maps were obtained within each orientation (Figure 13.1b right). To obtain conspicuity maps, the feature maps were summed across scales within each orientation level and normalised. The resultant four maps were then added together, and normalised once again (Figure 13.1c right).

13.2.2.2 The saliency map and winner-take-all neural network

The conspicuity maps are combined into the final input *S* to the saliency map (Figure 13.1d) by averaging at a predefined scale (a scale of 3 was used in the present simulations). Rather than directing attention to the largest value in *S*, the saliency map is modelled as a dynamical neural network of leaky *integrate-and-fire* neurons. Neurons of this form integrate their input and generate a spike when a voltage threshold is reached. This is illustrated in Figure 13.1 inset along with the subthreshold time course of neuron voltage (see Koch, 1998 for a general introduction to modelling neuron behaviour).

Neurons in the saliency map receive excitatory inputs only from the corresponding unit in *S*, and therefore act independently, while the potential of saliency map units increases faster for more salient locations in *S* (Figure 13.1 inset). Each unit in the saliency map in turn excites its corresponding winner-take-all neuron (Figure 13.1e), again evolving independently until one of them reaches the threshold (the same for all units) and fires (Figure 13.1 inset). The location of this spiking unit defines the winning location (Figure 13.1f). Although not utilised in the present simulations, the *n* most salient locations can also be identified. A process of local inhibition prevents the focus of attention from returning to the same location on successive runs. For details of how this is achieved please refer to Walther & Koch (2006).

13.2.3 Quantifying performance

In my earlier attempt to model movement-based signal detection, I presented signal and noise inputs separately to previously trained neural network models and could therefore determine network performance because input class was known (Peters & Davis, 2006). In the present approach, the signal is embedded within noise as part of a single input structure. To assess performance here, I needed to determine whether the winning location represented image motion generated by the lizard, or if attention is drawn to another region of the image characterised by plant motion. I achieved this by comparing the winning location for each frame in sequences featuring plant motion (signal and noise) with output from the respective frame in the corresponding static background sequence (signal only).

Rather than determining if the winning locations from each are identical, I compared the winning location from the signal and noise image with a wider region of interest (ROI) in the corresponding signal-only sequence (Figure 13.2). The ROI is computed during saliency analysis and provides an estimate of the spatial extent of the salient motion (Figure 13.2g). As Walther & Koch (2006) describe, quantifying this region involves stepping back through the conspicuity and feature maps (Figure 13.2c, d). The conspicuity map that contributed most to the activity of the winning location is identified, and in turn, the individual feature map that contributed most to the conspicuity map at this same location (Figure 13.2e). Within the winning feature map, a contiguous region of high activity containing the winning location is determined within a defined range of the winning location's activity (Figure 13.2f). By using a wider target area, I am assuming that the winning location and activity within the ROI are from the same source. If the winning location for a signal and noise input is contained within the ROI of the respective signal-only sequence, I assume that attention is correctly drawn to the lizard's movement and given a score of 1; otherwise the score is set to 0.

Two performance measures were obtained. As a measure of *response probability*, I calculated the proportion of frames in a sequence in which the winning location was contained within the wider ROI. My second measure of performance was *response time*. I located the first frame in which the winning location was contained within the wider ROI and divided this by the number of frames in the sequence to yield a relative time within the sequence.

13.3 Results and discussion

I have grouped the presentation of simulation results according to the three questions introduced above that prompted the modelling work. I first considered the relative effectiveness of signals as environmental conditions worsened, caused by increased plant movement from negligible (condition 1) to significant (condition 26). Here, only sequences where the lizards signalled at the same depth plane as plants were considered,

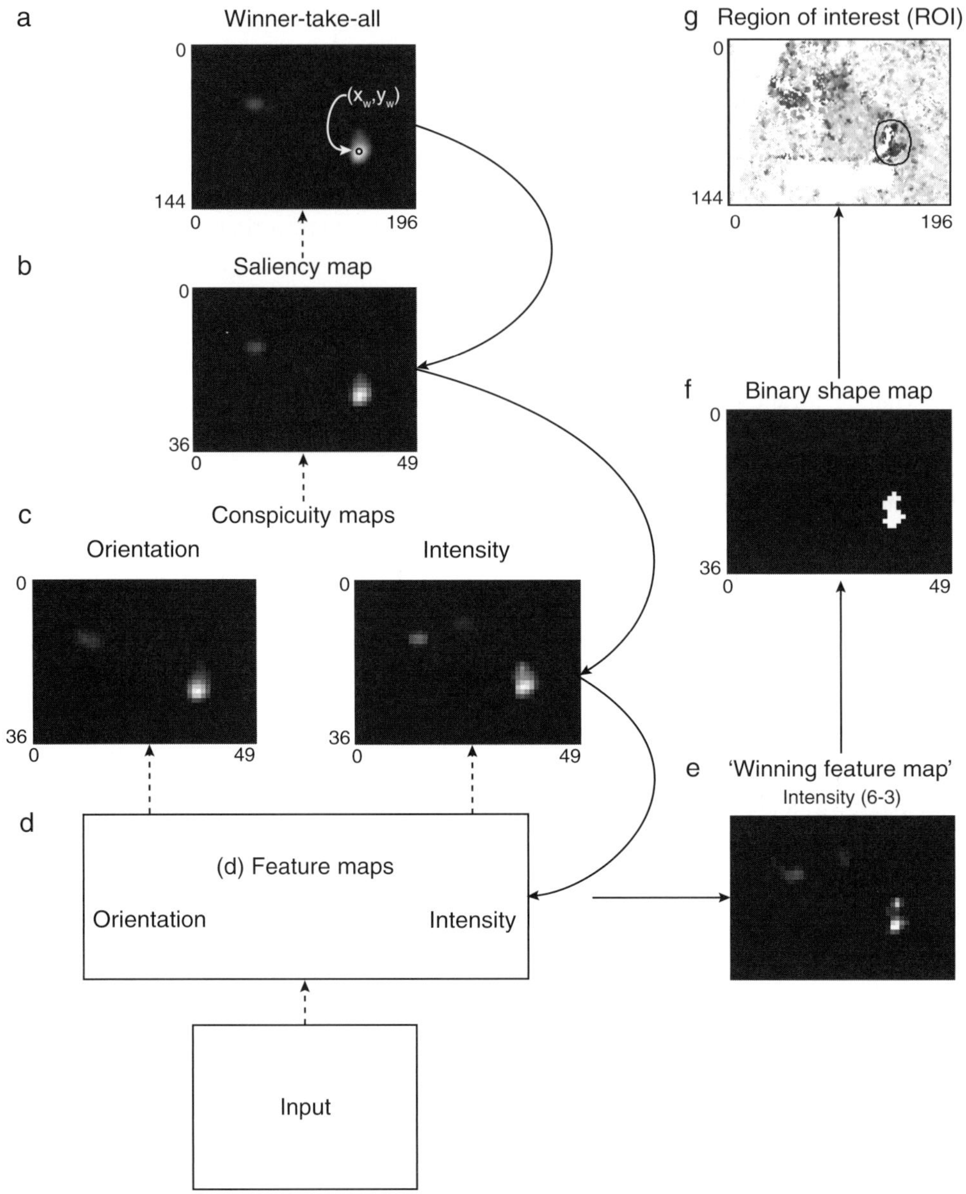

Figure 13.2. Schematic illustration describing the definition of a wider region of interest (ROI) in an input sequence used to compare performances across conditions. This is adapted from the description presented by Walther & Koch (2006) for defining the extent of a proto-object in an image. The left side of the figure from bottom-to-top reflects the schematic presentation in Figure 13.1, with the forward sequence of events represented by dashed arrow lines. The procedure for determining the ROI is given by solid arrow lines as follows and involves stepping back through the various maps. After identifying the winning location (a), the conspicuity map (c) that contributed most to the activity of the winning location in the saliency map (b) is identified, in this case the intensity conspicuity map. In turn, the individual feature map (d) that contributed most to this winning location is identified, which in this example, was defined by a centre at scale 6 and surround at scale 3 (e). A contiguous region of activity above a defined value that included the winning location was determined (f). This ROI is shown in (g) rescaled to the original size and overlaid onto the original angular speed input matrix.

which would be most typical of the circumstances facing *A. muricatus* receivers. I also consider separately the tail flick and display components that, under normal circumstances, occur consecutively in the same natural signal sequence. The next block of simulations addressed whether moving away from plants (and closer to the viewer) makes detection easier, particularly in adverse signalling conditions. A final set of simulations considered whether adjusting signalling speed could facilitate detection. In view of the preliminary nature of the work, and the small sample size (n = 3 lizards), I have not undertaken formal statistical analyses, leaving the data in raw (and/or summarised) form.

13.3.1 Signal efficacy and increased plant motion noise

The simulations demonstrated that worsening environmental conditions reduced the probability of detection, although the exact fate of a given exemplar across the 26 plant movement conditions varied between the sites (Figure 13.3). When overlaid onto *B. integrifolia* all three tail-flick signals were very effective in calm conditions, but their effectiveness decreased as environmental conditions worsened, reaching (near to) zero at environmental condition 16 (Figure 13.3a, middle). Display signals fared a little better for all three exemplars (Figure 13.3a, right). These signals remained more effective in a greater range of conditions, and indeed displays by two of the three lizards remained moderately effective up to environmental condition 25. Signal efficacy was notably highest when overlaid against *L. parviflorus* (Figure 13.3b). Compared with *B. integrifolia*, there was a less rapid decline to zero for tail flick and display signals, although the display again appeared to be marginally more robust than tail flicks in the worst environmental conditions. Relative performance at the remaining sites suggested signal efficacy would be extremely compromised. In light to moderate conditions, less than half of each tail flick sequence attracted attention when viewed against *A. longifolia* (Figure 13.3c, middle); they were generally ineffective in more adverse conditions. Displays faired a little better, exhibiting a gradual decline in performance as conditions worsened (Figure 13.3c, right). The masking effect of plant motion appeared strongest when signals were viewed against *L. longifolia*. The probability of response showed only minor variation with changing environmental conditions for both tail flicks and displays (Figure 13.3d, middle and right respectively).

The frames at which detection was first achieved, relative to signal duration, partially mirrored response probability data (Figure 13.4). Early detection occurred for tail flicks and displays against *B. integrifolia* in calm to moderate conditions, with longer reaction times for both signal types as environmental conditions worsened (Figure 13.4a). Rapid detection of signals was found in most environmental conditions for signals viewed against *L. parviflorus* (Figure 13.4b). In contrast, reaction times for signals viewed against *A. longifolia* (Figure 13.4c) and *L. longifolia* (Figure 13.4d) again suggested that detection at these sites is generally more difficult; as before, displays faired better when viewed against *A. longifolia* with detection times relatively quick in calm conditions, gradually getting later with worsening environmental conditions.

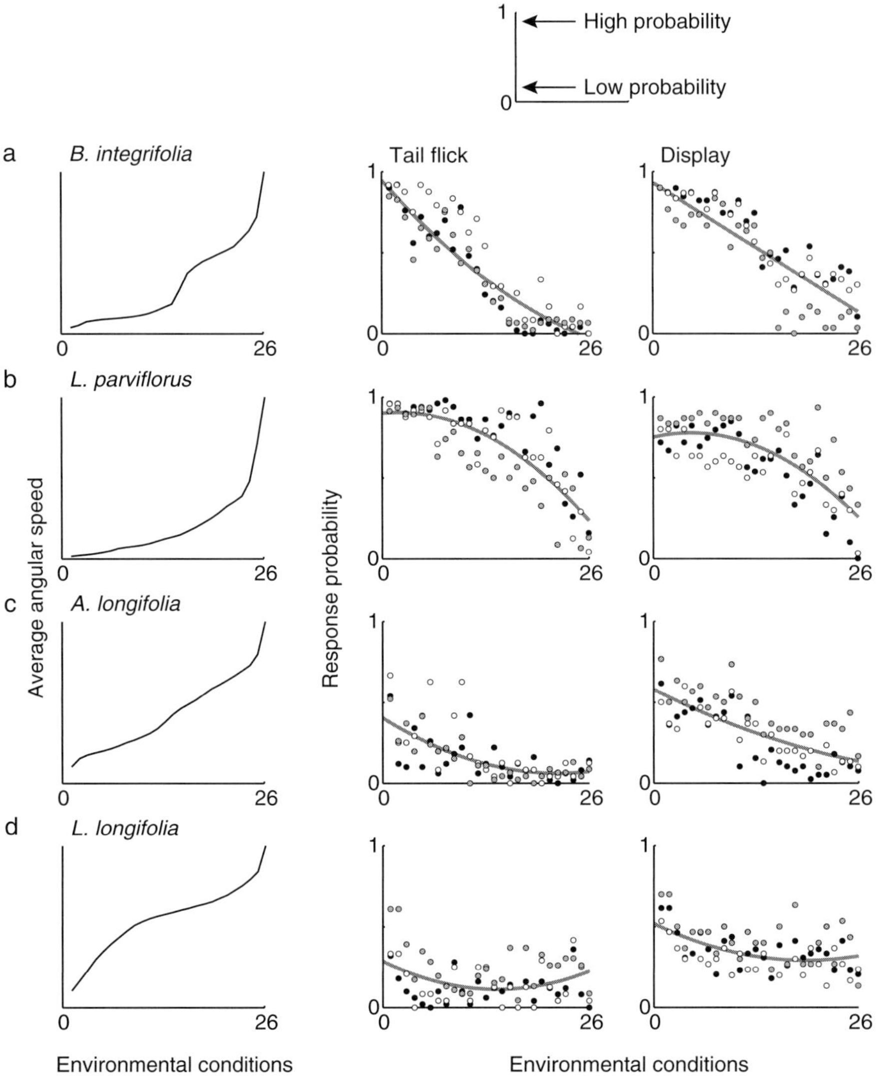

Figure 13.3. Summary of the relative efficacy of lizard signals performed at the same depth plane as plants as defined by response probabilities. Data are shown for each of the three lizards and by plant type: (a) *Banksia integrifolia*, (b) *Leucopogon parviflorus*, (c) *Acacia longifolia*, (d) *Lomandra longifolia*. Left column – Relative average angular speeds collapsed across space and time, of the 26 plant sequences and scaled within the range [0,1] by dividing each value by the maximum. Middle column – Probability of responding correctly to each frame of the tail-flick sequence for each lizard as a function of environmental conditions. Right column – Probability of responding correctly to each frame of the 'display' sequence for each lizard as a function of environmental conditions. Thick grey lines represent a quadratic fit to all data in a given plot.

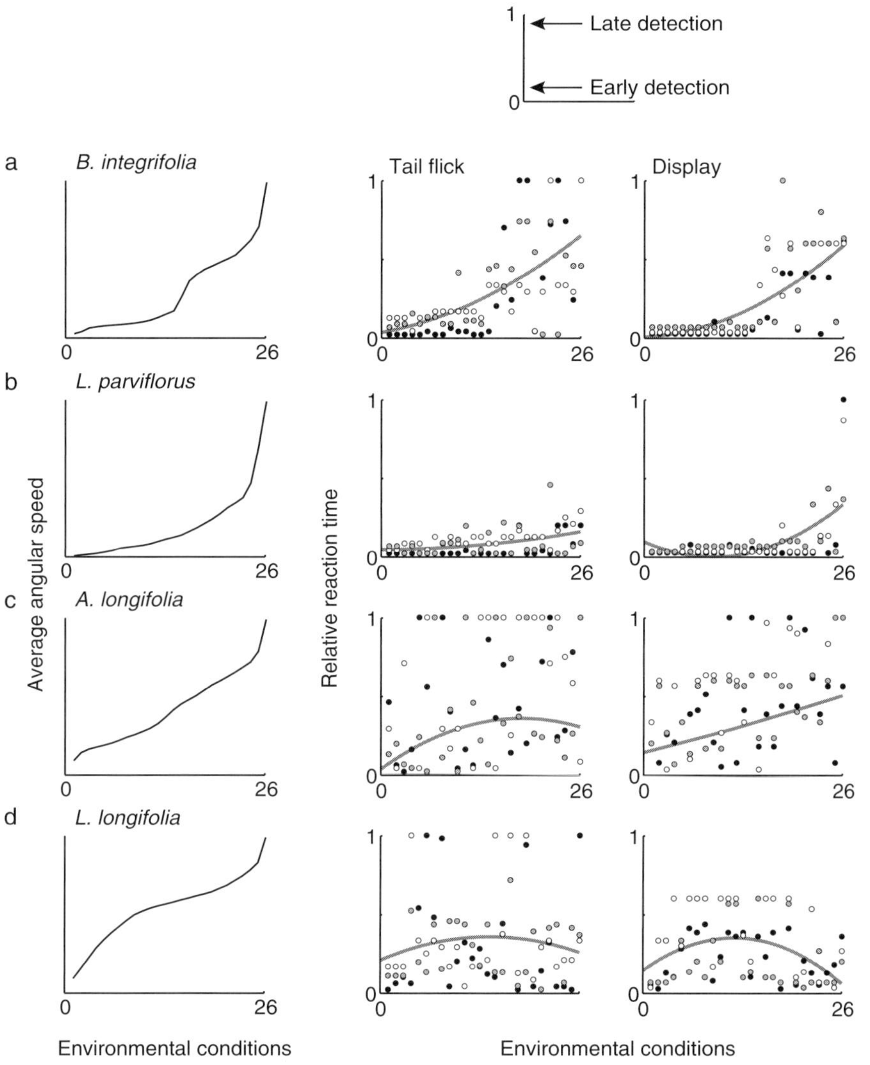

Figure 13.4. Summary of the relative efficacy of lizard signals performed at the same depth plane as plants as defined by reaction time. Data are shown for each of the three lizards and by plant type: (a) *Banksia integrifolia*, (b) *Leucopogon parviflorus*, (c) *Acacia longifolia*, (d) *Lomandra longifolia*. Left column – Relative average angular speeds collapsed across space and time, of the 26 plant sequences and scaled within the range [0,1] by dividing each value by the maximum. Middle column – Time to first correct detection of the lizard in noise relative to sequence duration for tail flicks, shown as a function of environmental conditions for each lizard. Right column – Time to first correct detection of the lizard in noise relative to sequence duration for 'displays', shown as a function of environmental conditions for each lizard. Thick grey lines represent a quadratic fit to all data in a given plot.

These simulations clearly indicated that the effectiveness of a given signal would depend critically on the environmental conditions at the time of signalling as well as the type of microhabitat in which the signal is performed. As suggested previously (see Peters *et al.*, 2008), the display portion of *A. muricatus*' signal appears, surprisingly, to be more robust to variation in signalling conditions than the introductory tail flicks that precede it. The spatiotemporal properties of signalling clearly overlap with those of wind-blown plants in more adverse conditions, regardless of motor pattern. One strategy to improve signal efficacy might be to signal faster (as discussed below), but another is to signal for longer. The exemplars for each signal type used here were short, in the order of 1–2 s duration. Within this time frame, signals featured rapid movements of multiple body parts (displays) or 1–3 flicks of the tail (tail flicking). This drastically under sampled typical occurrences in the case of tail flicking, which can extend to 120 s duration. Under natural circumstances, extremely variable wind conditions can shift the relative masking effect of plant motion from severe to negligible in a short period of time. Signallers could take advantage of this by extending signal duration, relying on environmental conditions to eventually become more conducive to reliable reception; longer duration signalling is indeed what *A. muricatus* lizards do in such circumstances (Peters *et al.*, 2007). Repeated flicks of the tail are presumably less costly than the repeated movements characterising displays in terms of energetic expenditure and increased conspicuousness to predators. I am currently investigating both of these issues in separate work.

13.3.2 Signal efficacy and signaller–plant distance

The effect of signaller–plant distance on response probability for each signal and plant combination is presented in Figure 13.5. Only three relative distances are shown for *L. parviflorus* (Figure 13.5b) and *L. longifolia* (Figure 13.5d). In order to scale the lizard to be twice the original size (see Section 13.2.1), a value greater than one was required that made the lizard too large for the frame. To facilitate inspection of the results, I have grouped environmental conditions into four categories: calm (conditions 1–7), light (8–13), breezy (14–19) and windy (20–26). The values shown are the mean (± SE) of the three lizard exemplars across the relevant environmental conditions. Comparing performance at the same depth plane as plants (averaged from Figure 13.3; filled circle, solid line), with each of the other profiles shows the benefit of moving away from plants. The results indicated, though not formally tested, an interaction between environmental conditions, plant microhabitat and signal type. First, no efficacy benefit for moving away from plants was observed for both signal types viewed against *L. parviflorus* (Figure 13.5b). Viewed against *B. integrifolia*, improvements are moderate for tail flicks in calm and light conditions, but more pronounced at greater distances in breezy and windy conditions (Figure 13.5a left). In contrast, there was little benefit for displays viewed against *B. integrifolia* (Figure 13.5b right). Both tail flicks and displays showed moderate improvements in all conditions when viewed against *A. longifolia*, showing incremental improvements with each distance (Figure 13.5c). Steady improvements were also found

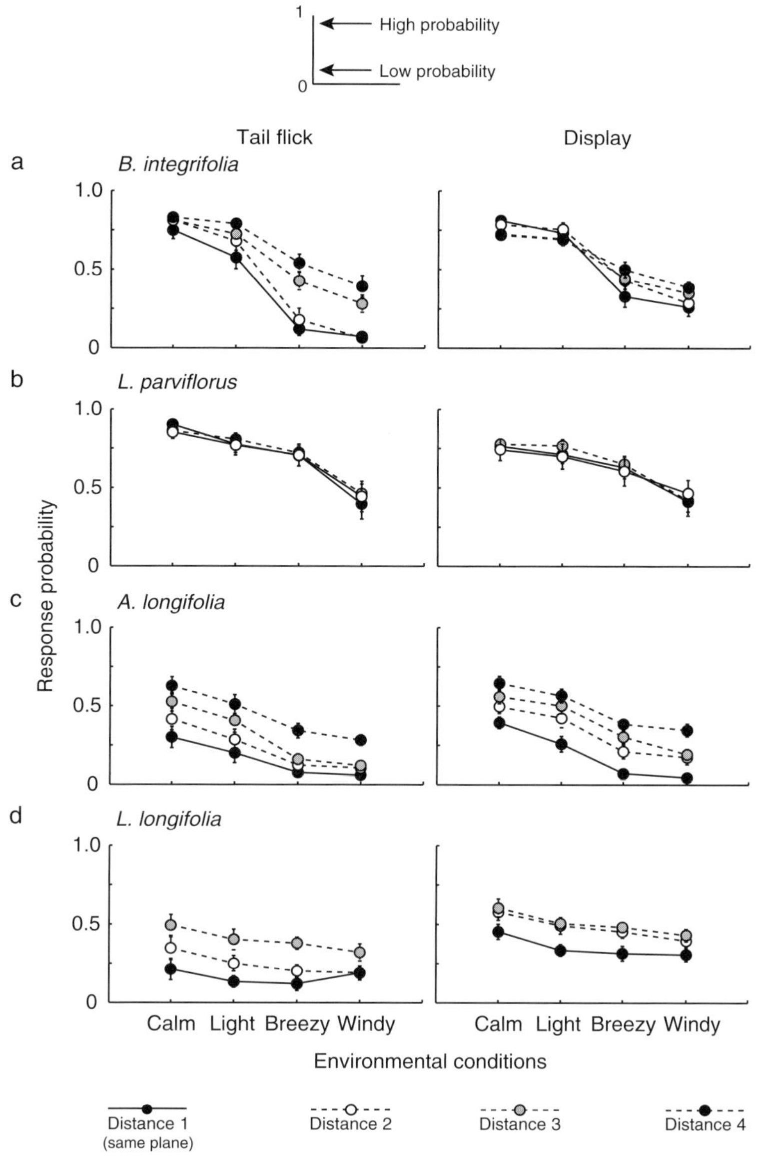

Figure 13.5. Mean (± SE) response probabilities as a function of environmental conditions grouped into four categories: 'calm' (environmental conditions 1–7), 'light' (8–13), 'breezy' (14–19) and 'windy' (20–26). Separate plots are shown for tail flicks (left) and displays (right) and for each plant species: (a) *Banksia integrifolia*, (b) *Leucopogon parviflorus*, (c) *Acacia longifolia*, (d) *Lomandra longifolia*. Separate lines represent performance at four signaller–plant distances starting from the same depth plane as plants (filled markers, solid line), followed by linearly increasing distances (dashed lines and marker as per legend). The fourth signaller–plant distance for *L. parviflorus* and *L. longifolia* was excluded from analysis (see text for details).

for tail flicks viewed against *L. longifolia* (Figure 13.5d left), although the benefit for displays of moving away from plants was only evident for the first new distance, with the next step resulting in no further improvement (Figure 13.5d right).

Changing signaller–plant distances also influenced relative reaction time (Figure 13.6), with the results being a mirror image of response probabilities. Signals that have relatively high response probabilities also seem to be detected quickly (e.g. compare Figure 13.6b and Figure 13.5b). One slight deviation from the respective performance data was observed for displays viewed against *B. integrifolia* (Figure 13.6a). Although response probabilities in the breezy and windy conditions did not change much with signaller–plant distance (Figure 13.5a), the reaction times showed a notable improvement (Figure 13.6a). Furthermore, although response probability to tail flicks viewed against *L. longifolia* improved with increasing signaller–plant distance (Figure 13.5d), particularly in calm and light conditions, this was not reflected in reaction times (Figure 13.6d), where the improvements were seen only in the move to the first new distance.

Obscured in Figure 13.5 was individual variation between lizards. Although I deliberately selected signals that were likely to be highly conspicuous, there were subtle differences between the relative performances of different exemplars. As an illustration, I have presented the difference between each new distance and the initial distance for each lizard in Figure 13.7, shown separately for tail flicks and displays viewed against *B. integrifolia*. While the tail flicks of two of the lizards showed general improvements across each condition (Figure 13.7b, A and C), improvements for the third were seen only in more adverse conditions (Figure 13.7b, B). Similarly, the display performed by lizard 3 (Figure 13.7c, C) showed reasonable improvements in adverse conditions, whereas the other two did not. Although the differences were not striking, I would expect that consideration of signals from a larger sample size would reveal greater individual variation. It may be possible, therefore, to identify which signal characteristics lead to an efficacy benefit from greater signaller–plant distances; alternatively, we might identify the characteristics that ensure a signal is robust to variation in location and environmental conditions.

13.3.3 Does signalling faster help?

The above simulations indicated that changing environmental conditions should affect signal efficacy. Adjustments to signalling strategies are necessary but contrasting strategies from different lizard species about the nature of these adjustments warrants further investigation. As a starting point, I make use of the modelling procedures described in this chapter to consider whether signalling faster makes displays more salient, particularly in adverse signalling conditions.

A new set of input sequences was generated that featured a modified signal overlaid onto a subset of the *B. integrifolia* sequences. I imported into Final Cut Express software (Apple Inc.) the tail flick and display sequences for one of the three exemplars. Using built-in software functions, I created replicas of the initial sequences that were twice as fast as the original. These new signal sequences were overlaid onto half of the *B. integrifolia*

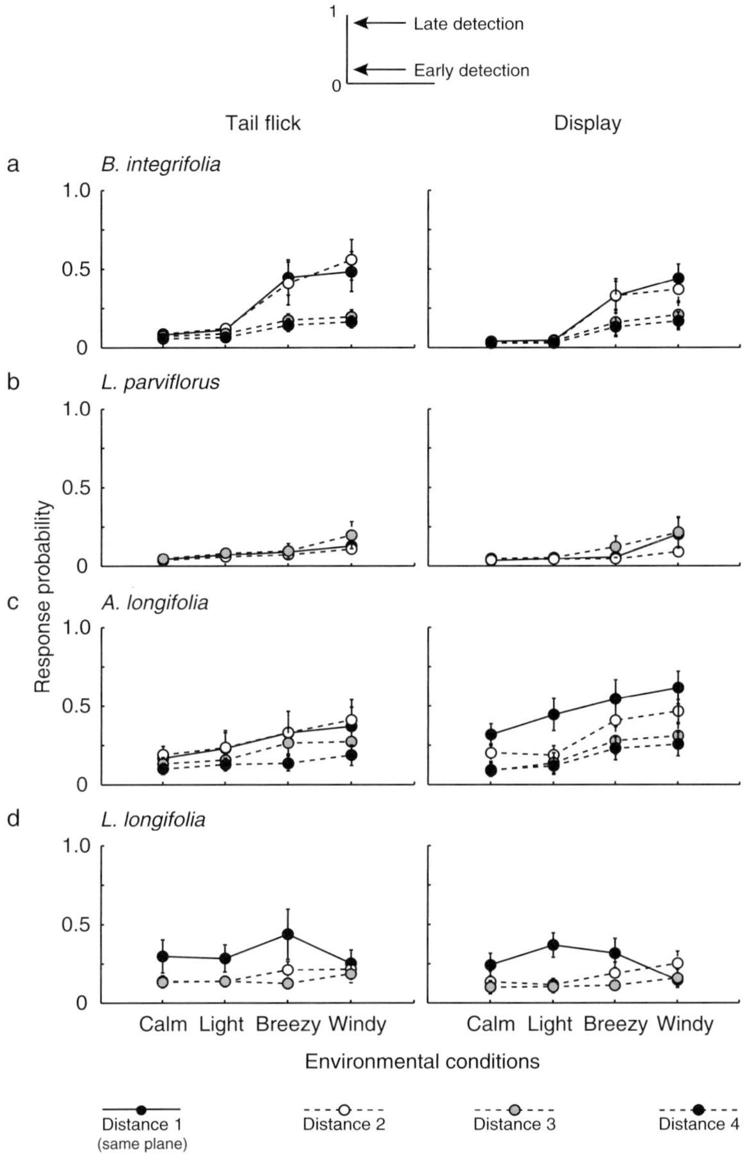

Figure 13.6. Mean (± SE) reaction times as a function of environmental conditions grouped into four categories: 'calm' (environmental conditions 1–7), 'light' (8–13), 'breezy' (14–19) and 'windy' (20–26). Separate plots are shown for tail flicks (left) and displays (right) and for each plant species: (a) *Banksia integrifolia*, (b) *Leucopogon parviflorus*, (c) *Acacia longifolia*, (d) *Lomandra longifolia*. Separate lines represent performance at four signaller–plant distances starting from the same depth plane as plants (filled markers, solid line), followed by linearly increasing distances (dashed lines and markers as per legend). Values represent time to first correct detection of the lizard in noise relative to sequence duration. The fourth signaller–plant distance for *L. parviflorus* and *L. longifolia* was excluded from analysis (see text for details).

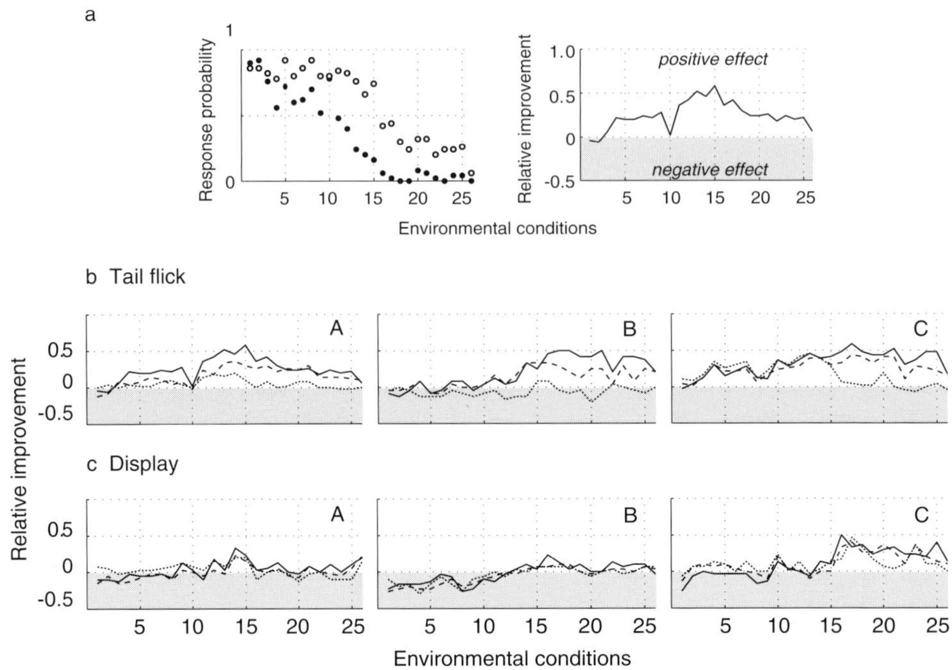

Figure 13.7. The effect on response probability of changing signaller–plant distance shown for each exemplar viewed against *Banksia integrifolia*. (a) Left – Response probability profiles at distance 1 (filled circles) and 4 (open circles) as a function of environmental conditions for one lizard, signal type and plant species. Right – Point-by-point subtraction of the two profiles shows the relative benefit of moving away from plants is greater in the mid-range of environmental conditions. Difference profiles for (b) tail flick and (c) displays shown separately for each lizard (A, B, C). Lines represent the relative effect of moving away from being at the same depth plane as plants: distance 2–1 (dotted line), distance 3–1 (dashed line) and distance 4–1 (solid line).

sequences (the odd-numbered environmental conditions). The faster signals, however, were also shorter in duration than their originals. Consequently, I created multiple signal + noise replicates (n = 6) within each environmental condition by shifting the starting position of the signal sequence within the plant sequence as shown in Figure 13.8a. As before, I varied the scale of the lizard sequences to simulate varying signaller–plant distances and created sequences with static backgrounds at each distance. Saliency analysis was carried out on the new set of 632 sequences in the same way as the preceding simulations.

The simulated efficacy benefits of faster signalling speeds are presented in Figure 13.8b–e, where the range of probabilities for the faster speed (shaded) are compared with outcomes at the original speed (closed circles). When signaller and plants were at equal depth planes (Figure 13.8b), modest improvements were observed for tail flicks under light to moderate environmental conditions, but there was no advantage to signalling faster in more adverse conditions. There was also no clear advantage for faster display speeds under all

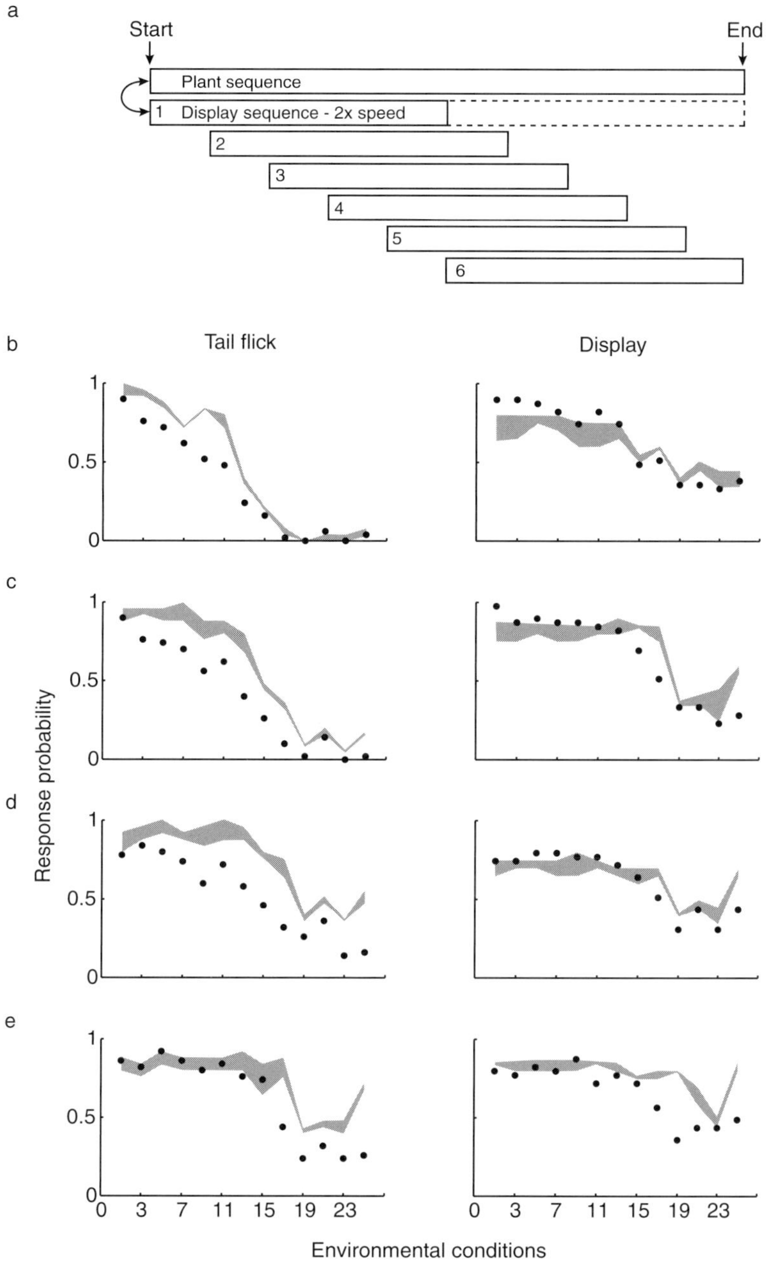

Figure 13.8. Results of simulations considering the efficacy benefit of faster signalling at the site featuring *Banksia longifolia*. (a) Display sequences were digitally modified to create sequences at twice the speed of original signals (see text for details). As the new, faster versions were half the duration of the original sequences, offsetting the starting position of the displays relative to the plant sequence created six replicate sequences spanning the original plant sequence. The effect of faster signalling speeds at each of the four signaller–plant distances, shown for the same depth plane as plants (b), followed by increasing distances 2, 3 and 4 (c–e respectively), is shown for tail flicks (left) and displays (right). A sub-sample of environmental conditions was used in these simulations represented by the odd numbered conditions (1–25). Closed circles represent response probabilities at the original speed, and the shaded region depicts the range of probabilities at the faster speeds across replicates and lizards.

conditions. The probability of detecting faster tail flicks increased relative to original speeds in most environmental conditions at the next two signaller–plant distances (Figure 13.8b, c), while the effectiveness of displays remained relatively unchanged. At the final signaller–plant distance (Figure 13.8d), the benefit of faster signalling is evident only in the more adverse range of signalling conditions for both tail flicks and displays.

These simulations offer a potential explanation for signal diversity. Signalling at the same depth plane as plants in adverse conditions, *A. muricatus* would not benefit from signalling faster regardless of the type of motor pattern. Longer duration signalling, possibly to wait for lulls in conditions that change the masking effect of plant motion, might be most beneficial; a light structure like a tail is more suited to this strategy than repeated whole-body signalling. *Anolis* lizards, on the other hand, are generally more separated from surrounding plants and may indeed be best served by increasing signalling speed, without needing to adjust signal duration. An exciting recent finding in *A. gundlachi* suggested that lizards achieve this by the facultative addition of a rapid series of 4-legged push-ups (Ord & Stamps, 2008).

13.4 General discussion and outlook

Models of signal evolution suggest that effective signals are ones that ensure reliable detection and efficient transfer of information (e.g. Endler & Basolo, 1998). Differences between microhabitats have been shown to contribute strongly to variations in signal structure across and within species (Slabbekoorn & Smith, 2002), while moment-to-moment changes in environmental conditions can prompt individual animals to modify aspects of their signals (Brumm & Todt, 2002). Crucial to our understanding of these systems has been knowledge of the sensory and brain mechanisms that govern perception, which show when and how the structure of signal and noise overlap. This type of knowledge is lacking for signals defined by movement. Recent findings suggested that movement-based signals are also adjusted in response to changing environmental conditions (Ord *et al.*, 2007; Peters *et al.*, 2007), but understanding the variation between species requires further investigations. As a starting point, the simulations presented in this chapter quantify the way movement-based animal signals and motion noise might overlap, where and when we can expect signals to diverge in response to sensory constraints, and offer explanations for signal diversity.

While I am confident that the modelling presented herein is informative and valuable, I recognise that I have taken a large-scale descriptive approach that does not consider the underlying complexity of motion vision. In many ways, it is likely that the complexity of the mechanisms underlying motion vision has restricted efforts within animal behaviour/behavioural ecology to understand the sensory constraints on movement-based signal design and evolution. Sensory mechanisms are important, of course, and will be of considerable use when understood in greater detail. Until this time, neural network models such as the approach presented herein can be informative without necessarily capturing the computations performed by the brain. The present simulations have provided quantitative predictions that can be tested empirically, for example, in controlled experiments in captivity and/or the field using mechanised ('robot') playbacks (e.g. Ord & Stamps, unpublished; Peters, 2008). The modelling presented herein has provided a way to refine our empirical strategies; in turn, I expect that the empirical data will help to refine the modelling approach. The present strategy permitted control over parameters believed to be important (plant type, wind conditions and spatial geometry), and as Phelps (2007) suggested, it now should be possible to uncouple the components of the prevailing theories of signal evolution, such as the Sensory Drive model (Endler & Basolo, 1998), as it pertains to signals of this type. There is also no reason that the present approach could not be utilised for understanding other related tasks such as detecting prey items or looming predators, and indeed, what happens when multiple functionally important motion cues are presented concurrently.

Regardless of the predictive utility in its present implementation, there are certainly areas to address in the future, either as modifications to the existing approach or with the adoption of a new one. The key issues at this stage concern how motion is represented in the input sequences and the suitability of saliency analysis in general. I discuss briefly both of these issues in the rest of this chapter.

Walther (2006) described a way of incorporating motion as a feature in the saliency model, making use of the multi-scale representation of the visual information (Gaussian pyramid). By the author's own admission, however, the suggested approach would have difficulty detecting small objects moving quickly. As this might have been relevant to the detection task I required of the model, I chose to calculate motion using gradient detectors independently of the saliency model. If included, however, I suspect that attention may be drawn to the lizard due to other features, irrespective of motion cues, because of the artificial nature of the composite sequences. Recorded under very different circumstances, the lizard and the perch stand out from the background in a manner that would not be the case in nature, where these lizards remain relatively cryptic (until motion destroys their camouflage). This is not to say that factors such as brightness and contrast are unimportant for motion signal detection. They are indeed important components of displays (Leal & Fleishman, 2002), which can enhance detection of moving objects (Persons *et al.*, 1999; Fleishman & Persons, 2001). By quantifying the velocity field first I ensured that attentional focus in my simulations was determined solely on motion cues. Of course there are alternative procedures for quantifying image motion, each having their relative strengths and weaknesses, as well as degrees of biological plausibility. For example, Borst (2007) has demonstrated that gradient detectors perform poorly in conditions characterised by low signal-to-noise ratios relative to alternatives such as Reichardt detectors. It is possible that my use of gradient detectors may have inadvertently contributed to the saliency model's poor performance in detecting the signal in adverse conditions. A comparison of motion analysis techniques is therefore warranted.

The second issue concerns the use of saliency analysis per se. There are certainly many neural network models for motion-based scene segmentation, based on optic flow fields (e.g. Cesmeli & Wang, 2000; Cirrincione & Cirrincione, 2003) or pattern recognition and object-tracking principles (e.g. Meier & Ngan, 1998). What models such as these did not offer, to my knowledge, was a convenient way to predict which regions of the segmented scene would attract attention. The notion of a salience map guiding the focus of attention has considerable support in the literature (see Fecteau & Munoz, 2006), and the availability of the Saliency Toolbox for Matlab (Walther & Koch, 2006) provided a suitable starting point for my investigations. Extensions to the saliency map approach highlight the importance of relevance (Fecteau & Munoz, 2006). Top-down influences such as behavioural relevance would be an intriguing addition to the current approach. There are, however, alternative approaches to location-based selection that do not make use of a saliency map. In one type of model, for example, attention is considered to be an emergent property of slow, competitive interactions that work in parallel across the visual field rather than a restricted region (see Desimone & Duncan, 1995). This idea has been implemented in neural network models using phase oscillators in which the attended location arises from synchronous oscillations among cell assemblies (Wu & Guo, 1999; see also Wang, 1999; Corchs & Deco, 2001).

In conclusion, the neural network modelling approach that I have adopted here is preliminary, but has helped to represent the task facing receivers in a changing environment. Importantly, it has begun to generate the kind of data needed to explain the diversity we see in movement-based animal signals. In view of the predictive utility, it represents an important improvement on my earlier effort (Peters & Davis, 2006). However, the approach does have room for improvement. Some of the issues to address would be relatively straightforward; other issues are more difficult to address and may represent a small challenge to researchers with a comprehensive background in computational neuroscience.

13.5 Appendix

A supplementary figure illustrating the processes involved in generating input sequences, as well as representative video clips used in the simulations described in this chapter are available online at http://richard.eriophora.com.au/simulations/efficacychapter.html.

Acknowledgements

This work was supported by funding from the Australian Research Council (DP0557018), with additional funding provided by the Centre for Visual Sciences at the Australian National University and the ARC Centre of Excellence in Vision Science. Thanks to Jan Hemmi and Jochen Zeil for useful discussions about aspects of the work described, and to Jan Hemmi, Ajay Narendra, Terry Ord and an anonymous referee for useful comments on an earlier draft of this chapter. I am also grateful to Colin Tosh and Graeme Ruxton for their invitation to submit a chapter for consideration in their book.

References

Adelson, E., Anderson, I., Bergen, I., Burt, I. & Ogden, I. 1984. Pyramid methods in image processing. *RCA Engineer* **29**, 33–41.

Borst, A. 2007. Correlation versus gradient type motion detectors: the pros and cons. *Phil Trans R Soc B* **362**, 369–374.

Brumm, H. & Todt, D. 2002. Noise-dependent song amplitude regulation in a territorial songbird. *Anim Behav* **63**, 891–897.

Cesmeli, E. & Wang, D. 2000. Motion segmentation based on motion/brightness integration and oscillatory correlation. *IEEE Trans Neural Netw* **11**, 935–947.

Cirrincione, G. & Cirrincione, M. 2003. A novel self-organizing neural network for motion segmentation. *Appl Intelligence* **18**, 27–35.

Cogger, H. G. 1996. *Reptiles and Amphibians of Australia*. Reed Books.

Corchs, S. & Deco, G. 2001. A neurodynamical model for selective visual attention using oscillators. *Neural Netw* **14**, 981–990.

Costermans, L. 2005. *Native Trees and Shrubs of South-eastern Australia*. Reed New Holland.

Desimone, R. & Duncan, J. 1995. Neural mechanisms of selective visual attention. *Annu Rev Neurosci* **18**, 193–222.

Endler, J. A. & Basolo, A. L. 1998. Sensory ecology, receiver biases and sexual selection. *Trends Ecol Evol* **13**, 415–420.

Fecteau, J. & Munoz, D. 2006. Salience, relevance, and firing: a priority map for target selection. *Trends Cogn Sci* **10**, 382–390.

Fleishman, L. J. 1986. Motion detection in the presence or absence of background motion in an *Anolis* lizard. *J Comp Physiol A* **159**, 711–720.

Fleishman, L. J. & Persons, M. 2001. The influence of stimulus and background colour on signal visibility in the lizard *Anolis cristatellus*. *J Exp Biol* **204**, 1559–1575.

Hannah, P., Palutikof, J. & Quine, C. 1995. Predicting windspeeds for forest areas in complex terrain. In *Wind and Trees* (ed. M. Coutts & J. Grace), pp. 113–129. Cambridge University Press.

Itti, L. & Koch, C. 2000. A saliency-based search mechanism for overt and covert shifts of visual attention. *Vision Res* **40**, 1489–1506.

Itti, L., Koch, C. & Niebur, E. 1998. A model of saliency-based visual attention for rapid scene analysis. *IEEE Trans Patt Anal Mach Intelligence* **20**, 1254–1259.

Koch, C. 1998. *Biophysics of Computation: Information Processing in Single Neurons*. Oxford University Press.

Koch, C. & Ullman, S. 1985. Shifts in selective visual attention: towards the underlying neural circuitry. *Human Neurobiology* **4**, 219–227.

Leal, M. & Fleishman, L. J. 2002. Evidence for habitat partitioning based on adaptation to environmental light in a pair of sympatric lizard species. *Proc R Soc Lond B* **269**, 351–359.

Meier, T. & Ngan, K. 1998. Automatic segmentation of moving objects for video object plane generation. *IEEE Trans Circuits Syst Video Technol* **8**, 525–538.

Ord, T. J., Peters, R. A., Clucas, B. & Stamps, J. 2007. Lizards speed up visual displays in noisy motion habitats. *Proc R Soc Lond B* **274**, 1057–1062.

Ord, T. J., Peters, R. A., Evans, C. S. & Taylor, A. J. 2002. Digital video playback and visual communication in lizards. *Anim Behav* **63**, 879–890.

Ord, T. J. & Stamps, J. A. 2008. Alert signals enhance animal communication in 'noisy' environments. *Proc Nat Acad Sci USA* **105**, 188300–188305.

Persons, M. H., Fleishman, L. J., Frye, M. A. & Stimphil, M. E. 1999. Sensory response patterns and the evolution of visual signal design in Anoline lizards. *J Comp Physiol A* **184**, 585–607.

Peters, R. 2008. Environmental motion delays the detection of movement-based signals. *Biol Lett* **4**, 2–5.

Peters, R. A., Hemmi, J. M. & Zeil, J. 2007. Signalling against the wind: modifying motion signal structure in response to increased noise. *Curr Biol* **17**, 1231–1234.

Peters, R. A., Hemmi, J. M. & Zeil, J. 2008. Image motion environments: background noise for movement-based animal signals. *J Comp Physiol A* **194**, 441–456.

Peters, R. A., Clifford, C. W. G. & Evans, C. S. 2002. Measuring the structure of dynamic visual signals. *Anim Behav* **64**, 131–146.

Peters, R. A. & Davis, C. J. 2006. Discriminating signal from noise: recognition of a movement-based animal display by artificial neural networks. *Behav Process* **72**, 52–64.

Peters, R. A. & Evans, C. S. 2003. Design of the Jacky dragon visual display: signal and noise characteristics in a complex moving environment. *J Comp Physiol A* **189**, 447–459.

Peters, R. A. & Ord, T. J. 2003. Display response of the Jacky dragon, *Amphibolurus muricatus* (Lacertilia: Agamidae), to intruders: a semi-Markovian process. *Austral Ecol* **28**, 499–506.

Phelps, S. M. 2007. Sensory ecology and perceptual allocation: new prospects for neural networks. *Phil Trans R Soc B* **362**, 355–367.

Slabbekoorn, H. & Smith, T. 2002. Habitat-dependent song divergence in the little greenbul: an analysis of environmental selection pressures on acoustic signals. *Evolution* **56**, 1849–1858.

Walther, D. 2006. *Interactions of Visual Attention and Object Recognition: Computational Modeling, Algorithms, and Psychophysics*, pp. 147. PhD thesis. California Institute of Technology.

Walther, D. & Koch, C. 2006. Modeling attention to salient proto-objects. *Neural Netw* **19**, 1395–1407.

Wang, D. 1999. Object selection based on oscillatory correlation. *Neural Netw* **12**, 579–592.

Wood, C. 1995. Understanding wind forces on trees. In *Wind and Trees* (ed. M. Coutts & J. Grace), pp. 133–164. Cambridge University Press.

Wu, Z. & Guo, A. 1999. Selective visual attention in a neurocomputational model of phase oscillators. *Biol Cybern* **80**, 205–214.

Part IV

Methodological issues in the use of simple feedforward networks

14

How training and testing histories affect generalisation: a test of simple neural networks

Stefano Ghirlanda and Magnus Enquist

14.1 Introduction

This paper deals with a general issue in the study of animal behaviour that we call *path dependence*. The expression refers to the fact that different histories of experiences (*paths*) may at first seem to produce the same behavioural effects yet reveal important differences when further examined. For instance, two training procedures may establish the same discrimination between two stimuli yet produce different responding to other stimuli, because the two paths have produced different internal states within the animal. There are several reasons why path dependence is an important issue. First, it comprises many phenomena that can provide stringent tests for theories of behaviour. Second, path dependence is at the root of several controversies, for instance whether animals encode absolute or relative characteristics of stimuli (Spence, 1936; Helson, 1964; Thomas, 1993) or whether learning phenomena such as backward blocking and un-overshadowing imply, in addition to basic associative learning, stimulus–stimulus associations or changes in stimulus associability (Wasserman & Berglan, 1998; Le Pelley & McLaren, 2003; Ghirlanda, 2005).

In this paper we use a simple neural network model of basic associative learning (Blough, 1975; Enquist & Ghirlanda, 2005) to show how path dependence can arise from fundamental properties of associative memory. The model has two core components: (1) distributed representations of stimuli based on knowledge of sensory processes and (2) a simple learning mechanism that can associate stimulus representations with responses. We consider examples of path dependence in experiments on generalisation (or 'stimulus control'). These consist of a training phase in which animals are trained to perform a specific response to several stimuli, and a test phase in which responding to a set of stimuli is recorded. The test stimuli often lie on a 'stimulus dimension' such as light wavelength or object size, so that generalisation is often described as a response gradient over the dimension. In generalisation experiments, path dependence appears as differences in the shape of generalisation gradients; different paths correspond to different training or testing procedures. We show that the model accounts for the following

Modelling Perception with Artificial Neural Networks, ed. C. R. Tosh and G. D. Ruxton. Published by Cambridge University Press.

phenomena (see below for details): lack of peak shift after 'errorless discrimination learning', decrease of peak shift during extinction testing, and the shift of generalisation gradients toward the average of the test stimuli (a kind of range effect).

Although we consider laboratory experiments, it is hardly necessary to note that path dependence exists in the wild as well, where paths are a consequence of environmental events rather than being arranged by an experimentalist.

14.2 Model

14.2.1 The neural network

A nervous system can be seen as a flexible structure that can be programmed to generate almost any behaviour, e.g. relationships between stimuli and responses. Concretely, such programming includes both the formation of the neural network and its pattern of connectivity and the adjustment of connections between cells in the network. Neural network models provide an understanding of how such processes can ultimately produce the behaviour we see in animals (Arbib, 2003; Enquist & Ghirlanda, 2005). Previous work has shown that simple neural network models of associative learning can reproduce many fundamental findings of learning and generalisation (Blough, 1975; Ghirlanda, 2005; Enquist & Ghirlanda, 2005). Here we use a standard feedforward network with one array of input nodes connected directly to one output node (there are no hidden nodes). Stimuli are modelled as eliciting graded patterns of activity in the array of input nodes. We write Si the activity induced in input node i by stimulus S ($i = 1,\ldots, N$). The input nodes are connected to the output node by weighted connections, the weight attached to node i being W_i. The strength or likelihood of responding to S is assumed to be an increasing function of the weighted sum r_s:

$$r_s = \sum_i w_i S_i \tag{1}$$

14.2.2 Learning

We model learning by the so-called δ rule (McClelland & Rumelhart, 1985), a simple case of gradient descent algorithms (Haykin, 1999) that was first derived by Widrow & Hoff (1960) and introduced to animal learning theory by Blough (1975), based on previous work by Rescorla & Wagner (1972). At each stimulus presentation, the algorithm prescribed as change ΔW_i in weight W_i according to

$$\Delta W_i = a(\lambda - r_S)S_i \tag{2}$$

where λ is the maximum value that responding to S can attain given the applied reinforcer, and a mainly regulates the speed of learning (Widrow & Stearns, 1985). Equation (2) is capable, through repeated applications, of establishing a different response to each of many stimuli, provided the corresponding patterns of activity satisfy certain requirements ('linear separability'). We refer to the literature for technical details and other applications to behaviour theory (Blough, 1975; Widrow & Stearns, 1985; McClelland & Rumelhart, 1986; Rumelhart & McClelland, 1986; Haykin, 1999; Enquist & Ghirlanda, 2005). Weights are assumed to start from a value of zero (drawing weights at random from a distribution symmetrical around zero would lead to the same conclusions).

14.2.3 Model stimuli and generalisation tests

The results below do not depend on the precise details of how stimuli are modelled, as long as the following general properties hold: (1) input node activity is positive; (2) each stimulus corresponds to a graded pattern of activity in input nodes; (3) physically more similar stimuli correspond to more similar patterns of activity; (4) higher intensity of stimulation corresponds to higher input node activity. Figure 14.1 illustrates the kind of model stimuli we use in practice. Dimensions such as sound frequency, light wavelength or spatial position can be modelled by translating an activity profile like the one in Figure 14.1b over the input array of the network. This simple scheme captures the empirical observation that, a stimulus change along these dimensions causes a change in the pattern of receptor activity but not in total

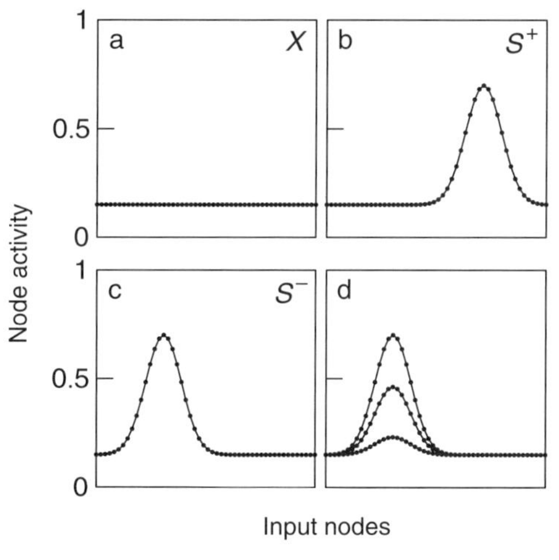

Figure 14.1. Examples of network input patterns used in this paper. The network input nodes are considered as a one-dimensional sense organ and each panel depicts the pattern of activity corresponding to particular stimuli. (a) Background stimuli are modelled as causing low activity in all input nodes (e.g. a dimly illuminated Skinner box); (b) a particular stimulus, say S^+, causes the activity of some input nodes to rise above background levels; different nodes are affected to different extents. (c) A distinct stimulus, say S^-, produces a pattern of activity of similar shape, but which peaks at a different position of the input array; (d) variation in stimulus intensity is modelled, following sensory physiology, by keeping the same pattern of activity but raising or lowering its overall level. See Appendix for details.

activity. Conversely, stimulus intensity dimensions are modelled by increasing or decreasing input node activity without changing which input nodes are stimulated (Figure 14.1d). We refer to Ghirlanda & Enquist (2003) and Enquist & Ghirlanda (2005) for further details. Once a stimulus dimension has been denned, a generalisation test is modelled simply by presenting the network with some stimuli from the dimension and recording the corresponding network output.

14.2.4 Network analysis

Neural networks can be analysed through a variety of tools (Haykin, 1999), such as formal mathematics, computer simulations and visualisation techniques that highlight some aspects of network organisation and functioning. In this paper we simulate the training and testing phases of generalisation experiments and we analyse the resulting network by plotting the weight array. The latter will usually contain both positive and negative weights, and the output to a particular stimulus will depend on the degree to which the corresponding pattern of activity overlaps with positive and negative weights. This, as we shall see, allows us to understand how the network responds to stimuli based on graphical representations of weight arrays and stimulus activity patterns.

14.3 Results

14.3.1 Errorless discrimination learning

The first step in most experimental studies of generalisation is to train animals to discriminate between several stimuli. The experimenter wishes, for instance, to have animals respond to a given stimulus, called S^+, and to ignore another one, called S^-. This may be achieved by instrumental conditioning, whereby responses to S^+ are 'reinforced', e.g. with food, whereas responses to S^- are not (Pearce, 1997). Details of training may vary, but it is most common to first train the desired response to S^+ and then to introduce S^-. The animal will usually respond to S^- in the first stages of discrimination training (especially if S^- is similar to S^+) but if such responses are never reinforced the animal will respond less and less to S^-. In practice, discrimination training is continued until a criterion is met such as 'three times more responding to S^+ than to S^-' or 'no responses to S^- in a 10-minute period'.

When a generalisation test is performed after discrimination learning, one may find that the stimulus that elicits most responses is not S^+, but a stimulus that is displaced away from S^+ so as to be more different from S^- (Figure 14.2, dotted line in left panel). Since its discovery by Hanson (1959), this phenomenon, called the 'peak shift', has fuelled extensive research to understand how a stimulus that was never reinforced (and, in many cases, never experienced) could be more powerful in eliciting a response than a reinforced stimulus (Mackintosh, 1974). It is known at least since the pioneering work of Blough (1975) that simple network models can reproduce the peak shift phenomenon. We

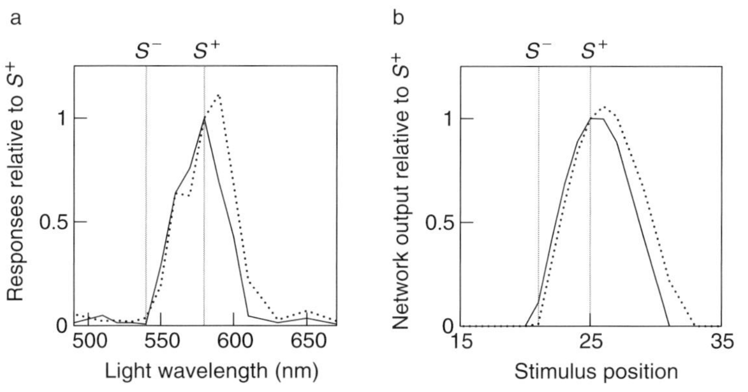

Figure 14.2. 'Errorless' discrimination learning. (a) Generalisation gradients from two groups of pigeons trained to solve a discrimination between monochromatic lights with a standard procedure (dotted line) or 'errorless learning' (see text; data from Terrace, 1964). Only the gradient from the former group shows a peak shift. (b) A neural network simulation of generalisation after standard (dotted line) vs. 'errorless' (continuous line) learning yields similar results.

will briefly review the mechanism below, referring to Enquist & Ghirlanda (2005) for further discussion. The dotted line in the right panel of Figure 14.2 shows a peak shift obtained from a neural network simulation of a discrimination experiment. For technical details regarding this and all other simulations in the paper we refer to the Appendix.

The left panel in Figure 14.2 also plots a generalisation gradient without a peak shift (continuous line), although the same S^+ and S^- have been used in training. The difference is that training did not follow the standard procedure outlined above but an alternative one, 'errorless discrimination learning', developed by Herbert Terrace at the beginning of the 1960s. In errorless discrimination learning S^- is introduced gradually rather than abruptly. In the experiment in Figure 14.2, for instance, Terrace (1964) trained a discrimination between a 580 nm monochromatic light (S^+) and a 540 nm light (S^-) beginning with a very faint 580 nm light, whose intensity was progressively increased until an intensity equal to S^+ is reached. The name 'errorless learning' derives from the fact the animal responds very little to S^- throughout training. Intuitively, this happens because, at any given moment, S^- is very similar to previously unreinforced stimuli (including, at the start of training, the experimental background) and thus has only a small probability of eliciting a response. Additionally, initial S^- presentations are so brief that the animal is effectively prevented from responding.

Terrace's finding that errorless learning prevents the peak shift has been replicated a few times but no agreement exists as to its causes (Purtle, 1973). Our aim is to explore what insight can be gained by simulating errorless learning with neural networks. We mimic Terrace's procedure by starting with a model S^- of low intensity (low activation of network input nodes) and progressively increasing its intensity, as shown for instance in Figure 14.1d. The resulting generalisation gradient (continuous line in Figure 14.2, right) peaks on

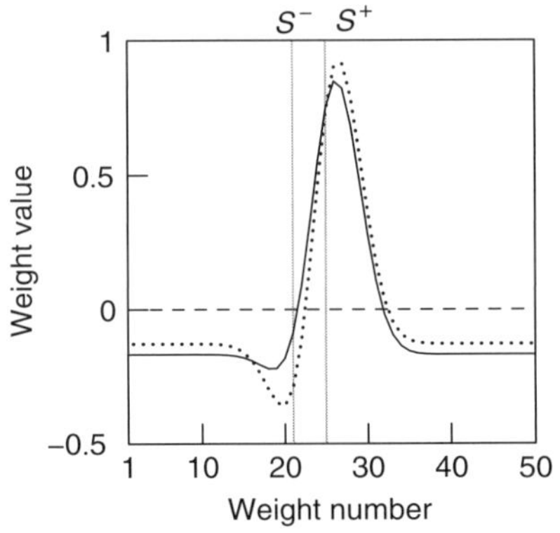

Figure 14.3. Network weights developed after simulations errorless discrimination training (continuous line) or standard discrimination training (dotted line). These weights produce the generalisation gradients in Figure 14.2b.

S^+ in agreement with Terrace's empirical result. To understand why this happens, we plot in Figure 14.3 the weight values obtained after both standard and errorless training. After standard training (dotted line) both positive and negative weight develop, associated respectively with the parts of the input array most activated by S^+ and S^-. When peak shift occurs, maximum responding is observed for stimuli that are close to S^+, but more distant from S^- than S^+ itself. Such stimuli retain most of S^+'s ability to excite nodes with positive weight while activating nodes with negative weights significantly less, which results in a more favourable balance between excitation and inhibition. During errorless learning, on the other hand, the input nodes most stimulated by S^- develop very small negative weights (continuous line). Thus the gains of departing from S^- cannot offset the losses caused by departing from $S+$. Interestingly, this explanation is consistent with Terrace's suggestion that errorless learning results in little inhibition being associated with S^-.

The reason why the weights develop as shown in Figure 14.3 can be understood by imagining what happens in the initial phases of training. The low intensity S^- used at the start of errorless learning is very similar to background stimuli to which responding is low (e.g. the dark response key in Terrace's experiment). To ensure that responding to such an S^- be low, therefore, it is sufficient to adjust the weights a little. On the other hand, at the beginning of standard training S^- is an intense stimulus not unlike S^+ and thus produces 'errors' in the form of a high network output, while the desired response is a low output. The learning algorithm must thus decrease responding to S^- considerably, which is achieved by attaching negative weight to input nodes most stimulated by S^-.

14.3.2 Disappearance of peak shift in extinction

The outcome of a generalisation test can be affected by different aspects of the testing procedure, e.g. its duration and what stimuli are used. The reason is, of course, that

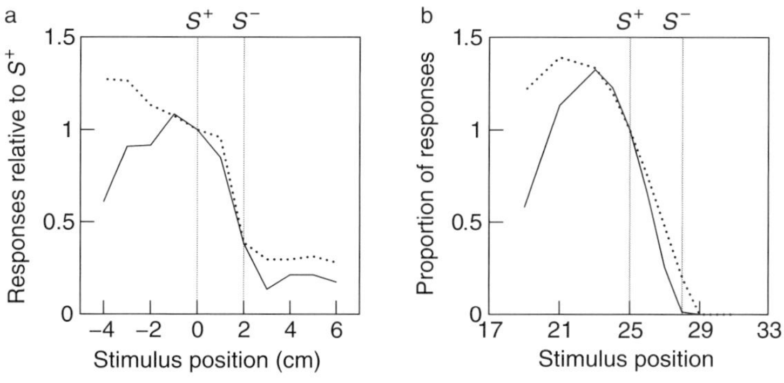

Figure 14.4. Disappearance of peak shift during generalisation testing in extinction in animal data (a, from Cheng *et al.*, 1997, experiment III; pigeons were trained to discriminate two small squares 2 cm apart on a computer screen, and tested with stimuli varying in horizontal position) and in the network model (b). Dotted lines represent the gradient just after training; continuous lines the gradient after testing. The empirical gradients are built using, respectively, the first and last few responses to each stimulus during testing in extinction (blocks 1 and 4 in Cheng *et al.*, 1997). Simulation parameters have been set to approximate the empirical post-discrimination gradient (dotted line in a), characterised by about three times more responding to S^+ than to S^-.

animals continue to learn during a test. Thus test results are not, as one would like, simply the result of probing the animal, but are partly due to learning caused by probing itself. The most common testing paradigm is testing 'in extinction', i.e. by unreinforced presentation of test stimuli. This causes a generalised decrease in responding and can also change the shape of the generalisation gradient. An interesting finding that we consider here is the reduction of peak shift during testing in extinction (Figure 14.4a; see also Purtle, 1973).

To model this finding we teach the network a discrimination between S^+ and S^-, then run a first generalisation test, which shows a peak shift (Figure 14.4b, dotted line). We then continue to test mimicking the extinction procedure, i.e. we apply the δ rule after each stimulus presentation with a low target value (low A in Eq. 2). In the generalisation gradient produced after many such presentations we find a greatly reduced peak shift. Network weights at the beginning and end of extinction testing are shown in Figure 14.5, where it is apparent that extinction testing has reduced the difference between positive and negative weights that underlies peak shift. Testing in extinction has also reduced the absolute values of the weights, which results in a general decrease in network output that parallels the decrease in responding as observed in experiments.

14.3.3 Range and frequency effects

Any set of stimuli may be, in principle, used in a generalisation test. It is most common to use a range of evenly spaced stimuli roughly centred around the training ones, with the

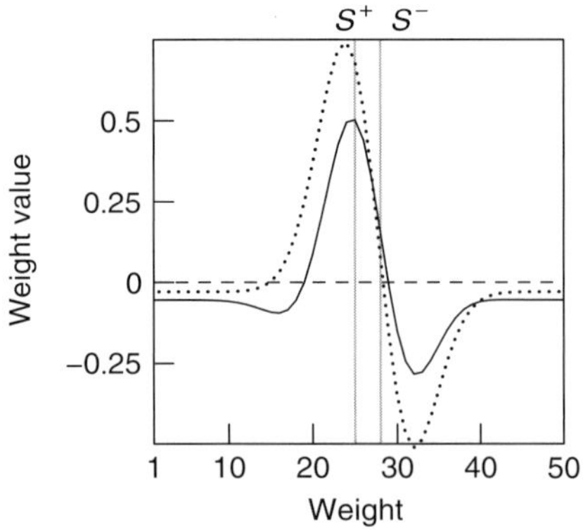

Figure 14.5. Weights at the beginning (dotted line) and end (continuous line) of generalisation testing in extinction. These two sets of weights produce the generalisation gradients in Figure 14.4b.

aim of getting an unbiased picture of generalisation (with only partial success, as seen above). Different kinds of tests, however, have been used specifically to study how generalisation is affected by post-training experiences. The most common manipulations include presenting some test stimuli more often than others and using only stimuli within a restricted range (reviewed in Thomas *et al.*, 1992). The changes in generalisation gradients brought about by such procedures may be collectively labelled 'range and frequency effects' and have generated considerable debate about underlying memory processes (Spence, 1936; Helson, 1964; Parducci, 1965; Thomas, 1993; Sarris, 2003).

One common finding is that extensive testing causes a 'central tendency effect' whereby the generalisation gradient appears shifted toward the middle of the stimulus range used in the test (Figure 14.6a; reviewed by Thomas *et al.*, 1992). It is easy to test the network in these conditions, simply running tests in extinction with different ranges of stimuli. The outcome of such tests is indeed a central tendency effect (Figure 14.6b). Figure 14.7 shows how testing with different ranges of stimuli modifies the weight array: probing with a particular stimulus range causes a shift in the weight array toward the middle of the range.

Our results suggest that at least some range and frequency effects may arise from simple mechanisms of associative learning, while current thinking often appeals to more complex memory processes (cf. 'adaptation level' theory and 'frequency-range' theory; Helson, 1964; Parducci, 1965; Thomas *et al.*, 1992). The need for such additional processes is partly inferred from the belief that range and frequency effects are virtually absent in non-human animals, although it is possible to find examples in the animal literature (reviewed by Thomas, 1993; Sarris, 2003). Perhaps the relative ease with which range and frequency effects appear in humans may just follow from the fact that humans learn faster (i.e. the effect of testing is seen even in relatively short tests).

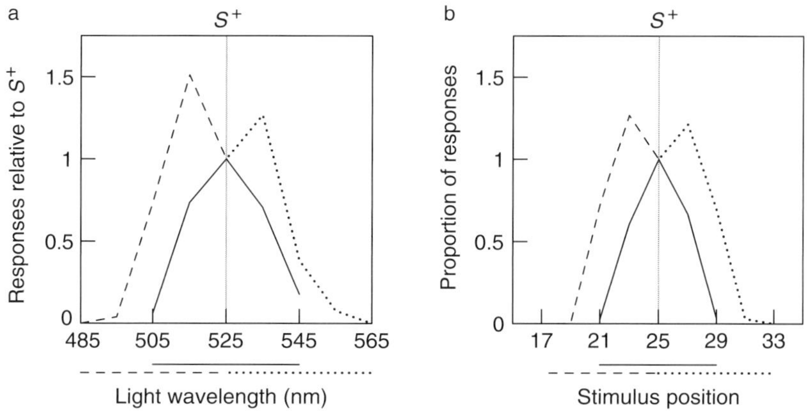

Figure 14.6. Central tendency effect. (a) Generalisation gradients obtained in extinction with different ranges of test stimuli, indicated by the lines below the graph in matching style, after identical training to respond to S^+. The gradient peak appears displaced toward the centre of the test range (data from Thomas & Jones, 1962, asking humans to identify a 525 nm light). (b) Network simulation of the same experiment, showing a similar central tendency effect.

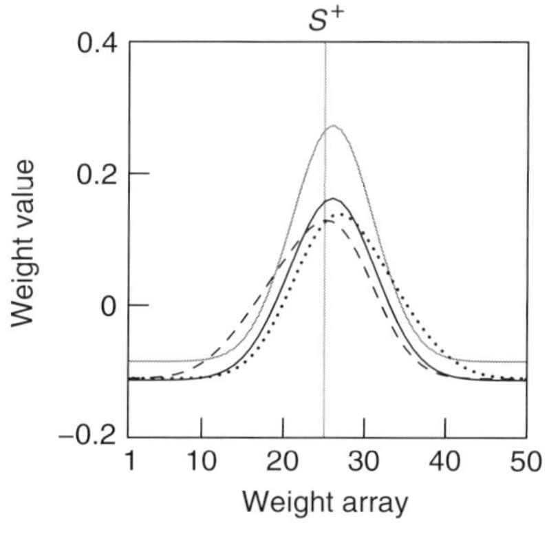

Figure 14.7. Weight arrays for the simulations in Figure 14.6b. The grey line represents weights just after training, the other lines shows how weights have changed after testing in extinction with three stimulus ranges. In addition to a general decline cause by the extinction procedure we see a shift in the pattern of weights such that the largest weights move toward the centre of the probed stimulus range.

14.4 Discussion

In this exploratory study we have used simple neural networks to test the hypothesis that path dependence phenomena arise from basic mechanisms of associative learning in distributed memory systems. We have shown that a simple network model of learning can reproduce three particular findings: the effect of errorless learning and of extinction testing on peak shift and the central tendency effect. We chose to consider these findings for several reasons. A first one is that they provide a true test of neural network models, which were developed to account for different phenomena. We stress that we have used, without modification, a very basic model, essentially Blough's (1975) model with the addition that stimulus representations be built with knowledge of relevant sensory processes (Ghirlanda & Enquist, 1999; Enquist & Ghirlanda, 2005).

A second reason is that the considered phenomena have been known for many decades, yet their theory is still unsatisfactory. The effects of errorless learning and extinction testing have been repeatedly considered in the peak shift literature (Purtle, 1973; Mackintosh, 1974), but theory is limited to verbal arguments such that animals may learn 'from the experience of being tested' (Prokasy & Hall, 1963, quoted by Purtle, 1973). Range and frequency effects have received considerable attention (Thomas *et al.*, 1992; Thomas, 1993; Sarris, 2003) but the extent to which they can be accounted for in terms of simple associative learning is still unknown. The main theoretical difficulty posed by all these phenomena, and by path dependence in general, is that they require us to track the cumulative effect of sequences of experiences with many stimuli. This is difficult in most models (to put it mildly) and reveals one crucial advantage of neural networks: the ability to simulate *arbitrary sequences of experiences* and to get predictions about responding to *any stimulus that can be received.* Simple neural networks are also amenable, in some cases, to mathematical analysis, although we have not pursued this approach here (see Haykin, 1999; Enquist & Ghirlanda, 2005).

In conclusion, neural networks provide a very natural framework to study path dependence and thus increase our knowledge of how experiences shape behaviour. Neural network models are already a promising account for a large body of behavioural phenomena (Enquist & Ghirlanda, 2005), and including path dependence would contribute to a unified picture of how nervous systems bring about behaviour.

14.5 Appendix: Simulation details

14.5.1 Stimuli

Stimuli are modelled as Gaussian activity profiles raising over a baseline activity level (Figure 14.1). If we consider a stimulus of intensity I and value x along a dimension such as sound frequency, spatial position or light wavelength, the activity of input node i is

$$S_i(I,x) = b + Ie^{-(x-x_i)^2/2}\sigma_i^2 \tag{3}$$

where b is a baseline activity, x_i is the stimulus value to which input node i is most sensitive and σ_i regulates how quickly node activity drops when x departs from xi. According to Eq. 3, $S_i(I,x)$ can range from a minimum value of b (when x is very different from x_i) to a maximum of $b+I$ (when $x=x_i$). The relationship between physical intensity I and receptor activation is usually nonlinear in nervous systems, but it is not necessary to take this into account here (often the relationship is approximately linear if intensity is

Table 14.1. *Values of stimulus parameters used in simulations (Eq. 3).*

Simulation	b	I	σ
Errorless learning			
$S+$ and standard group S^-	0.2	0.5	3
Errorless group S^-	0.2	$0.5(t/T)^{5\dagger}$	3
Extinction testing	0.2	0.5	4
Central tendency	0.2	0.65	5

† T is the length of discrimination training.

measure logarithmically, such as in dB for sound). We also assume $x_i = i$. This means that receptor sensitivity spans evenly the physical dimension in consideration (or the part of it probed in an experiment) and is an approximation to what happens in most nervous systems. We refer to Enquist & Ghirlanda (2005) for further details on stimulus modelling.

The exact values of stimulus parameters used in simulations are given in Table 14.1. These values have been chosen to approximate the particular empirical gradients that we decided to model, but the general phenomena (e.g. that peak shift recedes during extinction testing) are rather insensitive to exact parameter values. It is unsurprising that different parameter values had to be used since the simulations refer to different stimulus modalities and species. An interesting potential development, to be pursued in further research, would be to tie simulation parameters more firmly in sensory physiology.

14.5.2 Training and testing

All experiments considered here establish an initial discrimination between S^+ and its absence, or background stimuli (modelled as in Figure 14.la). We include such a training in our simulations, followed in the simulations on errorless learning and decrease of peak shift in testing by discrimination training between S^+ and S^-. The latter also included nonreinforced presentation of background stimuli. We used $\lambda = 1$ for reinforced stimuli and $\lambda = 0.05$ for nonreinforced ones. The only exception to the latter was the simulation in Figure 14.4, for which $\lambda = 0.2$ was used to approximate the actual discrimination ratio observed in the experiment.

The length of each stage of discrimination training was set to 5000 iterations of Eq. 2 in the text (with $\alpha = 0.05$). In each iteration one of the stimuli to be learned was presented to the network. This ensured that network output to training stimuli was within a few per cent of the required output. Extinction testing was simulated as 5000 nonreinforced stimulus presentations; this was enough to yield significant gradient changes but not so much as to cause complete extinction of responding.

References

Arbib, M. A. 2003. *The Handbook of Brain Theory and Neural Networks*. 2nd Edn. MIT Press.

Blough, D. S. 1975. Steady state data and a quantitative model of operant generalization and discrimination. *J Exp Psychol Anim Behav Process* **104**(1), 3–21.

Cheng, K., Spetch, M. L. & Johnson, M. 1997. Spatial peak shift and generalization in pigeons. *J Exp Psychol Anim Behav Process* **23**(4), 469–481.

Enquist, M. & Ghirlanda, S. 2005. *Neural Networks and Animal Behavior*. Princeton University Press.

Ghirlanda, S. 2005. Retrospective revaluation as simple associative learning. *J Exp Psychol Anim Behav Process 31, 107–111.*

Ghirlanda, S. & Enquist, M. 1999. The geometry of stimulus control. *Anim Behav* **58**, 695–706.

Ghirlanda, S. & Enquist, M. 2003. A century of generalization. *Animl Behav* **66**, 15–36.

Hanson, H. 1959. Effects of discrimination training on stimulus generalization. *J Exp Psychol* **58**(5), 321–333.

Haykin, S. 1999. *Neural Networks: A Comprehensive Foundation*. 2nd Edn. Macmillan.

Helson, H. 1964. *Adaptation-level Theory*. Harper & Row.

Le Pelley, M. E. & McLaren, I. P. L. 2003. Learned associability and associative change in human causal learning. *Q J Exp Psychol* **56B**(1), 68–79.

Mackintosh, N. J. 1974. *The Psychology of Animal Learning*. Academic Press.

McClelland, J. & Rumelhart, D. 1985. Distributed memory and the representation of general and specific information. *J Exp Psychol Gen* **114**(2), 159–188.

McClelland, J. L. & Rumelhart, D. E., eds. 1986. *Parallel DistributedProcessing: Explorations in the Microstructure of Cognition, Vol. 2*. MIT Press.

Parducci, A. 1965. Category judgment: a range-frequency model. *Psychol Rev* **72**(6), 407–418.

Pearce, J. M. 1997. *Animal Learning and Cognition*. 2nd Edn. Psychology Press.

Prokasy, W. F. & Hall, J. F. 1963. Primary stimulus generalization. *Psychol Rev* **70**, 310–322.

Purtle, R. B. 1973. Peak shift: A review. *Psychol Bull* **80**, 408–421.

Rescorla, R. A. & Wagner, A. R. 1972. A theory of Pavlovian conditioning: Variations in the effectiveness of reinforcement and nonreinforcement. In *Classical Conditioning II: Current Research and Theory* (ed. A. H. Black & W. F. Prokasy), pp. 64–99. Appleton-Century-Crofts.

Rumelhart, D. E. & McClelland, J. L., eds. 1986. *Parallel Distributed Processing: Explorations in the Microstructure of Cognition, Vol. 1*. MIT Press.

Sarris, V. 2003. Frame of reference models in psychophysics: a perceptual-cognitive approach. In *Perception beyond Sensation* (ed. C. Kaernbach, E. Schröger & H. Müller), pp. 69–88. Lawrence Erlbaum Press.

Spence, K. 1936. The nature of discrimination learning in animals. *Psychol Rev* **43**, 427–449.

Terrace, H. S. 1964. Wavelength generalization after discrimination training with and without errors. *Science* **144**, 78–80.

Thomas, D. R. 1993. A model for adaptation-level effects on stimulus generalization. *Psychological Review* **100**(4), 658–673.

Thomas, D. R. & Jones, C. G. 1962. Stimulus generalization as a function of the frame of reference. *J Exp Psychol Gen* **64**(1), 77–80.

Thomas, D. R., Lusky, M. & Morrison, S. 1992. A comparison of generalization functions and frame of reference effects in different training paradigms. *Percept Psychophys* **51**(6), 529–540.

Wasserman, E. A. & Berglan, L. R. 1998. Backward blocking and recovery from overshadowing in human causal judgement: the role of within-compound associations. *Q J Exp Psychol* **51B**(2), 121–138.

Widrow, B. & Hoff, M. E. J. 1960. Adaptive switching circuits. In *IRE WESCON Convention Record*, Vol. **4**, pp. 96–104. IRE.

Widrow, B. & Stearns, S. D. 1985. *Adaptive Signal Processing*. Prentice-Hall.

15

The need for stochastic replication of ecological neural networks

Colin R. Tosh and Graeme D. Ruxton

15.1 Introduction

Artificial neural networks are increasingly being used by ecosystem, behavioural and evolutionary ecologists. A particularly popular model is the three-layer, feedforward network, trained with the back-propagation algorithm (e.g. Arak & Enquist, 1993; Ghirlanda & Enquist, 1998; Spitz & Lek, 1999; Manel *et al.*, 1999; Holmgren & Getz, 2000; Kamo *et al.*, 2002, Beauchard *et al.*, 2003). The utility of this design (especially if, as is common, the output layer consists of a single node) is that for a given set of input data, the network can be trained to make decisions, and this decision apparatus can subsequently be applied to inputs that are novel to the network. For example, an ecosystem ecologist with a finite set of ecological, biochemical and bird-occurrence data for a river environment can train a network to produce a predictive tool that will determine the likelihood of bird occurrence through sampling of the environment (Manel *et al.*, 1999). Or in behavioural and evolutionary ecology, a network can be trained to distinguish between a 'resident animal' signal and 'background' signals, and subsequently used to determine how stimulating a mutant animal signal is, and hence, how signals can evolve to exploit receiver training (Kamo *et al.*, 2002). Reasons for the popularity of the back-propagation training method (Rumelhart *et al.*, 1986) include its computational efficiency, robustness and flexibility with regard to network architecture (Haykin, 1999).

Despite the sensitivity of many other nonlinear modelling methods to variation in initial system state (Scott, 1999), and the inherent tendency of ecologists to 'replicate', detailed investigation of the nature of variation in network properties following stochastic replication of training data and/or starting weight composition is rarely reported in ecological applications using the aforementioned network design. Treatment of replication in published studies varies greatly, from cases of apparent nonreplication (a single sequence of training data and a single starting weight array) to studies where genuine, stochastic variation of starting weights and/or training data is clearly evident. It is also common for research procedures to be reported in insufficient detail to be able to gauge treatment with respect to replication of training. Even in studies where considerable effort

Modelling Perception with Artificial Neural Networks, ed. C. R. Tosh and G. D. Ruxton. Published by Cambridge University Press.

has been made to replicate training procedures, the precise nature of resultant variation in network properties is not usually a significant theme in presentation of results. Hence, it is not always clear whether the variation in network properties that might result from, for example, training networks with different samples of the same variables from the same environment, or random sampling of the same set of environmental data, is trivial, or can lead to substantial differences in the predictive properties of networks.

The issue of stochastic replication of artificial neural networks is also important from more than a technical view point. Ghirlanda and Enquist (Chapter 14, this volume) term the phenomenon in which different histories of experiences (paths) may at first seem to produce the same behavioural effects yet reveal important differences when further examined, 'Path Dependence'. Using a two-layer neural network with a single output node and the so-called δ training rule, the authors demonstrate the path dependence can contribute to such phenomena as lack of peak shift after 'errorless discrimination learning', decrease of peak shift during extinction testing, and the shift of generalisation gradients toward the average of the test stimuli. Thus, examination of the properties of individual neural networks that differ subtly in training regime may also inform on fundamental behavioural phenomena and also potentially relate to individual differences in the behaviour of real organisms.

As part of ongoing work into the sensory basis of predator–prey interactions, we have investigated the consequences of stochastic variation in training regime, resulting from random sampling of the same underlying statistical distribution, for the predictive properties of a three-layer, feedforward, back-propagated neural network. As in Ghirlanda and Enquist (Chapter 14, this volume), we show that the consequences of subtle variation in the training data set can be far from trivial, and can lead to networks with quite different, discreet, predictive properties.

15.2 Methods

15.2.1 The neural network and training procedure

We are using three-layer, feedforward neural networks as caricatures of the sensory surface/interneuron/sensory map system found in a range of taxa from flies to primates (Ashley & Katz, 1994; Cohen & Newsome, 2004; Shipp, 2004). The first layer consists of 20 nodes, and subsequently 'bottlenecks' into a hidden layer of 10 nodes (neural bottlenecks are often seen as nerves leave sensory organs). Architecture then differs from the common single-node output architecture, and the 20-node architecture of the input layer is reconstructed to represent an isomorphic spatial map. All layers are fully connected, hyperbolic tan (tanh) activation functions (Krakauer, 1995; Ghirlanda & Enquist, 1998) are used throughout, and training rate is set at 0.2.

During training we presented networks with 4000, 20-element vectors consisting of '–1s' (representing 'empty space') and '1s' (an 'object), whose relative numbers per vector followed a normal random distribution with mean = 10 and SD = 2. Positioning within vectors followed a uniform random distribution. Starting weight arrays were constructed from a uniform random distribution between –1 and 1. Fifty-two such training input/weight array sets were created, and each used to train a network over a maximum of 1000 epochs. The task of the back-propagation training algorithm was to reconstruct each input pattern in the output layer. Epoch for termination of training was chosen according to an early stopping procedure (Hecht-Nielsen, 1990) in which a 4000 × 20 test input array was created as above, error in reconstructing its constituent vectors calculated after each epoch, and stopping point defined as the epoch where error was at a minimum. Weight updating was sequential and input vectors were presented in the same order in each epoch.

15.2.2 Object targeting

We are particularly interested in how object targeting by organisms is affected by the number of distracter objects (this is sometimes investigated under the guise of the 'confusion effect'; e.g. Landeau & Terborgh 1986). Following training, we therefore chose an arbitrary network position ('position 10' in a 2D network input, if the reader wishes to visualise the procedure) into which a target object of value = 1 was input, and the number of distracter objects (also = 1) varied from 0 to 19 (empty space again = –1). For each number of distracters, positioning around the target object was varied 50 times according to a uniform random distribution, and targeting accuracy gauged as the proportion of occasions in which an object (gauged as output > 0.9) was reconstructed at *output* position 10 (the equivalent position in the spatial map to the input unit stimulated by the target object). The same test arrays as just described were applied to all trained networks (hence the following results are not due to stochastic variation in *test* data).

15.2.3 Characterising network error-weight surfaces

We initially suspected the discrete network predictive states that resulted from stochastic replication of training (see the results section) to be artefactual, due to settling of the back-propagation algorithm in different local minimal in the network error-weight hyperspace (Haykin, 1999). We therefore characterised error-weight surfaces for one training data-starting weight set that terminated in each of the predictive states. For each of these four training data-starting weight sets, fourteen random walks in weight space were initiated from both the end and start point of training. Each walk started by randomly changing each element in the weight arrays by + or –3 and subsequently determining total error in reconstructing the appropriate 4000 training vectors, measured using the total summed, squared measure employed within the back-propagation algorithm (see Rumelhart *et al.*, 1986). This was repeated 50 times in order to characterise the region immediately surrounding the start and end points of training. Incremental change to weight elements was then increased to + or –10 and the process repeated a further 50 times in order to characterise the wider weight space, and finally increments were increased to + or –50 and the process repeated a further 50 times, to characterise gross weight space. Thus each walk consisted of 150 samplings of weight space. As sampling of gross weight space was relatively coarse, the complete sampling procedure of 28 walks was repeated 8 times for each training data-starting weight set. While this is a thorough sampling of weight space (33 600 samples for each unique surface) we still cannot entirely exclude the possibility of extremely narrow error minima, especially in the area of gross sampling. The 400 weight values in each vector of the set of 4200 samples of weight space (28 walks of 150 samples each) were reduced to a single points in 2D space using Sammon mapping (Sammon, 1969) implemented using the 'sammon' function of the SOM Toolbox for Matlab (http://www.cis.hut.fi/projects/somtoolbox/), run for 200 iterations with a step size of 0.2. This dimensional reduction procedure incrementally reduces the discrepancy between total Euclidian interpoint-distance in multi- and lower-dimensional space using a pseudo-Newton error minimisation method, and tends to associate points with similar weight compositions. It was used in the present application in preference to more familiar procedures such as PCA because the first two components in the latter method (and allied dimensional reduction procedures) usually captures only a small proportion of the original variation in data with high-order data sets such as those used presently. Following 2D mapping, surfaces were produced by Delauney triangulation and plotted against network training input reconstruction error.

15.2.4 Mapping the relationship between end-weight composition and network predictive properties

We further employed Sammon mapping to characterise weight arrays at the start and end of training for two of the distinct network predictive states obtained (those shown in Figure 15.1, A and B; results and conclusions presented applied to the other predictive patterns, but data were not shown to ease visualisation). Specifically, we wanted to know if similar predictive patterns arose from similar weight compositions, or if quite different weight compositions can produce the same predictive property.

15.3 Results and discussion

Figure 15.1 shows the object targeting vs. group size relationships obtained following 52 stochastic training runs. It can be seen clearly that stochastic variation in training procedure was sufficient to create at least four discrete network predictive states. Object targeting was either (A) 'Always accurate', (B) 'More accurate for smaller numbers of input objects', (C) 'More accurate for small and large numbers of input objects, less for intermediate sizes', or (D) 'More accurate for larger numbers of input objects'.

Generally, error-weight surfaces of representative training data-starting weight sets that terminated in each of the predictive states, consisted of a plateau with a single major error minimum, to the bottom of which the back-propagation algorithm lead weight

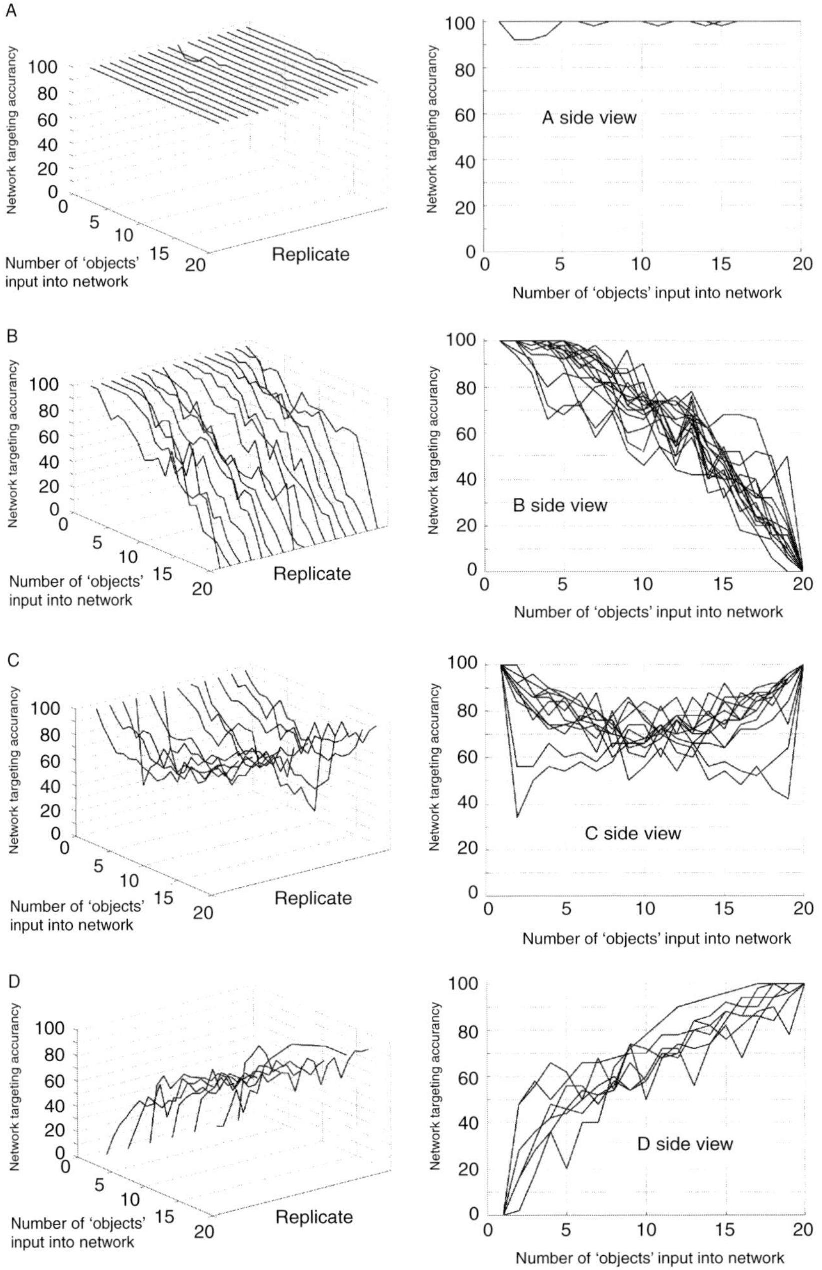

Figure 15.1. The predictive properties of 52 neural networks created through stochastic replication of input training data set and starting weight composition. Replicates with similar properties are presented together in each graph. Refer to the Methods section for the treatment of object targeting by neural networks, including the definition of 'Network targeting accuracy' (y-axis).

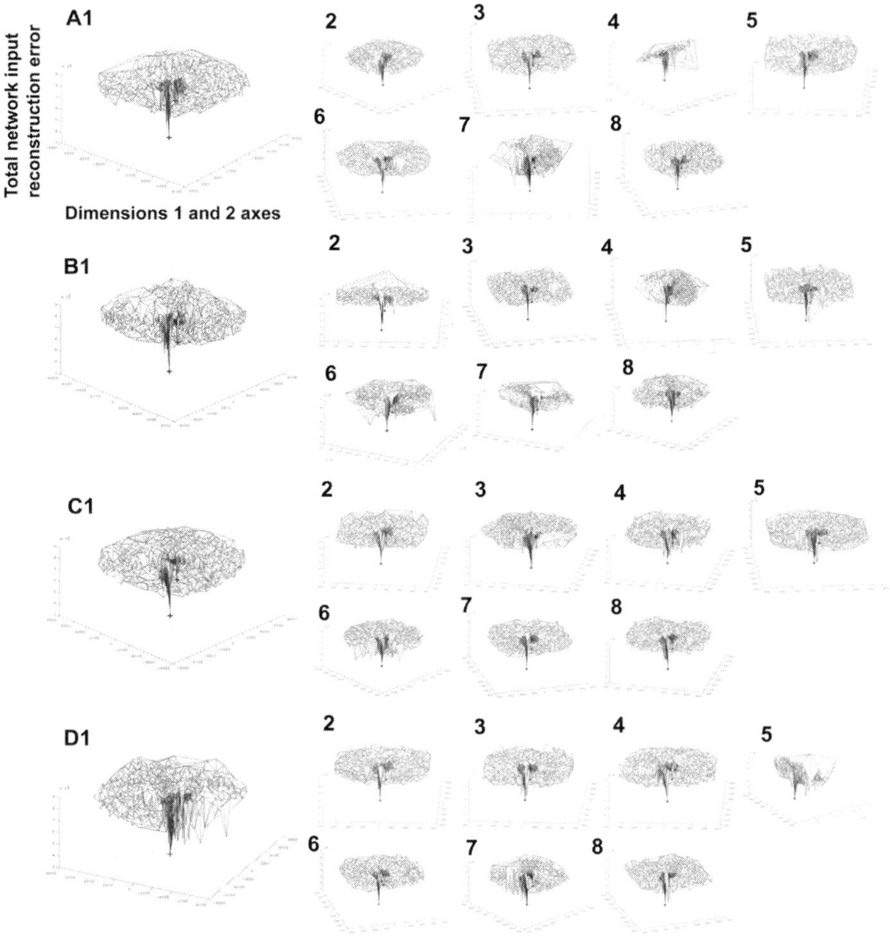

Figure 15.2. Neural network error surfaces for one example training data-starting weight set associated with each network predictive pattern illustrated in Figure 15.1. A, B, C, D refer to the pattern with corresponding letters in Figure 15.1. Multiple random walks were initiated from the start and endpoint of back-propagation training (shown by crosses on the surface), and surfaces produced by Sammon mapping of weight vectors, Delaunay triangulation of Sammon map values in two dimensions, and subsequent plotting against network training input reconstruction error. Numbers 1–8 refer to repeats of the random walk procedures for each training data-starting weight set. Further details can be found in the Methods section.

arrays (Figure 15.2). We found no example of a lower error value than the endpoint of training in any of the 33 600 sample points for each unique error surface, although there did appear to be examples of major alternative minima in some surfaces (Figure 15.2, A7, B5, C5–6, D1, D7, D8 for example). These regions should be viewed with caution, however, as they could be produced through misplacement of points in Sammon

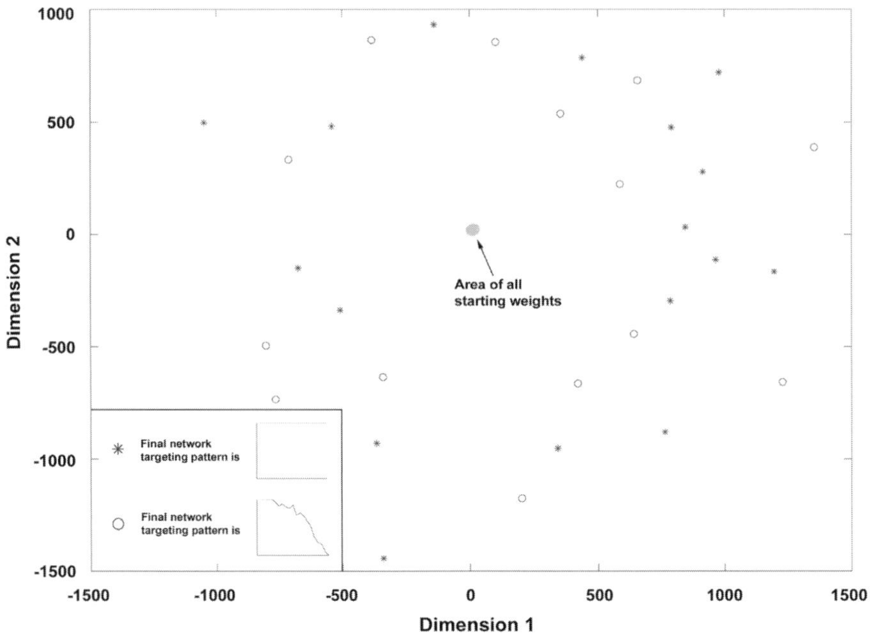

Figure 15.3. Sammon map of weight composition at the start and end of network training, for networks that converge to the predictive states shown in Figure 15.1, A and B. The fact that there is no pronounced same-symbol aggregation within the figure indicates that multiple networks with quite different weight array compositions can have the same predictive properties, and networks with relatively similar weight compositions can have quite different properties: the weight space appears to be a 'patchwork' of different network predictive states.

mapping. We conclude that the end weight values for networks predicting each of the object targeting patterns (Figure15.1) are major error minima (if not the principal minima) in the error surface and there is little evidence to suggest that the various network predictive patterns result from suboptimal convergence of networks into local regions of intermediate error reduction.

Figure 15.3 indicates that quite different weight compositions at the end of network training can produce the same predictive property and vice versa. This suggests that the weight space of this neural network can be considered something of a 'patchwork' of predictive states, and the eventual properties of a network will depend on the particular 'patch' the training algorithm converges on. The observation from Figure 15.3 that the system can converge to different weight compositions and predictive properties from similar starting weight arrays, indicates the primacy of stochastic variation of training input data rather that starting weights in the phenomena discussed. This was confirmed by training with the same starting weight arrays but different training sets: networks still converge to different predictive states (data not shown).

15.3.1 The need for stochastic replication of ecological neural networks

A point we wish to make from the preceding demonstrations is that a researcher investigating ecological or behavioural phenomena using the described system in a non-replicated fashion might lead to a quite different biological conclusion relative to the researcher who has stochastically replicated the system and fully investigated predictive properties therefrom. The stochastic variation in network training data that results from random sampling of the same statistical distribution, can lead to dramatic differences in network predictive states in three-layer, feedforward, backpropagated neural networks. Although the network we have used differs both architecturally and in training objective to the typical decision-making network used in ecology, pronounced effects on the behaviour of artificial neural networks consequent of subtle differences in training regime have also been demonstrated in a two-layer feedforward network with a single output node trained with the δ rule (Ghirlanda and Enquist, Chapter 14, this volume) and a five-layer feedforward network with a single output node trained with a genetic algorithm (C. Tosh and G. Ruxton, unpublished). In the latter system networks were trained to be very specialised with regard to choice of plant-like resource objects projected onto the input layer, and varied randomly in starting weight composition and the position and order of objects projected onto the input layer. When the ability of networks to choose an appropriate resource object while being 'distracted' by different numbers of inappropriate objects was analysed, some networks showed decreasing discrimination accuracy with increasing numbers of distracters and some showed increasing discrimination. Thus, path-dependent effects may be observed in a variety of artificial neural network systems. We do not think that ecologists and behavioural biologists who have applied networks with similar designs without a full appreciation of path-dependent effects should be unduly alarmed by our results, however. We suspect that many conscientious researchers in these disciplines do in fact replicate networks and analyse the range of resultant properties routinely in research procedures (presumably excluding this as a significant source of variation in their system), prior to presenting work for publication. We do, however, encourage researchers who have not considered the variation in neural network properties that can arise from stochastic replication of training procedures, to include such replication as a matter of course in research procedures. Procedures that ecologists might consider include obtaining multiple training data sets from the same environment, random subsampling from the same data set prior to network training, and random ordering of input vectors from the same data set for presentation to the network. These procedures may involve considerable additional effort on the part of the researcher, and may make results 'more' or 'less interesting' for subsequent publication, but workers who do not apply such procedures should be aware that the research results they present may represent only one of a range of possible predictive network states.

What of the biological significance of path-dependent effects? Ghirlanda and Enquist (Chapter 14, this volume) show that subtle variation in an artificial neural network training regime can contribute to important phenomena in animal behaviour such as lack of peak shift after 'errorless discrimination learning', decrease of peak shift during extinction testing, and the shift of generalisation gradients toward the average of the test stimuli. Could path-dependent effects also contribute to inter-individual variation in behaviour? Certainly individuals of a species from the same population will experience a similar set of stimuli during neural network training, but with subtle differences in the temporal order and content of stimulus sets. Could such variation contribute to behavioural syndromes, the consistent differences in the behavioural tendencies of individuals of the same species (Sih *et al.*, 2004)? While speculative, the demonstration that many artificial neural networks are inherently sensitive to subtle variation in training regime indicates that path dependence will be a useful framework to tackle such questions.

Acknowledgement

This work was funded by the UK Biotechnology and Biological Sciences Research Council (Grant No: BBS/B/01790).

References

Arak, A. & Enquist, M. 1993. Hidden preferences and the evolution of signals. *Phil Trans R Soc B* **340**, 207–213.

Ashley, A. A. & Katz, F. N. 1994. Competition and position-dependent targeting in the development of the *Drosophila* R7 visual projections. *Development* **120**, 1537–1547.

Beauchard, O., Gagneur, J. & Brosse, S. 2003. Macroinvertebrate richness patterns in North African streams. *J. Biogeogr* **30**, 1821–1833.

Cohen, M. R. & Newsome, W. T. 2004. What electrical microstimulation has revealed about the neural basis of cognition. *Curr Opin Neurobiol* **14**, 169–177.

Ghirlanda, S. & Enquist, M. 1998. Artificial neural networks as models of stimulus control. *Anim Behav* **56**, 1383–1389.

Haykin, S. 1999 *Neural Networks: A Comprehensive Foundation*. Prentice Hall.

Hecht-Nielsen, R. 1990. *Neurocomputing*. Addison-Wesley Publishing Company.

Holmgren, N. M. A. & Getz, W. M. 2000. Evolution of host plant selection in insects under perceptual constraints: a simulation study. *Evol Ecol Res* **2**, 81–106.

Kamo, M., Ghirlanda, S. & Enquist, M. 2002. The evolution of signal form: effects of learned versus inherited recognition. *Proc R Soc B*. **269**, 1765–1771.

Krakauer, D. J. 1995. Groups confuse predators by exploiting perceptual bottlenecks: a connectionist model of the confusion effect. *Behav Ecol Sociobiol* **36**, 421–429.

Landeau, L. & Terborgh, J. 1986. Oddity and the confusion effect in predation. *Anim Behav* **34**, 1372–1380.

Manel, S., Dias, J. M., Buckton, S. T. & Ormerod, S. J. 1999. Alternative methods for predicting species distribution: an illustration with Himalayan river birds. *J Appl Ecol* **36**, 734–747.

Rumelhart, D., Hinton, G. & Williams, R. 1986. Learning internal representations by error propagation. In *Parallel Distributed Processing: Explorations in the Microstructure of Cognition. Vol. 1: Foundations* (ed. D. Rumelhart & J. McClelland), pp. 318–363. MIT Press.

Sammon, J. W. 1969. A nonlinear mapping for data structure analysis. *IEEE T Comput* **C-18**, 401–409.

Scott, A. 1999. *Nonlinear Science*. Oxford University Press.

Shipp, S. 2004. The brain circuitry of attention. *Trends Cogn Sci* **8**, 223–230.

Sih, A, Bell, A. M., Johnson, J. C. & Ziemba, R. E. 2004. Behavioral syndromes: an integrative overview. *Q Rev Biol* **73**, 241–278.

Spitz, F. & Lek, S. 1999. Environmental impact predication using neural network modelling. An example of wildlife damage. *J Appl Ecol* **36**, 317–326.

16

Methodological issues in modelling ecological learning with neural networks

Daniel W. Franks and Graeme D. Ruxton

16.1 Introduction

A key attribute of all but the simplest organisms is an ability to modify their actions in the light of experience – that is to learn. This attribute allows individuals to adapt to rapidly changing environments. Learning is a fundamental aspect of animal behaviour (Barnard, 2003). One aspect of animal behaviour where learning has been particularly extensively studied is food gathering (see recent reviews by Adams-Hunt & Jacobs, 2007; Sherry & Mitchell, 2007; Stephens, 2007), and it is this aspect that we will focus on. We use the term *ecological learning* to describe an organism learning about its environment.

Neural network models are being used increasingly as effective tools for the description and study of animal behaviour (see Enquist & Ghirlanda, 2005 for a review). There are many different techniques that can be used to model animal learning, with Bayesian approaches being one such example. However, with the desire of taking advantage of neural networks' ability to generalise, neural networks have also been used to model stimulus learning in animals, and have even been used to examine the difference between neural network predators that evolve or learn (for example, see Kamo *et al.*, 2002). In this paper we focus solely on the use of neural networks to represent ecological learning (such as a predator learning and generalising over prey) and argue that there are fundamental differences between the way neural network models are generally trained and the way organisms learn. We further argue, and show, that this means that neural network models can be used to study ecological learning only with very great care.

Most neural networks used in ecology and evolution are treated as black boxes that are used to find the optimal or evolutionarily stable strategy for organisms to play in given situations. Generally, the focus is on the outcome (the final strategy) rather than on the mechanisms and trajectory by which this final state was reached. This paper represents a first step in a programme aimed at exploring whether the use of neural networks can be extended to provide insights into these mechanisms as well as to the final behavioural state. Our motivation for this is the study of predatory responses to chemically defended prey that signal their defence with bright displays (such as the yellow and black stripes of

Modelling Perception with Artificial Neural Networks, ed. C. R. Tosh and G. D. Ruxton. Published by Cambridge University Press.

many wasps). Predatory responses are shaped both by evolution and by previous experience. That is, predators often have innate unlearned biases for and against attacking individuals with certain appearance traits, but these unlearned responses are modified in the light of an individual's positive and negative experiences with individuals that they attack. Individual predators learn over their lifetime (ecological learning, as we defined it above), and populations of individuals experience evolutionary selection both on unlearned biases, and of aspects of how they learn. A neural-network model is ideally suited to representation of the decision making of the predators, but in order to be useful, we must be able to represent the interlinked effects of two very different processes of neutral network weights: evolution over generational timescales, and ecological learning on the timescale of an individual's attacks on specific prey items.

Representation of evolution in a population of predators is relatively simple to encode. Many neural network models used to study animal behaviour relate to change in behaviour over generational timescales rather than over the lifetime of an individual. That is, they describe the evolution of animal behaviour, and the network weights are optimised according to some predefined fitness measure. For example, the network weights might be coded in an artificial genome and evolved using a genetic algorithm. In a genetic algorithm individuals with a low fitness are replaced with fitter individuals each generation. Individuals inherit a combination of their parents' network weights and occasionally weights are mutated. Thus, training a network with a genetic algorithm is often considered as studying the evolution of a stimulus response. However, the weights are generally fixed within the lifetime of an individual, and so the behaviour of an individual is unchanging throughout its lifetime, and learning is excluded from the model. Such models would often be greatly enhanced in their flexibility if they could represent ecological learning as well as evolutionary change. Genetic algorithms can be used as an optimisation technique to represent learning within realistic time frames (e.g. Haas, 2006), in the same way that back-propagation can, requiring multiple chromosomes per learning individual. Our arguments therefore apply equally to the use of genetic algorithms to represent learning as to back-propagation and other learning techniques. Hence, one aim of our paper is to present a means of representing ecological learning of a neural network in a way that can later be combined with conventional means of representing evolution, in a way that captures the essential components of each mechanism, preserves the essential separation of timescales of the two processes, and allows the interplay between the two to be studied.

Some attempt has been made in the computer science literature to model both learning and evolution in neural networks (Ackley & Littman, 1991; and see Toquenaga *et al.*, 1995 for an application). The research focuses on implementing unsupervised learning by evolving desired network outputs, and examining its effect on network fitness. However, the fidelity of the network learning to ecological learning is not considered. In addition, the effect of the learning on stimulus generalisation is unexplored. In contrast, here we are interested in a methodological question: can neural networks appropriately model real world ecological learning?

We focus on whether neural network training algorithms produce learning rates and generalisation gradients that are an appropriate analogue to their ecological counterpart. We first capture what we mean by 'ecological learning' by scrutinising the important features of animal lifetime learning and their effect on animal behaviour. We then discuss our attempts at developing networks that behave appropriately and capture the important differences between evolution and learning.

16.2 What do we mean by ecological learning?

By ecological learning we simply refer to a situation where an organism learns to adapt to its ecology during its lifetime. For example, a predator might start naïve with respect to what are (and what are not) appropriate items to include in its diet and, based on experience throughout its lifetime, learn which food sources to consume and which to avoid. We focus on the effect of the learning process at the level of the animal, rather than the lower-level processes (for example, at a neurological level) that allow such learning to occur. Here we are trying to identify exactly what properties are required from neural networks for them to be good models of animal learning, so that we then have a basis with which to assess neural network learning behaviour. An important aspect of this is for us to tease apart the differences between evolution and learning as means of inducing change in animal behaviour. The main difference between the two is the timescale at which they operate. Learning occurs at the level of the individual and within the individual's lifetime, whereas evolution occurs at a higher level (i.e. evolution cannot occur in a sole individual) and over many generations. Of course the mechanisms allowing learning will themselves be under evolutionary pressure and so within-lifetime learning and between-generation evolution are inextricably linked. Another difference between ecological learning and evolutionary change is that there is a separation of the adaptation and evaluation phases in evolution (i.e. the animal's fitness is assessed at the end of its lifetime), whereas there is no such separation in ecological learning (i.e. the animal evaluates after each experience). That is, with ecological learning change and the driver of change are intermingled: a forager eats a prey item, which changes its perception of food availability in its environment, which in turn influences the subsequent food choices of the forager. In evolutionary change, the feedback is not as immediately responsive, since change can only occur when a new generation is produced. Hence, for animals with fixed behaviour and no ability to learn, each feeding experience has no effect on the next; however, the aggregate of these experiences has an effect on the final fitness of the individual.

Ecological learning can be very quick. Indeed, sometimes only a single aversive experience is required for foragers to learn that potential prey of a given type should be avoided in future (Shettleworth, 1998). However, such one-trial learning is the extreme, and normally a larger number of experiences are required before foragers learn to avoid unpalatable or toxic prey. Generally, this number is between 5 and 50 (e.g. Skelhorn & Rowe, 2005, 2006a, 2006b; Ham *et al.*, 2006; Ihalainen *et al.*, 2007). The number of prey items an individual will consume in its lifetime will vary greatly between species, but for the insectivorous and granivorous birds used in the experiments above the number of food items consumed in a lifetime could easily have been hundreds of thousands.

Thus, there would be a big timescale difference in a neural network model of ecological learning and a neural network model of evolution. Some might argue that this timescale difference does not matter and evolution can be thought of as analogous to learning and that the only real difference is that evolution takes longer to find the solution than learning. However, whether change in behaviour occurs solely by evolution or by a combination of evolution and learning can result in important consequences for the evolution of the stimuli. Let us take our predator example and apply it to modelling the evolution of warning signals in prey. Let us assume that a prey species that is unprofitable to the predator begins with a conspicuous colouration, along an appearance dimension that defines the degree of crypsis versus degree of conspicuousness. Predators are modelled as neural networks with the prey appearance as the input and the attack probability as the output. Note that predators' responses to prey are best described as propensity to attack, and not as dichotomous outcomes, as predators come into the world naïve and gradually adjust their attack rates over time (still occasionally sampling prey they have learned are defended) (Ruxton *et al.*, 2004). If we model evolving predators, and predators begin with a tendency to attack, it will take many contacts with individual prey for predators to begin to avoid the prey species. In the meantime the prey species would have quickly evolved to become cryptic since too many conspicuous individuals would perish in the time that predators are naïve and any mutant individuals that are more cryptic would fixate in the population, even if it meant being mimicked by profitable cryptic prey (Franks & Noble, 2004). If we model predators that give the desired properties of ecological learning, then predators would be able to quickly learn to avoid the unprofitable species. Thus, far fewer prey individuals would be sacrificed during the predators' naïve stage, giving the unprofitable species a chance to stay conspicuous. There are therefore important implications following from whether we model animals as evolving or learning. To give another example, the evolution of Müllerian mimicry between defended prey species is generally understood to be driven by the costs to these prey caused by predators learning that they are unpalatable (Ruxton *et al.*, 2004). It is desirable, then, for us to develop a concept of ecological learning that captures the essentials of this process and can be applied universally to modelling problems involving animal learning, or the interplay of learning of evolution. Later in this article we suggest such a concept.

16.3 Why neural networks as they are commonly implemented are not good models of ecological learning

To judge whether neural networks are good models of ecological learning we need to assess whether or not they capture the properties of ecological learning that we have discussed. Kamo *et al.* (2002) present a model intended to examine the differences in signal evolution when neural network receivers evolve and when they learn. They use a genetic algorithm to represent evolution and back-propagation to represent ecological learning. It is therefore essential that the difference between evolution and ecological learning is captured. Kamo *et al.* (2002) state that '. . . in studies of signal evolution we are not interested in the learning process per se. It is enough that receivers, after experiences with a given sender signal, react realistically to mutant signals: in other words, we need realistic models of generalisation.' Kamo *et al.* take an important first step in showing the difference in stimulus evolution as a result of differences in generalisation between evolution and learning. However, we have argued above that aspects of the learning process in addition to generalisation can have an important impact on signal evolution (particularly in terms of mutant survival rates).

Superficially, back-propagation appears to be an ideal candidate for allowing neural networks to model ecological learning. Back-propagation is the most commonly used algorithm for training a multi-layer feedforward neural network to produce desirable responses to a range of inputs (Enquist & Ghirlanda, 2005). However, the word 'learning' often attached to the end of the word 'back-propagation' is not intended to infer algorithmic fidelity to ecological learning. We now discuss fundamental properties of ecological learning that are not captured by standard backpropagation. We return to the issue of timescale of adaptation. We have discussed how receivers such as predators often learn the ecologically correct response to a stimulus after just 5–50 experiences. Capturing this property with back-propagation is a problem; even on very simple discrimination problems the training data (e.g. collections of stimuli and the correct response) have to be presented to the network thousands of times before the network learns the correct responses (even if the response is simply an attack probability; see our Results section). Of course, the amount of computational time the algorithm takes does not matter, what matters is that number of training samples required is typically hundreds of times greater than we desire for modelling ecological learning. Put simply, real animals learn effectively after a much smaller number of discrete experiences (e.g. attacks on different prey individuals for the predators discussed previously) than a neural network subject to back-propagation.

Our second criterion for effective representation of biological learning is that there is no separation between the training and evaluation phases. If a predator, for example, experiences an unprofitable prey item then the probability of it attacking the same prey type again may be lowered in time for its next experience. The standard back-propagation algorithm works as follows. First, the training set (e.g. a set of stimuli and the appropriate response) is batch-processed by the network and an error is calculated (neural networks do not have to be batch processed, but it is the most common method). Next, a back-propagation iteration is performed to adjust the network's weights and reduce the error. This process is repeated until a predefined condition is met (e.g. until the network has learnt the correct responses). Thus, back-propagation focuses on non-incremental learning tasks where the training data are arranged a priori and learning stops when the data have been processed. This approach clearly separates the training and evaluation phases in a manner that is analogous to neither evolution nor ecological learning. There is no separation of training and evaluation in the natural world. Each and every prey item that a predator decides to eat or decides not to eat will have an impact (no matter how small) on its ultimate fitness.

What if we modify the approach so that we perform a back-propagation iteration after each prey experience (as is often done when using neural networks to model behaviour)? The network weights only adjust by a small amount each back-propagation iteration and does not adjust the response to the stimuli fast enough to change the response on the next encounter more than negligibly. In nature, an experience with a single stimulus can change the receiver's future response to the stimulus dramatically. For example, a single aversive experience can make a predator much more wary of similar stimuli in future (Shettleworth, 1998). Thus, referring back to our predator example, it would affect the subsequent probability of re-sampling that stimulus on the next encounter. This would, in turn, affect the speed of learning for that stimulus and affect the response gradient for similar stimuli. In terms of predators, this would affect the attack inhibition generalisation gradient.

16.4 The suitability of different weight optimisation algorithms

16.4.1 Methods

We performed an analysis of neural networks with different training algorithms in an attempt to find a method that captures the key properties of ecological learning. The receiver (for example, the predator) is represented by a fully connected three-layer artificial feedforward neural network with a single input node, four hidden nodes and a single output node. Network weights are initialised with randomly selected values being drawn (independently for each node) from a uniform distribution between –1.0 and +1.0. The response to a given stimulus (for example, a prey item's appearance) is found by feeding the value of the sampled stimulus into the network's input. This stimulation in the input node is fed through the network, and the corresponding value given by the output layer is taken as the response.

We tested the following algorithms which, based on extensive literature searching, seemed to offer the potential to represent ecological learning: *back-propagation*, *Rprop* (Riedmiller & Braun, 1993), *quickprop* (Fahlman, 1988) and the *Associative Reward-Penalty algorithm* (Mazzoni *et al.*, 1991; Barto, 1995). Our tests showed that the Associative Reward-Penalty algorithm was the fastest (i.e. the network trained effectively after the smallest number of experiences) and so it receives our focus for illustrating the results. With this algorithm inputs may be continuous and output units are binary stochastic elements with p_i, the probability of firing of the ith unit, being defined by

$$p_i = g\left(\sum_{j=1}^{M} w_{ij}x_j\right),$$

where the jth unit provides input x_j to the ith unit via the connection w_{ij}, M is the number of inputs to the unit (always 1 in our study), and $g(x)$ is the binary sigmoid function $g(x) = 1/(1 + \exp[-x])$. During weight updating the reinforcement signal is calculated from the error between the actual and desired output as $r = 1 - \varepsilon$ with

$$\varepsilon = \left\{ \frac{1}{K} \sum_{k=1}^{K} |x_k^* - x_k| \right\}^{1/n},$$

where k indexes the K output units in the network (always 1 in our study), x_k^* is the desired output of the kth unit in the output layer, x_k is its actual output, and n is a constant. Weights are then updated according to

$$\Delta w_{ij} = qr(x_i - p_i)x_j + \lambda q(1 - r)(1 - x_i - p_i)x_j,$$

where x_i is the output of the ith unit of the network, and q and λ are constants. The network's feedforward process is stochastic and gives a response of either one or zero. The probability of giving a one or zero is optimised by the weight update process.

A learning iteration is performed on the network after it is presented with a training sample. The network is trained on n samples selected randomly with equal probability from two different stimuli. Unless otherwise stated, one stimulus has a value of 0.2 and an optimal network response of 0.9, and the other stimulus has a value of 0.8 and an optimal network response of 0.1. This task has been used previously in Ghirlanda & Enquist (1998). The learning task may appear to be overly simplistic, as might the network structure with only one input node. However, we are interested in studying continuous-valued inputs, and purposely use a simple case to avoid the possibility of low learning speed as the result of an overly complicated task. A one-dimensional input also allows us to examine network generalisation in a graphical manner. We allow the network multiple hidden nodes to avoid the possibility of slow learning as the result of the inability of the network to solve the problem.

16.4.2 Results

This type of algorithm does not allow the network to learn quickly enough to meet our ecological learning requirements. Figure 16.1 shows that the network is still very naïve after 90 iterations. The network only begins to respond well to each stimulus after hundreds of iterations (Figure 16.2). We used $\lambda = 0.01$ and $p = 0.5$ for the results shown in the graphs. However, the results for our tasks of discriminating between continuous inputs are qualitatively robust to variations of λ, q and to variations in the random seed. Hence, we find no commonly used training method that provides a realistic analogy to ecological learning. In each case, they require an unrealistically high level of separate experiences in order to learn appropriate behaviour. In the next section, we present a method, based on multiple back-propagation-iterations per experience, which can provide an appropriate model of such learning. The purpose of this paper is to examine the

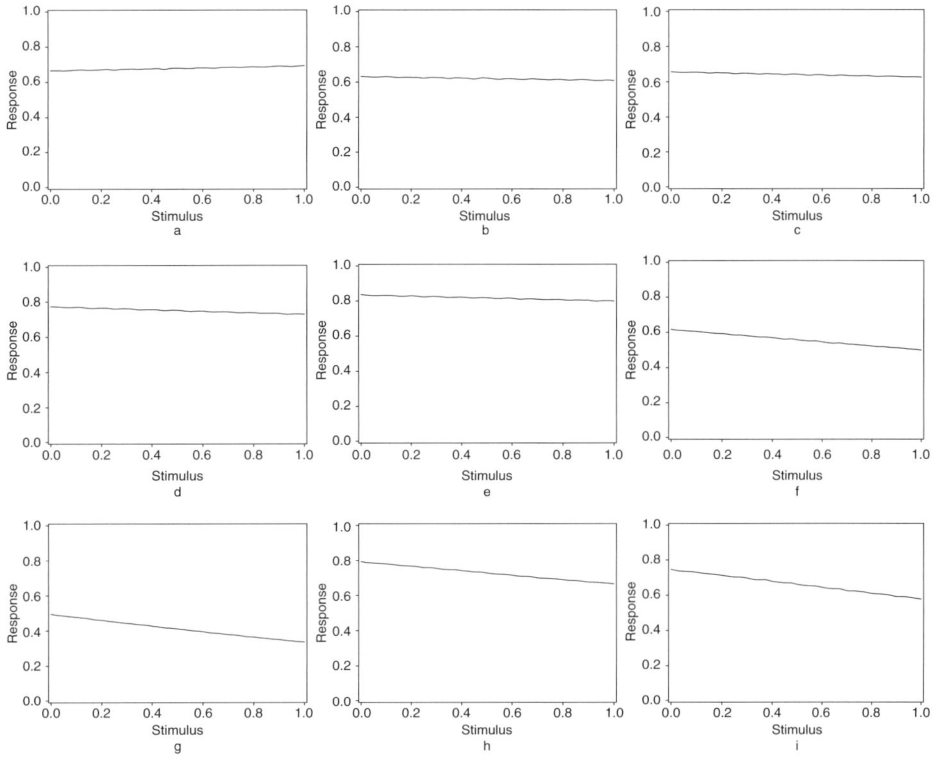

Figure 16.1. Network response gradients after training ($\lambda = 0.01$ and $p = 0.5$) after the following number of iterations/learning experiences: (a) 10, (b) 20, (c) 30, (d) 40, (e) 50, (f) 60, (g) 70, (h) 80, (i) 90. Networks are sequentially trained with a stimulus of 0.2 and a stimulus of 0.8. The correct response to the stimulus of 0.2 is 0.9 and the correct response to the stimulus of 0.8 is 0.1. Each response data-point represents the proportion of times (over 10 000 feedforwards) the network gives a response of 1.0. After 40 experiences (20 with each stimulus) the network's response has little improved. Even after 90 experiences the network gives an inadequate response to both stimuli.

appropriateness of different artificial neural network training algorithms to the modelling of ecological learning, and not simply to present a network-training algorithm. However, the next section demonstrates one way that one of the problems (the need for too many training examples) might begin to be addressed with minor modifications to currently used algorithms.

16.5 A neural network with some properties analogous to ecological learning

In this section we present an algorithm that can allow a neural network to learn in the correct timeframe for ecological learning. We reiterate, however, that the purpose of this

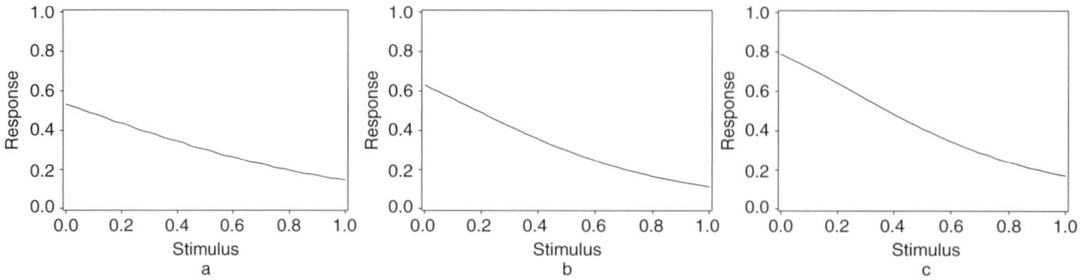

Figure 16.2. Network response gradients after training (with $\lambda = 0.01$ and $p = 0.5$) after the following number of iterations/learning experiences: (a) 200, (b) 300, (c) 400. Networks are sequentially trained with a stimulus of 0.2 and a stimulus of 0.8. The correct response to the stimulus of 0.2 is 0.9 and the correct response to the stimulus of 0.8 is 0.1. Each response data-point represents the proportion of times (over 10 000 feedforwards) the network gives a response of 1.0. After 400 experiences (200 with each stimulus) the network begins to give a reasonable (although still highly suboptimal) response to each stimulus.

article is not to present a new algorithm, but to highlight methodological design issues related to the common practice of using feedforward neural networks to model ecological learning.

16.5.1 Methods

Back-propagation is performed on the network after it is presented with a training sample. The network is trained on n samples selected randomly with equal probability from two different stimuli. Unless otherwise stated, one stimulus has a value of 0.2 and an optimal network response of 0.9, and the other stimulus has a value of 0.8 and an optimal network response of 0.1. Typically, in previous studies using back propagation, after feeding a sample through the network, a single back-propagation iteration is performed. We deviate from this by performing r back-propagation iterations after each learning experience, and recalculate the error after each iteration.

Specifically, back-propagation takes place as follows:

1. The inputs to the network are fed forward through the network to the output.
2. The error is calculated for the output neuron:

$$\delta_k = \psi'(p_k)(d_k - p_k).$$

3. The weights to the output node are changed:

$$\Delta w_{ki} = a\delta_k p_i.$$

4. The errors are calculated for the hidden neurons in the previous layer:

$$\delta_k = \psi'(p_k) \sum_j \delta_j w_{jk}.$$

5. The hidden weights are changed:

$$\Delta w_{ki} = \alpha \delta_k p_i.$$

6. Steps 4 and 5 are repeated until the input layer is reached.

Steps 1–6 were repeated r times for each stimulus, before the next stimulus was presented. The transfer function used on the input to each node in the hidden and output layers is a simple sigmoid function performed on the sum of the node's weighted inputs

$$\psi(x) = \frac{1}{1 + \exp(-x)}$$

d is the desired output, p is the actual output, α is the learning rate, k is the focal node, i points to nodes in the previous layer and j points to nodes in the next layer. We set the learning rate to $\alpha = 1$.

16.5.2 Results

With multiple iterations per training experience the network can learn near-optimal responses very quickly; with $r = 30$ the network can learn a near-optimal response in ~40 experiences (Figure 16.3). This method does not separate evaluation and learning phases and meets our timeframe requirement for ecological learning for the given task (although it may be affected by the type of task, and the task complexity). The number of back-propagation iterations per learning experience will need to be adjusted (and tested for overtraining) for any modelling task as it will depend on the task complexity, the number of inputs and the size of the network, among other parameters. Note that this technique can be used on a number of different training algorithms, but we select back-propagation for simplicity and because it is very widely used.

The response of the network during the early learning phase is different with this method than with standard back-propagation and its variants. The generalisation gradient is shifted dramatically after each early experience instead of gradually adjusting (Figure 16.4). Following a negative experience the network responds with a low response to all stimuli, and after a positive experience the network responds with a high response to all stimuli. For the example in Figure 16.4, the network is presented with positive and negative stimuli alternatively in a systematic manner to illustrate the effect. This effect may be analogous to some instances of biological learning; at first the predator is entirely naïve and does not realise that individual prey items differ along stimuli range and that this variation contains useful information. Thus, initially the response is flat across the stimuli range as this is the quickest end state the network can find with so little experience, and only after experience with several individuals differing in their stimulus does it begin to understand how this stimuli range can be utilised in its decision making (i.e. it learns to differentiate between instances on the basis of the levels of stimuli presented by each). Similarly the large changes in behaviour in response to the last individual attacked

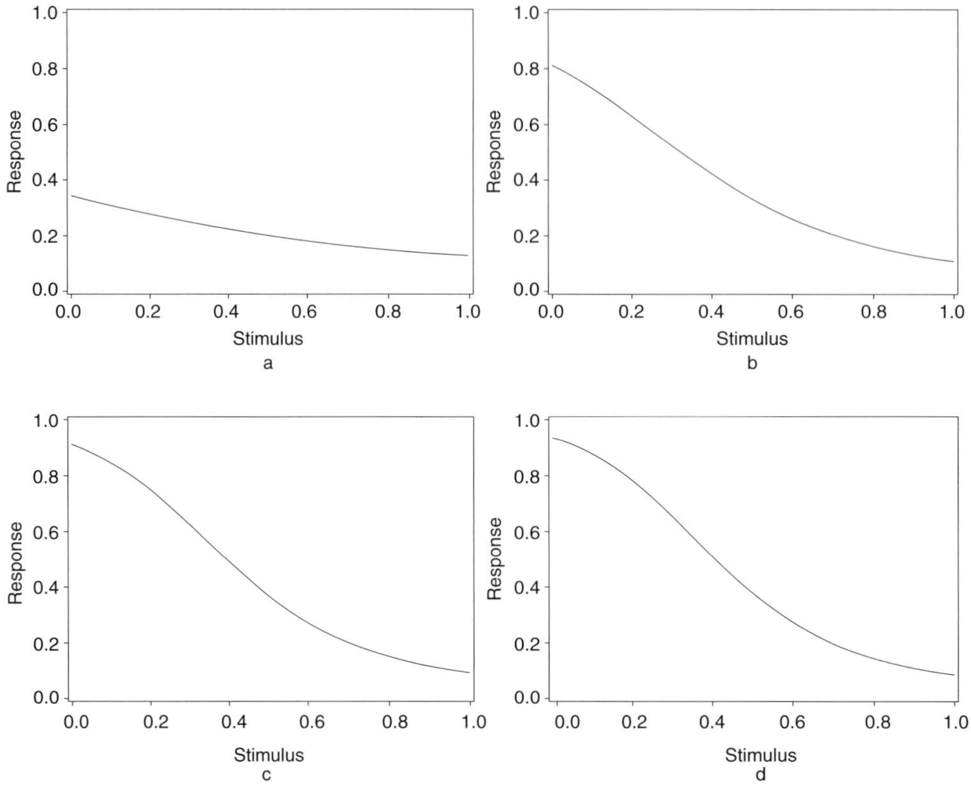

Figure 16.3. Generalisation gradients showing the network output (response) produced from a given input (stimulus). The sampled stimuli train the network to output 0.9 for an input of 0.2 and to output 0.1 for an input of 0.9. The network topology comprised a single input node, a single output node and learning rate $a = 1$. Network weights were initialised with values selected from a random uniform distribution between –1.0 and +1.0. The number of training experiences for each case is (a) 10, (b) 20, (c) 30, (d) 40. The network learns near-optimal responses after less than 40 experiences. For all cases $r = 30$.

can be seen as an effect of extreme naïvety with the predator not yet having the experience to associate previous experiences with each other, and hence being very strongly swayed by the last. However, over time the association is drawn and the predator integrates over a number of experiences that it now appreciates are related, and is less swayed by its most recent experience. Analogous to ecological learning by real animals, these initial effects lessen in the model as the predator becomes more experienced.

16.6 Discussion

We argue that there are fundamental differences between the way neural network models are generally trained and the way that animals learn. For example, evidence shows that animals learn correct responses to stimuli after a significantly smaller number of separate

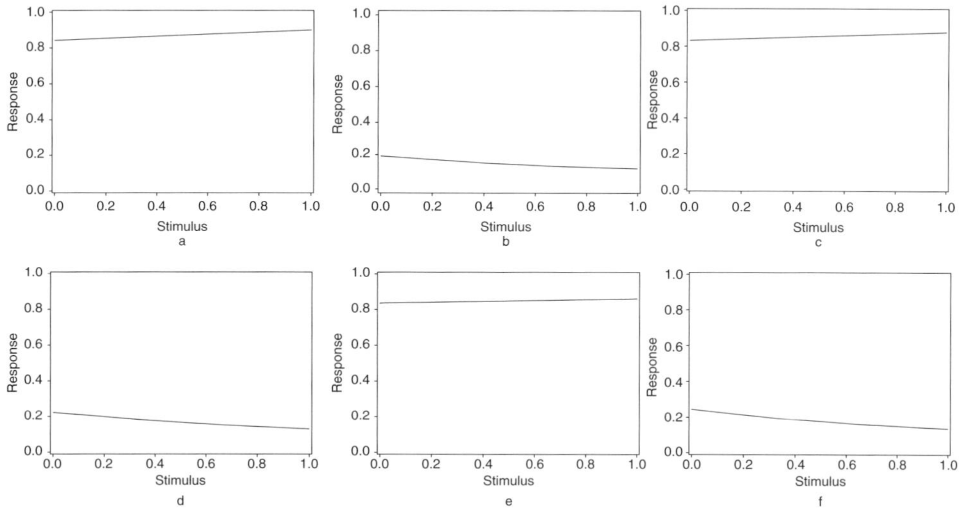

Figure 16.4. Generalisation gradients showing the network output (response) produced from a given input (stimulus) for the early part of the learning phase after naïvety. The sampled stimuli train the network to output 0.9 for an input of 0.2 and to output 0.1 for an input of 0.9. The network topology comprised a single input node, a single output node and learning rate $a = 1$. Network weights were initialised with values selected from a random uniform distribution between –1.0 and +1.0. The number of training experiences for each case was (a) 1, (b) 2, (c) 3, (d) 4, (e) 5, (f) 6. For all cases $r = 30$. During the early learning phase the network switches between extremes: following a negative experience the network responds with a low response to all stimuli, and after a positive experience the network gives a high response to all stimuli. This effect lessens as the predator becomes more experienced. For the example we show here the network is presented with positive and negative stimuli alternately in a systematic manner to illustrate the effect.

experiences than do artificial feedforward neural networks as they are typically modelled. This means that neural network models can be used to study ecological learning, but we must pay great care and attention to their fidelity to their biological analogue for the particular context of the study. Although there are different methods of implementing ecological learning, we focus on the use of neural networks in this study. Note that we are not arguing against a straw man – models exist that use back-propagation to represent ecological learning. Unfortunately, such models are not faithful to real ecological learning, which could have an effect on the predicted evolution of the stimuli. The purpose of this article is not to add to the machine learning literature, but to highlight important methodological considerations when choosing to use feedforward neural networks to represent ecological learning. In this article we have outlined what we mean by ecological learning and show how feedforward neural networks, as they are typically used, may – in some aspects – be inadequate for modelling ecological learning.

A key part of ecological learning is that individuals only receive reinforcement if they have sampled a stimulus. For example, upon a predator encountering a prey item, it has to decide whether or not to attack. If it chooses to attack then it will find out about that prey item and receive either a positive (if the prey item is profitable) or negative (if the prey item is unprofitable) reinforcement signal. However, if it chooses not to attack then it learns nothing. That is, if it chooses not to attack then it does not subsequently discover whether that particular prey individual was profitable or unprofitable and so it does not get a chance to evaluate whether its decision was correct. Thus, individuals that are encountered but not attacked do not contribute to the predator's learning. This important aspect of ecological learning has been captured in our model by only training the network when it selects to sample a stimulus. Note however, that the decision to forgo attacking an encountered prey item will have consequences for the fitness of the individual (loss of potential nutrients, avoidance of dangerous toxins) over and above consequences of the individual forgoing the opportunity to learn from sampling that particular prey individual.

Empirical work could be stimulated by our findings. A great deal has been made of the ability of neural networks to predict similar generalisation curves to those observed in (experienced) real animals (see Enquist & Ghirlanda, 2005 for an overview). However, the development of such generalisation through learning from previous experience has not currently been compared between model and reality. Such comparisons would powerfully test the ability of neural networks to mimic the decision making of real animals. The combination of the methodology presented here to represent ecological learning with more conventional techniques (such as genetic algorithms) to represent evolutionary processes may provide the ideal methodological framework for the study of the adaptive value of learning. In the context of interest to us, foraging and food choice by predators, the classically considered advantage of learning is that it allows rapid tracking of temporally varying circumstances. The disadvantage of learning in this context is that some sampling of unpalatable or otherwise defended individuals is required before learning is complete. This sampling potentially wastes time and energy that could more usefully be invested in alternative palatable prey types. There is also the risk that the sampled unpalatable prey may injure or even kill the predator.

We show that a minor but influential modification to the back-propagation training technique can offer a network that meets the criteria for ecological learning under the conditions tested. The modification is trivial and is not intended as an addition to the machine learning literature. It is simply used to illustrate how such problems may start to be overcome, and to illustrate the difference between the initial behaviour and our desired behaviour. Although the network is still learning from the same number of training samples as with the unmodified algorithm, it is responding correctly after each simulated experience in the modified case (and thus its response is only considered for its first learning iteration with each new stimuli). Another technique that may be explored to examine whether the network learns with the correct number of samples for a particular task is to tune the parameters of a learning algorithm. For example, Enquist & Ghirlanda (2005) present simulations of rats learning a T-maze, using the Associative Reward–Penalty algorithm, with learning speeds identical to the ones in the experimental data (roughly 30 experimental sessions). Thus, learning speeds are clearly variable and dependent on the type of problem, the network architecture (and whether the input needs to be binary or continuous; as binary inputs may be much easier to learn, although not appropriate for every modelling situation) and on particular algorithm parameters. Our point is that we need to be sure in each case that if we wish to model ecological learning, then the network should work with the same number of samples as the modelled animals.

Why might we want to have a neural network capable of ecological learning? Such a network can be used to study stimulus evolution in response to learning and, along with the use of a genetic algorithm, to re-examine the difference between learning and evolution using neural networks. The advantage of using a neural network (and the reason they are widely used in ecological modelling) is that they automatically generalise over stimuli in a realistic manner. It can also be used to further study the way that learning and evolution interact, such as how learning might guide evolution (Baldwin, 1896). Of course, there are other methods of modelling learning and generalisation that might be more suitable, depending on the modelling task (see e.g. Balogh & Leimar, 2005).

Our intuition suggests that learning might at first seem an unlikely strategy for predators. Wasps, for example, seem to look the same and to deliver the same toxic defence year after year, so why should each generation of birds have to learn about wasps? But we have to be careful about this argument. Let us say that a given predator species had a genetically determined innate aversion to wasps such that they never sampled them. If wasps then lost their defence and became palatable, then those fixed-strategy predators could not take advantage of this newly available food source until such time as there was a generic change in the population. This could happen relatively rapidly if there was an existing genetic polymorphism in the predator population such that some predator individuals did not have the aversion and willingly consumed wasps. Such a polymorphism seems unlikely since those predators without the aversion would presumably have been out-competed by those with the aversion in previous generations (when wasps were defended). Thus, in order to take advantage of the newly available prey type, the fixed-strategy predators would have to rely upon the correct mutation happening along. These predators may have to wait a very long time for the mutation, during which time competing predatory species with a learning strategy would be taking advantage of the new food source, and potentially thus out-competing the fixed behaviour predators. But why would the wasps evolve to abandon their defence? Presumably their defence is costly, and if the fixed strategy predator is their only predator (or even just a significant part of their predation threat), then prey that invest in defences may be at a competitive disadvantage compared with their conspecifics that invest instead in other functions. Thus, if the predators are selected to entirely ignore the prey, then the prey will be selected to abandon their defence and become profitable foraging prospects to predators. However, it may be difficult for the fixed-behaviour predators to respond quickly to the lowering of the prey defences. Such problems do not arise for the learning predators for two reasons: (1) their sampling means that there is always some predation pressure on the prey, and this will reduce the propensity for strong changes within the prey population in level of defence, and (2) if the prey do change, then the predators can respond quickly.

Hence, there appear to be circumstances where learning about potential food is more adaptive than fixed rules under evolutionary control. However, careful quantitative modelling of predator–prey interactions will be required to carefully delineate these circumstances (for a general abstract model see Hinton & Nowlan, 1987). If neural networks are to be used to explore the interaction between learning and evolution, then it is important that we think about the methodological issues raised here.

Acknowledgements

Thanks to Colin Tosh and Donald DeAngelis for suggestions.

References

Ackley, D. & Littman, M. 1991. Interactions between learning and evolution. In *Artificial Life II. Studies in the Sciences of Complexity* (ed. C. G. Langton, C. Taylor, J. D. Farmer, and S. Rasmussen). Addison-Wesley.

Adams-Hunt, M. M. & Jacobs, L. F. 2007. Cognition for Foraging. In *Foraging: Behaviour and Ecology* (ed. D. Stephens, J. Brown & R. Ydenberg), pp. 105–140. Chicago University Press.

Baldwin, M. J. 1896. A new factor in evolution. *Am Naturalist* **30**, 441–451.

Balogh, A. C. V. & Leimar, O. 2005. Müllerian mimicry: an examination of Fisher's theory of gradual evolutionary change. *Proc R Soc B* **272**, 2269–2275.

Barnard, C. 2003. *Animal Behaviour*. Prentice Hall.

Barto, A. J. 1995. Reinforcement learning. In *The Handbook of Brain Theory and Neural Networks* (ed. M. Arbib), pp. 804–809. MIT Press.

Enquist, M. & Ghirlanda, S. 2005. *Neural Networks and Animal Behavior*. Princeton University Press.

Fahlman, S. E. 1988. Faster-learning variations on back-propagation: an empirical study. *Proceedings of the 1988 Connectionist Models Summer School*. Morgan-Kaufmann.

Franks, D. W. & Noble, J. 2004. Warning signals and predator-prey coevolution. *Proc R Soc B* **271**, 1859–1866.

Ghirlanda, S. & Enquist, M. 1998. Artificial neural networks as models of stimulus control. *Anim Behav* **56**, 1383–1389.

Haas, B. A. 2006. Speciation by perception. *Anim Behav* **72**, 139–146.

Ham, A. D., Ihalainen, E., Lindstrom, L. & Mappes, J. 2006. Does colour matter? The importance of colour in avoidance learning, memorability and generalisation. *Behav Ecol Sociobiol* **60**, 482–491.

Hinton, G. E. & Nowlan, S. J. 1987. How learning can guide evolution. *Complex Syst* **1**, 495–502.

Ihalainen, E., Lindstrom, L. & Mappes, J. 2007. Investigating Mullerian mimicry: predator learning and variation in prey defences. *J Evol Biol* **20**, 780–791.

Kamo, M., Ghirlanda, S. & Enquist, M. 2002. The evolution of signal form: effects of learned versus inherited recognition. *Proc R Soc B* **269**, 1765–1771.

Mazzoni, P., Andersen, R. A. & Jordan, M. I. 1991. A more biologically plausible learning rule for neural networks. *Proc Natl Acad Sci US A* **88**, 4433–4437.

Riedmiller, M. & Braun, H. 1993. A direct adaptive method for faster backpropagation learning: the RPROP algorithm. *Proceedings of the IEEE International Conference on Neural Networks*.

Ruxton, G. D., Sherratt, T. N. & Speed, M. P. 2004. *Avoiding Attack: The Evolutionary Ecology of Crypsis, Warning Signals and Mimicry*. Oxford University Press.

Sherry, D. F. & Mitchell, J. B. 2007. Neuroethology of foraging. In *Foraging: Behaviour and Ecology* (ed. D. Stephens, J. Brown, and R. Ydenberg), pp. 61–104. Chicago University Press.

Shettleworth, S. J. 1998. *Cognition, Evolution and Behaviour*. Oxford University Press.

Skelhorn, J. & Rowe, C. 2005. Frequency-dependent taste-rejection by avian predation may select for defence chemical polymorphisms in aposematic prey. *Biol Lett* **1**, 500–503.

Skelhorn, J. & Rowe, C. 2006a. Avian predators taste-reject aposematic prey on the basis of their chemical defence. *Biol Lett* **2**, 348–350.

Skelhorn, J. & Rowe, C. 2006b. Prey palatability influences predator learning and memory. *Anim Behav* **71**, 1111–1118.

Stephens, D. W. 2007. Models of information use. *In Foraging: Behaviour and Ecology* (ed. D. Stephens, J. Brown & R. Ydenberg), pp. 31–60. Chicago University Press.

Toquenaga, Y., Kajitani, I. & Hoshino, T. 1995. Egrets of a feather flock together. *Artific Life* **1**, 391–411.

17

Neural network evolution and artificial life research

Dara Curran and Colm O'Riordan

17.1 Introduction

Neural networks have been employed as research tools both for machine learning applications and the simulation of artificial organisms. In recent times, much research has been undertaken on the evolution of neural networks where the architecture, weights or both are allowed to be determined by an evolutionary process such as a genetic algorithm. Much of this research is carried out with the machine learning and evolutionary computation community in mind rather than the artificial life community and as such, the latter has been slow to adopt innovative techniques which could lead to the development of complex, adaptive neural networks and in addition, shorten experiment development and design times for researchers.

This chapter attempts to address this issue by reminding researchers of the wealth of techniques that have been made available for evolutionary neural network research. Many of these techniques have been refined into freely available and well-maintained code libraries which can easily be incorporated into artificial life projects hoping to evolve neural network controllers.

The first section of this chapter outlines a review of the techniques employed to evolve neural network architectures, weights or both architectures and weights simultaneously. The encoding schemes presented in this chapter describe the encoding of multi-layer feedforward and recurrent neural networks but there are some encoding schemes which can (and have been) employed to generate more complex neural networks such as spiking (Floreano & Mattiussi, 2001; Di Paulo, 2002) and gasNets (Smith *et al.*, 2002) which are beyond the scope of this chapter. This section includes references to papers drawn from the machine learning, evolutionary computation and artificial life community. The next section discusses some recent work undertaken by artificial life researchers and their approaches to the evolution of neural networks.

17.2 The evolution of neural networks

The evolution of neural networks combines two artificial intelligence tools: evolutionary computation (in particular, genetic algorithms) and neural networks.

Modelling Perception with Artificial Neural Networks, ed. C. R. Tosh and G. D. Ruxton. Published by Cambridge University Press.

Genetic algorithms attempt to apply evolutionary concepts to the field of problem-solving, notably function optimisation, and have proven to be valuable in searching large, complex problem spaces.

Neural networks are highly simplified models of the working of the brain. A neural network consists of a combination of neurons (or nodes) and synaptic connections (or connection weights), which are capable of passing data through multiple layers. The end result is a system which is capable of generalisation, pattern recognition and classification.

In the past, algorithms such as back propagation have been developed to refine one of the principal components of neural networks: the connection weight. The approach has worked well, but is prone to becoming trapped in local maxima and is incapable of optimisation where problems lie in a multi-modal (i.e. the solution landscape contains more than one peak) or non-differentiable problem space.

Genetic algorithms and neural networks can be combined such that a population of neural networks compete with each other in a Darwinian 'survival of the fittest' setting. Networks which are deemed to be fit are combined and passed onto the next generation producing an increasingly fit population, so that following a number of iterations, an optimised neural network can be obtained without resorting to trial-and-error manual tweaking of the neural network architecture and weights.

In addition, evolutionary computation methods of generating neural networks are analogous to the organic evolution of simple nervous systems and thus provide a model of evolution that can be employed to study the origins of biological systems.

The remainder of this section outlines some of the characteristics of neural networks which have been evolved and some of the many possible solutions for representing a neural network in a format suitable to the genetic algorithm, i.e. transforming a network to a gene code or chromosome.

17.2.1 Layers of evolution

The types of evolution of neural networks can be classified according to the goals behind such evolution. Some schemes have proposed the evolution of the weights, starting with a fixed architecture, while others have suggested that the architecture is more important. Other approaches have included the evolution of transfer functions and learning rules. Perhaps the most potentially interesting area for new research is the combination of such techniques. In the following sections, some of the most popular evolutionary frameworks are discussed, including some which combine more than one aspect.

17.2.2 Evolution of weights

The evolution of weights assumes that the architecture of the network remains static. This implies some degree of pre-processing on the part of the human designers and is generally employed when some ideas exist pertaining to the correct structure of the neural network.

One of the primary motivations for using evolutionary techniques to establish the weighting values rather than traditional gradient descent techniques, such as back-propagation, lies in the inherent problems associated with gradient descent approaches. Back-propagation in particular can become easily trapped in local maxima. Furthermore, it is difficult for a back-propagation algorithm to find the optimal solution if the function being optimised is multi-modal or nondifferentiable (Sutton, 1986). In addition, it has been shown that back-propagation is sensitive to the initial condition of the neural network causing additional problems (Kolen & Pollack, 1991).

Evolutionary approaches such as genetic algorithms, however, are able to optimise functions in such environments and, furthermore, do not even require the function to be continuous. This is because genetic algorithms employ fitness functions which can be tailored to suit the problem at hand and are not restricted in the manner that neural networks are.

Much research has been completed on the evolution of connection weights (Montana & Davis, 1989; De Garis, 1990; Whitley *et al.*, 1990; Belew *et al.*, 1992; Branke, 1995; Chellapilla & Fogel, 1999; Sasaki & Tokoro, 1999). Another approach uses the global searching capability of genetic algorithms to search the broad area of the weight problem space, and then uses back-propagation as a local search to refine the weights (Belew *et al.*, 1992). This approach ensures that the networks evolved are very accurate, more accurate than they may have been had genetic algorithms been used in isolation.

17.2.3 Evolution of architectures

Work which addresses the evolution of architectures views structure as the defining characteristic of the neural network. The view is that once a suitable structure is found, an algorithm such as back-propagation can be used to find the correct weights. Prior to the evolutionary approach, techniques consisted of two basic operations: constructive and destructive.

Broadly speaking, the constructive method begins with a minimal network and successively adds nodes and connections until the network is capable of solving the desired problem with sufficient accuracy. The destructive method takes the mirror approach. It begins with an already functioning network and successively removes connections and neurons until the network is no longer able to solve the problem, at which point the last move is undone. There are clearly problems with these kinds of approaches, which are really a type of hill-climbing and are likely to become trapped in local maxima (Angeline *et al.*, 1994).

Research suggests that genetic algorithms may be a more suitable approach to this type of problem and evolution of neural network architectures has been largely successful for both feedforward and recurrent neural networks (Miller *et al.*, 1989; Kitano, 1990; Harp & Samad, 1991; Koza & Rice, 1991; Angeline *et al.*, 1994; Luke & Spector, 1996; Yao, 1999; Stanley & Miikkulainen. 2002).

17.2.4 Transfer functions

The transfer function for all neurons of a neural network is generally taken to be fixed, although some attempts have been made to allow its adaptation over generations (White, 1994; Yao & Liu, 1996). These schemes typically begin with a fixed proportion of transfer functions, such as sigmoidal or Gaussian, and allow the genetic algorithm to adapt to a useful combination according to the situation.

17.2.5 Learning rules

Neural networks have a number of learning rules which govern the speed and accuracy with which the network will train. Examples of these are the learning rate and momentum. These parameters can be difficult to assign by hand, and therefore make good candidates for evolutionary adaptation. Typically the parameters are encoded into the gene code of each network and allowed to evolve (Harp & Samad, 1991; Hochman *et al.*, 1996).

17.2.6 Simultaneous evolution

One of the most interesting areas of evolutionary neural networks is the combination of several schemes which simultaneously evolve different aspects of the networks. One of the most important is the combination of architecture and weight evolution (De Garis, 1990; Koza & Rice, 1991; Maniezzo, 1993; Zhang & Muhlenbein, 1993; Gruau, 1994; White, 1994; Richards *et al.*, 1997; Hussain & Browse, 1998; Yao, 1999). The advantage of combining these two basic elements of a neural network is that a completely functioning network can be evolved without the need to specify the number of nodes in a network, the connections between the nodes or the weights of each connection. It might be advantageous to simultaneously evolve more neural network features, thus leading to more efficient and accurate results. However, it is not clear whether the increase in complexity of the resulting encoding scheme would be offset by a marked improvement in performance. Typical convergence times for combined approaches run into days, and the addition of more complexity might make the problem intractable.

17.2.7 Comparison of approaches

A number of different criteria were selected to evaluate each approach. The first of these is complexity. Empirical results indicate that more complex approaches to neuro-evolution do not necessarily provide an improvement in performance (Koehn, 1996). Therefore, it seems logical to favour simpler approaches to the expense of complex ones.

The second evaluation criterion is the expressiveness of each approach. The expressiveness of an approach indicates the flexibility of neural networks that can be generated by the evolutionary process, in terms of both connection weights and architecture. In other words, expressiveness represents the types of networks that can be evolved. A very expressive encoding allows the evolutionary process great freedom in selecting more complex network architectures while a poorly expressive encoding constrains the evolutionary algorithm by restricting the type of network that can be evolved.

We have attempted to classify each approach according to these criteria in as objective a manner as possible based on previous research and our experiences.

Table 17.1 shows a comparison of each of the approaches outlined above according to these two criteria.

17.3 Encoding strategies

Crucial to the successful evolution of a neural network is the way its structure and/or weights are encoded into the chromosome used by the genetic algorithm. A great deal of research has been carried out in this area and as a consequence many systems exist. These can be divided into two main categories: direct and indirect encoding.

17.3.1 Direct encoding

Direct encoding is a strategy where some or all of a neural network's defining parameters, such as weight values, number of nodes, connectivity etc., are encoded into the gene code. Thus it is possible to recreate the exact neural network from the underlying genotype (Koehn, 1994).

17.3.1.1 Connectionist encoding

Connectionist encoding is concerned with mapping a neural network's connectivity to the gene code. In other words, connectionist encoding concentrates on the connections between the *weights* of a neural network and the network's *nodes*.

An early implementation of such an encoding scheme is that of Miller's Innervator (Miller *et al.*, 1989). Innervator has a fixed number of nodes and connectivity is denoted by a single bit. A matrix is then derived containing a full connectivity map of the neural network. The networks are selected according to their performance following several epochs of training.

Another approach to connectivity encoding is to encode the weights for each connection in the neural network. In the canonical genetic algorithm (Holland, 1975; Goldberg, 1989), chromosomes are defined as a binary string – so early approaches used a binary string to encode the real number weight values (Whitley *et al.*, 1990; Belew *et al.*, 1992).

This is the approach of GENITOR (Whitley *et al.*, 1990) where each connection and its corresponding weight are encoded into the gene code. Each weight in GENITOR is encoded as an 8-bit binary string, indexed by a single bit which is used to denote the presence or absence of the connection. The weight values are first evolved to an optimum, and then a pruning algorithm streamlines the neural network's architecture, by changing the index bit as required.

Table 17.1. *A comparison of each of the approaches outlined.*

Layers of evolution		
Approach	Complexity	Expressiveness
Evolution of weights (Montana & Davis, 1989; De Garis, 1990; Whitley *et al.*, 1990; Belew *et al.*, 1992; Branks, 1995; Chellapilla & Fogel, 1999; Sasaki & Tokoro, 1999).	**Simple**: Weights can be easily encoded into a genome and evolved using standard evolutionary computation techniques.	**Poor**: Network architectures are fixed and the scheme only evolves the weighting values of connections between nodes.
Evolution of architectures (Miller *et al.*, 1989; Kitano, 1990; Harp & Samad, 1991; Koza & Rice, 1991; Angeline *et al.*, 1994; Luke & Spector, 1996; Stanley & Miikkulainen, 2002).	**Quite simple**: The architecture of the network (connections and nodes) can be encoded into a graph-like structure and evolved using evolutionary computation techniques.	**Poor**: While evolving the topology of the network is important, weight values should also be considered.
Transfer functions (White, 1994; Yao & Liu, 1996).	**Quite simple**: Typically, transfer functions are evolved along with weight or architecture evolution.	**Mediocre**: Evolving a suitable transfer function can be seen as a valuable first step in the evolutionary process but is not expressive enough because it is generally combined with either weight or architecture. evolution.
Learning rules (Harp & Samad, 1991; Hochman *et al.*, 1996).	**Quite simple**: Learning rules can be encoded into a genome and evolved in a relatively simple manner.	**Mediocre**: Learning rule evolution is similar to transfer function evolution in that it provides a potentially useful first step by evolving the most adequate parameter setting. However, on its own, it does not provide enough expressiveness to evolve flexible neural networks.
Simultaneous evolution (De Garis, 1990; Koza & Rice, 1991; Maniezzo, 1993; Zhand & Muhlenbein, 1993; Gruau, 1994; White, 1994; Richards *et al.*, 1997; Hussain & Browse, 1998; Yao, 1999).	**Quite complex**: The genome must contain both weight and architecture information, making it more complex than previous approaches.	**Good**: Evolving both the architecture and weights of a neural network gives the evolutionary process more flexibility and is more likely to produce useful problem-solving neural networks.

There are obvious advantages to preserving the traditional binary approach. It is both very simple and extremely general. In addition, no new genetic operators need to be devised. However, converting a real number to a binary representation invariably means a loss of accuracy, unless very large strings are created. Large chromosomes are very often detrimental to the genetic algorithm's performance in term of processing time (Holland, 1975).

To combat this problem, Montana & Davis (1989) devised an encoding scheme which represents weights as real numbers. They also created a number of tailored genetic operators which are able to deal with the change. Another approach has been the use of integer fractions to denote the real number value (Gruau, 1995; Hochman *et al.*, 1996), as opposed to approximating the value by direct encoding.

17.3.1.2 Node-based encoding

Node-based encoding strategies concentrate on the number of neurons, or nodes, which should be used. While weight-encoding schemes assume that an architecture has already been designed, the construction of an efficient network structure is just as difficult and is therefore well suited to a genetic algorithm approach.

In Schiffmann *et al.*'s (1993) approach, a blueprint is used to describe the neural network's structure. Starting with the input node, each node is numbered in sequence and placed in a list. Then, each node in the list is taken in turn and the numbers of the connecting nodes from the previous network layers are placed before its entry in the list, forming a complete node mapping of the network. Crossover is implemented between nodes only, and several mutation operators are used to add or delete weak connections.

A similar system, GANNet, has been devised (White, 1994), but with a few restrictions – notably on the number of input nodes and the fact that only connections between adjacent layers are allowed.

17.3.1.3 S-expressions

The use of LISP Symbolic expressions has been adopted in the creation of an alternate encoding strategy. Each network is represented by a number of functions (representing nodes) and terminals. Rather than encoding the neural network structure as a list, Koza & Rice (1991) represent this as a parameter tree.

A number of operators can be used to define the network: arithmetic and weighting functions can be combined to create the weights of the network and the bias of a node can be altered via special processing nodes in the grammar tree.

The S-expressions approach to encoding neural networks results in quite streamlined gene codes which do not suffer from the same scalability problems as direct encoding. Crossover takes place at a sub-tree level, ensuring that learned portions of the network are not entirely disrupted.

17.3.1.4 Layer-based encoding

Layer-based encoding uses a chromosome which is subdivided into areas corresponding to the neural networks' layers. In the GENESYS system (Harp & Samad, 1991), each area has an identifying index, the number of nodes within it and a number of projector fields which specify a node's connectivity to the next layer. Mandischer (1993) modified the system and specifies a radius and density of connections for each layer. The radius describes the spread of connections to nodes in the given area, while the density indicates how many nodes in the layer are connected. Crossover is applied between layers and mutation alters the learning rate, momentum, radius, density and the number of nodes in a layer.

17.3.1.5 Marker-based encoding

Marker-based encoding is inspired by the structure of DNA in living organisms (Moriarty & Miikkulainen, 1995). In DNA, structures known as nucleotide triplets specify amino acids which make up a protein. Some triplets are given special 'marker' status, which allows them to denote the start and end of the protein definition.

Marker-based chromosomes are said to be circular, in that one end of the chromosome can be wrapped around to join the other. The start marker does not necessarily have to be placed at the start of the gene code because the algorithm reads the chromosome until such a marker is found. The scheme can be said to be complete, in that any gene code can be correctly converted into a functioning network. Marker-based encoding certainly allows more freedom in the definition of a neural network and the mutation and crossover operators can operate without restrictions.

17.3.1.6 Neuro-evolution of augmenting topologies (NEAT)

The NEAT system evolves both neural network structure and weights by incrementally increasing the complexity of a neural network (Stanley & Miikkulainen, 2002). A NEAT genotype is made up of connection genes, each describing connections between two nodes, including weight and node labels. In addition, layer information is also encoded into the genome. The main innovation of NEAT is that the system keeps track of the historical origin of each gene and ensures that genes that do not have common ancestors do not compete against each other. This means that NEAT eliminates the problem of competition between neural network architectures that are fundamentally different.

17.3.2 Indirect encoding

The indirect encoding strategies attempt to describe a neural network in terms of assembly instructions or recipes (Schiffmann *et al.*, 1993). While in the direct encoding approach, parameters of the network were explicitly present in the genetic code, with indirect encoding, only a method of assembling the network is present.

The main motivations for such a shift are size and modularity. While direct encoding schemes are, in general, quite straightforward to implement, they suffer from a lack of scalability – the more complex the network the more computationally expensive they become due to the size of the gene code in the genetic algorithm. This problem is less prevalent with indirect encoding methods.

17.3.2.1 Matrix re-writing

One of the first indirect encoding schemes proposed is that of Kitano's matrix rewriting (Kitano, 1990). The scheme is based around the connectivity matrix seen in direct encoding schemes (Miller *et al.*, 1989). It begins with a base 2×2 matrix and repeatedly applies generation rules to each non-terminal element in the matrix until all elements are terminals.

Generally a fixed number of rewriting steps is used to create the final matrix. These rewriting rules can be encoded into a gene code. In this case, each rule can correspond to four alleles on the chromosome corresponding to the first start rule matrix, containing A and B. Typically the rules defining the final re-writing step (i.e. to the binary stage) are predefined and do not play a part in the evolution of the rules.

Some good results have been reported for this scheme (Kitano, 1990). However, recent work (Siddiqi & Lucas, 1998) has shown that direct encoding can be at least as good as the matrix rewriting proposed here.

17.3.2.2 Cellular encoding

Cellular encoding, created by Gruau (1994, 1995), represents neural networks as grammar trees, i.e. the grammar describing the network is encoded as a tree. The building block of cellular encoding is the cell which represents a node in an ordered graph. Each cell has a reading head which reads the cellular code and acts upon the instructions therein. The cell manages internal variables which can govern its development or regulate neural network-related parameters such as weights or thresholds. The cell can be viewed as a Turing machine, only reading sections of the grammar tree instead of tape.

The development begins with a single cell known as the ancestor cell, which is connected to an input and an output cell. The cell's reading head is placed at the start of the cellular code (itself in the beginning) and executes the operator located there. Various operators exist which control division and bias modification.

Cellular encoding can be used to evolve both weights and architecture of a neural network (Gruau *et al.*, 1996). It compares quite favourably with direct encoding, in that while the cellular encoding takes longer to compute, the relative amount of effort required to achieve efficient neural networks make it attractive.

17.3.2.3 Edge encoding

A scheme similar to cellular encoding, edge encoding (Luke & Spector, 1996) grows network graphs using edges instead of nodes. While the cellular encoding approach evaluates each grammar tree node in a breadth-first search manner, thus ensuring parallel execution, edge encoding is designed to work using depth-first search.

The cellular encoding approach tends to create graphs with a large number of connections which must then be pruned using the CUT or CLIP operators. Edge encoding, using its depth-first search approach, favours graphs with fewer connections. However, the relative lack of connections does not necessarily imply a smaller genetic code – on the contrary, edge encodings often have larger gene codes than those produced by cellular encoding. This is because there is implicitly more information required to store edges in a network graph than is required when simply storing node information.

Luke & Spector (1996) argue that this is not too significant and that the real benefit of edge encoding is that modularity can be created through the development of building blocks resulting from their depth-first search approach.

17.3.2.4 Lindermayer-systems

The set of Lindermayer-systems are an encoding scheme based on the work of Lindermayer. They are based on a biological model where cells exchange information with their neighbours. They use a specialised grammar in which production rules are applied in parallel rather than sequentially, as seen in previous examples. Boers & Kuiper (1992) have used this model to generate neural networks. The rewriting rules take into account the relative position of each cell to determine whether the production rule should apply. All possible production rules which are applicable are applied immediately, rather than waiting for other portions of the graph to catch up.

The encoding method generates strings which are not always guaranteed to produce correct production rules. Therefore, error recovery operations were implemented to address this problem. The system was successful with the XOR problem and simple letter recognition tasks.

17.3.2.5 Growth encoding

Cangelosi *et al.* (1994) have argued for a more biological approach to the evolution of neural networks. They criticise the direct encoding mechanism for its lack of scalability and also for its lack of biological plausibility as it is unlikely that the entire nervous system of an organism is mapped out in detail in its genetic code.

Their work is based on an earlier project which was concerned with simulating the growth of synaptic connections between neurons in a neural network (Nolfi & Parisi, 1992). In the earlier work, the gene code contained information on the manner in which connections were allowed to grow from neurons. The network was mapped in 2D space, and those connections which had reached other nodes in the given time frame were considered valid; others were discarded.

Cangelosi takes the work further by taking this previous growing principle but also adding the possibility of nodes dividing and migrating in their 2D environment. The set of rules employed is similar to the rule rewriting schemes seen earlier but the application of the rules is radically different from previous approaches. The genetic code specifies the type of cell which is being represented, the number of divisions allowed per cell, the connection growing rate and the angle of growth.

17.4 Comparison of encoding schemes

The lack of theoretical foundations for the vast majority of neural network encoding schemes makes direct comparison between them difficult. While there are a large number of articles detailing individual encoding schemes, each is tested on different problem tasks, populations and parameters, making it impossible to compare the performance of each approach.

As before, a number of different criteria must be selected to evaluate each scheme. Given that empirical results suggest that more complex systems do not necessarily provide an increase in performance (Koehn, 1996), complexity becomes an undesirable element and can be used as an evaluation criterion for the schemes.

A second criterion for comparison is the expressiveness of each scheme (how flexibly and completely the scheme encodes the weights and architecture of neural networks). The expressiveness of a scheme indicates the flexibility of neural networks that can be generated by the evolutionary process, in terms of both connection weights and architecture.

Table 17.2 shows a comparison of direct and indirect encoding schemes according to these two criteria.

17.5 Case studies

The following case studies highlight recent work in Artificial Life research which employs evolutionary neural networks. The case studies were selected as a sample of Artificial Life research spanning a number of subjects which are typically of interest to Artificial Life researchers:

- Imitation.
- Navigation.
- Representation of the world.
- Social learning.

Of the four presented, only two use an encoding technique which simultaneously evolves the architecture and weights of the neural networks. This underlines the need for a more widespread acceptance of the many available encoding techniques within the Artificial Life community. Embracing these techniques will allow researchers to attempt more challenging experiments to simulate more complex behaviour and cognition.

17.5.1 Examining the effects of imitation on populations of evolutionary neural networks

Work by Borenstein & Ruppin (2003) examined the effect of allowing agents to imitate each other to solve a number of problem tasks of increasing difficulty. The work employed populations of neural networks which evolved using a genetic algorithm. Experiments compared the performance of neural networks using evolutionary learning and neural networks employing both evolutionary learning and learning by imitation.

The neural network architecture for each problem was defined by the experimenters and fixed for the duration of the experiments. The evolutionary algorithm allowed only the weights of each network to evolve using a simple direct encoding mechanism.

Table 17.2. *A comparison of direct and indirect encoding schemes.*

Direct encoding schemes		
Encoding scheme	Complexity	Expressiveness
Connectionist encoding (Whitley *et al.*, 1990).	**Simple:** can either be implemented as a binary connection map or as a direct encoding of weight values.	**Poor:** network architectures are fixed and the scheme only encodes connections between nodes.
Node-based encoding (Schiffman *et al.*, 1993; White, 1994).	**Quite simple:** genome represents a tree of node information.	**Poor:** can be used to evolve connection weights or architectures, but not both simultaneously.
S-expressions (Koza & Rice, 1991).	**Quite complex:** neural networks represented using LISP S-expressions. Results in quite streamlined genomes.	**Good:** the scheme can be made to encode both neural network weights and architecture.
Layer-based encoding (Harp & Samad, 1991).	**Complex:** encoding contains complex descriptions of connectivity between lists of layers. Specialist genetic operators are required.	**Poor:** weights values are typically not encoded.
Marker-based encoding (Moriarty & Miikkulainen, 1995).	**Quite simple:** neural networks are represented as strings, with markers delineating node information.	**Good:** encoding allows evolution of both architecture and weights for networks of any size.
NEAT (Stanley & Miikkulainen, 2002).	**Quite complex:** the historical origin of each gene must be maintained as it has a direct bearing on the selection and crossover process.	**Good:** encoding allows evolution of both architecture and weights for networks of any size.
Matrix re-writing (Kitano, 1990).	**Quite complex:** it isbased on a connectivity matrix that is manipulated using a number of rewrite rules encoded in the genome.	**Quite good:** the scheme can evolve both weights and architecture, but recent work has shown that direct encoding can be at least as good as matrix rewriting.
Cellular encoding (Gruau, 1994, 195).	**Quite complex:** neural networks are represented by grammar trees and a number of operators are used to encode/decode the genome. Computing time is significantly higher than direct encoding schemes.	**Good:** both weights and architecture of a neural network can be encoded.
Edge encoding (Luke & Spector, 1996).	**Quite complex:** this scheme is similar to cellular encoding but uses edges rather than nodes as its base unit for grammar trees.	**Good:** both weights and architecture of a neural network can be encoded.

Table 17.2. (*cont*)

Direct encoding schemes		
Encoding scheme	Complexity	Expressiveness
Lindermayer-systems (Boers & Kuiper, 1992)	**Complex:** neural networks are evolved using a grammar tree where production rules are applied in parallel. Error recovery operations must be included to guarantee validity.	**Good:** both weights and architecture of a neural network can be encoded.
Growth encoding (Cangelosi *et al.*, 1994).	**Complex:** neural networks are mapped to 2D space and connections between nodes are grown according to a number of rules, including number of divisions allowed per cell, the growing rate and the angle of growth.	**Good:** both weights and architecture of a neural network can be encoded.

While the encoding scheme employed was sufficient to allow the experimenters to obtain interesting results (particularly on the simpler problems), it is possible that using a more sophisticated encoding mechanism evolving both weights and architectures would have allowed the experimenters to obtain better results for the final, more complex problem.

17.5.2 Insect path integration

A recent adaptive behaviour paper by Haferlach *et al.* (2007) employed evolutionary neural networks to develop a path integration technique for robot navigation. Path integration is a navigation strategy employed by many animal species which explains the ability of many animals to return home via a direct route after a prolonged foraging expedition, even in an environment with few landmarks.

The authors chose to allow both the weights and architecture of each neural network to evolve using the marker-based encoding scheme (see Section 17.3.1). This allowed neural network structures to evolve free of many constraints and was capable of generating neural networks with complex topographies, although the size of networks was limited to 50 nodes.

The resulting neural networks were both compact and effective and the best of these was loaded onto a robot to test the evolved behaviour in a real-world situation. The experiment showed that the evolved neural network was robust enough to perform well even when exposed to noisy real-world environments.

17.5.3 Development of spatial representations in robots

Research by Gigliotta & Nolfi (2008) presented an evolutionary technique to allow robots to develop spatial representations of their environment and to self-localise themselves within this environment. Discriminating between different environmental features and to localise oneself in an environment are complex cognitive processes and form key components of animal navigation.

The experiments presented in the paper placed a robot in a maze environment where the robot was required to identify when it found itself in sections of the maze it had previously visited. The robot was controlled by a neural network which perceived the environment through a number of sensors and performed actions based on the output of three motor neurons.

Each neural network had a fixed architecture and the researchers employed a simple direct encoding technique to evolve the neural network weights. Although the results of the experiments showed that the robots were capable of evolving the desired behaviour, the fitness of the evolving neural networks over time shows a dramatic jump from very poor to very fit over five generations. It is therefore likely that the evolutionary process happened to stumble across a useful mutation which completely changed the dynamic of the neural network controller. Had the researchers employed a neural network encoding mechanism which evolved both weights and architecture simultaneously, it is likely that the evolutionary path would have been less difficult for the evolutionary algorithm.

17.5.4 The effects of cultural learning in populations of neural networks

Research undertaken by Curran & O'Riordan (2006) examines the effect of adding cultural learning to populations of neural networks that evolve using a genetic algorithm. Culture can be defined as the transmissions of nongenetic information from one generation to another. Cultural learning refers to the process whereby individuals within a population acquire knowledge from others.

Each neural network was allowed to evolve using marker-based encoding (see Section 17.3.1), where both weights and architecture of the neural network evolve simultaneously. A number of experiments were carried out examining the effects of cultural learning using three benchmark tasks: the 5-bit parity problem, the game of tic-tac-toe and the game of connect-four. Results showed that the addition of cultural learning promotes improved fitness and significantly increases both genotypic (the genetic make-up of individuals) and phenotypic (the behaviour of individuals) diversity in the population.

17.6 Chapter summary

This chapter has outlined the state of current research into the encoding of neural networks for the purposes of evolution by genetic algorithm. While a large number of different implementations exist, the basic dichotomy between direct and indirect encoding remains prevalent within the evolutionary neural network community and no system has become dominant. However, it must be said that direct encoding has a larger following than indirect encoding. This may be due to a number of factors including the relative ease of implementation and the intuitive mapping that such systems provide. As many of the neural network architectures that are evolved in the majority of experiments tend to be relatively small in scale, the advantages of indirect encoding in compressing the available information may not necessarily outweigh the disadvantages of difficult implementation and, in some cases, poor efficiency and execution time.

The chapter also presents a number of case studies of recent Artificial Life research employing evolutionary neural networks. Of these, only two employ an encoding scheme which simultaneously evolves both network weights and architectures. From this, it is clear that the Artificial Life community is not making full use of the available evolutionary neural network research which could lead to more complex simulation of behaviour and cognition.

References

Angeline, P. J., Saunders, G. M. & Pollack, J. P. 1994. An evolutionary algorithm that constructs recurrent neural networks. *IEEE Trans Neural Netw* **5**(1), 54–65.

Belew, R. K., McInerney, J. & Schraudolph, N. N. 1992. Evolving networks: using the genetic algorithm with connectionist learning. In *Artificial Life II* (ed. C. G. Langton, C. Taylor, J. D. Farmer & S. Rasmussen), pp. 511–547. Addison-Wesley.

Boers, E. J. W. & Kuiper, H. 1992. *Biological Metaphors and the Design of Artificial Neural Networks*. Master's thesis, Leiden.

Borenstein, E. & Ruppin, E. 2003. Enhancing autonomous agents evolution with learning by imitation. *Interdisciplin J Artific Intell Simul Behav* **1**(4), 335 – 348.

Branke, J. 1995. Evolutionary algorithms for neural network design and training. *Technical Report No. 322. University of Karlsruhe, Institute AIFB.*

Cangelosi, A., Nolfi, S. & Parisi, D. 1994. Cell division and migration in a 'genotype' for neural networks. *Network* **5**, 497–515.

Chellapilla, K. & Fogel, D. B. 1999. Evolving neural networks to play checkers without relying on expert knowledge. *IEEE Trans Neur Netw* **10**, 1382–1391.

Curran, D. & O'Riordan, C. 2006. Increasing population diversity through cultural learning. *Adapt Behav* **14**(4), 315–338.

De Garis, H. 1990. Genetic programming: building artificial nervous systems using genetically programmed neural network modules. In *Machine Learning: Proceedings of the Seventh International Conference* (ed. B. W. Porter & R. J. Mooney), pp. 132–139. Morgan Kaufmann.

Di Paolo, E. A. 2002. Evolving spike-timing dependent synaptic plasticity for robot control. In *EPSRC/BBSRC International Workshop: Biologically-inspired Robotics, The Legacy of W. Grey Walter, WGW2002.*

Floreano, D. & Claudio Mattiussi, C. 2001. Evolution of spiking neural controllers for autonomous vision-based robots. In *Evolutionary Robotics. From Intelligent Robotics to Artificial Life* (ed. T. Gomi), pp. 38–61. Springer.

Gigliotta, O. & Nolfi, S. 2008. On the coupling between agent internal and agent/environmental dynamics: development of spatial representations in evolving autonomous robots. *Adaptive Behav* **16**(2–3), 148–165.

Goldberg, D. E. 1989. *Genetic Algorithms in Search, Optimization and Machine Learning*. Addison-Wesley.

Gruau, F. 1994. *Neural Network Synthesis using Cellular Encoding and the Genetic Algorithm*. PhD thesis, Centre d'etude nucleaire de Grenoble, Ecole Normale Superieure de Lyon.

Gruau, F. 1995. Automatic definition of modular neural networks. *Adapt Behav* **3**(2), 151–183.

Gruau, F., Whitley, D. & Pyeatt, L. 1996. A comparison between cellular encoding and direct encoding for genetic neural networks. In *Genetic Programming 1996: Proceedings of the First Annual Conference*, pp. 81–89.

Haferlach, T., Wessnitzer, J., Mangan, M. & Webb, B . 2007. Evolving a neural model of insect path integration. *Adapt Behav* **15**(3), 273–287.

Harp, S. & Samad, T . 1991. Genetic synthesis of neural network architecture. In *Handbook of Genetic Algorithms* (ed. L. Davis), pp. 202–221. Van Nostrand Reinhold.

Hochman, R., Khoshgoftaar, T. M., Allen, E. B. & Hudepohl, J. P. 1996. Using the genetic algorithm to build optimal neural networks for fault-prone module detection. In *Proceedings of the Seventh International Symposium on Software Reliability Engineering,* pp. 152–162. IEEECS.

Holland, J. H. 1975. *Adaptation in Natural and Artificial Systems.* The University of Michigan Press.

Hussain, T. S. & Browse, R. A. 1998. Genetic encoding of neural networks using attribute grammars. In *CITO Researcher Retreat*, Hamilton, Ontario, Canada.

Kitano, H. 1990. Designing neural networks using genetic algorithm with graph generation system. *Complex Syst* **4**, 461–476.

Koehn, P. 1994. Combining genetic algorithms and neural networks: The encoding problem. Master's thesis, University of Erlangen and The University of Tennessee.

Koehn, P. 1996. Genetic encoding strategies for neural networks. In *Proceedings of Information Processing and Management of Uncertainty in Knowledge-Based Systems.*

Kolen, J. F. & J. B. Pollack, J. B. 1991. Back propagation is sensitive to initial conditions. *Adv Neural Inform Process Syst* **3**, 860 – 867.

Koza, J. R. & Rice, J. P. 1991. Genetic generation of both the weights and architecture for a neural network. In *International Joint Conference on Neural Networks, IJCNN-91*, Vol. II, pp. 397–404. IEEE Computer Society Press.

Luke, S. & Spector, L. 1996. Evolving graphs and networks with edge encoding: preliminary report. In *Late Breaking Papers at the Genetic Programming 1996 Conference Stanford University* (ed. J. R. Koza), pp. 117–124. Stanford University.

Mandischer, M. 1993. Representation and evolution of neural networks. In *Artificial Neural Nets and Genetic Algorithms. Proceedings of the International Conference at Innsbruck, Austria* (ed. R. F. Albrecht, C. R. Reeves & N. C. Steele), pp. 643–649. Springer.

Maniezzo, V. 1993. Searching among search spaces: Hastening the genetic evolution of feedforward neural networks. In *Artificial Neural Nets and Genetic Algorithms. Proceedings of the International Conference at Innsbruck, Austria* (ed. R. F. Albrecht, C. R. Reeves & N. C. Steele), pp. 635–643. Springer-Verlag.

Miller, G., Todd, P. M. & Hedge, S. U. 1989. Designing neural networks using genetic algorithms. In *Proceedings of the Third International Conference on Genetic Algorithms and Their Applications*, pp. 379–384.

Montana, D. J. & Davis, L. 1989. Training feedforward neural networks using genetic algorithms. In *Proceedings of the Eleventh International Joint Conference on Artificial Intelligence*, pp. 762–767. Morgan Kaufmann.

Moriarty, D. E. & Miikkulainen, R. 1995. Discovering complex othello strategies through evolutionary neural networks. *Connect Sci* **7**(3–4), 195–209.

Nolfi, S. & Parisi, D. 1992. Growing neural networks. *Technical Report.* Institute of Psychology, CNR Rome.

Richards, N, Moriarty, D., McQuesten, P. & Miikkulainen, R. 1997. Evolving neural networks to play Go. In *Proceedings of the 7th International Conference on Genetic Algorithms, East Lansing, MI.*

Sasaki, T. & Tokoro, M. 1999. Evolving learnable neural networks under changing environments with various rates of inheritance of acquired characters: comparison between Darwinian and Lamarckian evolution. *Artific Life* **5**(3), 203–223.

Schiffmann, W., Joost, M. & Werner, R. 1993. Application of genetic algorithms to the construction of topologies for multilayer perceptrons. In *Proceedings of the International Conference on Artificial Neural Networks and Genetic Algorithms*, pp. 675–682.

Siddiqi, A. & Lucas, S. 1998. A comparison of matrix rewriting versus direct encoding for evolving neural networks. In *Proceedings of the 1998 IEEE International Conference on Evolutionary Computation*, pp. 392 – 397.

Smith, T., Husbands, P., Philippides, A. & O'Shea, M. 2002. Neuronal plasticity and temporal adaptivity: Gasnet robot control networks. *Adapt Behav* **10**(3–4), 161–184.

Stanley, K. O. & Miikkulainen, R. 2002. Efficient evolution of neural network topologies. In *Proceedings of the 2002 Congress on Evolutionary Computation CEC 2002* (ed. D. B. Fogel *et al.*), pp. 1757–1762. IEEE Press.

Stanley, K. O. & Miikkulainen, R. 2002. Evolving neural networks through augmenting topologies. *Evol Comput* **10**(2), 99–127.

Sutton, R. S. 1986. Two problems with backpropagation and other steepest-descent learning procedures for networks. In *Proceedings of 8th Annual Conference of the Cognitive Science Society*, pp. 823–831.

White, D. W. 1994. *GANNet: A Genetic Algorithm for Searching Topology and Weight Spaces in Neural Network Design*. PhD thesis, University of Maryland College Park.

Whitley, D., Starkweather, T. & Bogart, C. 1990. Genetic algorithms and neural networks – optimizing connections and connectivity. *Parallel Comput* **14**(3), 347–361.

Yao, X. 1999. Evolving artificial neural networks. In *Proceedings of the IEEE*, pp. 1423–1447.

Yao, X. & Liu, Y. 1996. Evolving artificial neural networks through evolutionary programming. In *Proceedings of the 5th Annual Conference on Evolutionary Programming*, pp. 257–266. MIT Press.

Zhang, B. & Muhlenbein, H. 1993. Evolving optimal neural networks using genetic algorithms with Occam's razor. *Complex Syst* **7**(3), 199–220.

18

Current velocity shapes the functional connectivity of benthiscapes to stream insect movement

Julian D. Olden

18.1 Introduction

Ecological thresholds have long intrigued scientists, dating from the study of threshold effects for age-specific human mortality (Gompertz, 1825) to present-day investigations for biodiversity conservation and environmental management (Roe & van Eeten, 2001; Folke *et al.*, 2005; Huggett, 2005; Groffman *et al.*, 2006). Defined as a sudden change from one ecological condition to another, ecological thresholds are considered synonymous with discontinuities in any property of a system that occurs in nonlinear response to smooth and continuous change in an independent variable. Understanding ecological thresholds and incorporating them into ecological and socio-ecological systems is seen as a major advance in our ability to forecast and thus properly cope with environmental change (Carpenter, 2002; Rial *et al.*, 2004; Gordon *et al.*, 2008). Consequently, ecologists and economists continue to be extremely attracted to the idea that ecological thresholds may exist and can be used in a management context (Muradian, 2001).

Empirical studies of ecological thresholds are diverse and have grown in number in recent years (Walker & Meyers, 2004). In landscape ecology, a working hypothesis is the existence of critical threshold levels of habitat loss and fragmentation that result in sudden reductions in species' occupancy (Gardner *et al.*, 1987; Andrén, 1994; With & Crist, 1995). As the landscape becomes dissected into smaller and smaller parcels, landscape connectivity – referring to the spatial contagion of habitat – may suddenly become disrupted (With & Crist, 1995). In this case, habitat fragmentation leads to decreased landscape connectivity, thereby limiting organism dispersal that is essential for maintaining population viability (Kareiva & Wennergren, 1995; Fahrig, 2001). Critical thresholds, therefore, may also comprise so-called extinction thresholds (*sensu* Lande, 1987), which represent a transition point in the functionality of a landscape when a population is abruptly and unpredictably lost. Evidence for such critical thresholds to animal movement has been advanced by advocates of neutral landscape models (Gardner *et al.*, 1987; With & King, 1997), in particular the application of percolation theory used to describe how landscape structure affects the abundance, distribution and behaviour of organisms (McIntyre & Wiens, 1999).

Modelling Perception with Artificial Neural Networks, ed. C. R. Tosh and G. D. Ruxton. Published by Cambridge University Press.

Simulation experiments have been used extensively to predict whether critical thresholds to landscape connectivity exist and how this affects the movement of animals (With *et al.*, 1997). Empirical evidence for such thresholds, however, remains scarce and is limited almost exclusively to terrestrial insects (Wiens *et al.*, 1997; McIntyre & Wiens, 1999; Schooley & Wiens, 2003, 2005). Moreover, while it is clear that critical thresholds to animal movement are likely to depend on a number of factors, including dispersal ability (Keitt *et al.*, 1997), the movement 'rules' of an organism (Pearson *et al.*, 1996) and the specifics of the environment (Loxdale & Lushai, 1999), it is still unclear whether such hypotheses are supported by empirical data. Given differences in how species perceive and respond to landscape heterogeneity (With, 1994; Olden *et al.*, 2004a; Pe'er & Kramer-Schadt, 2008) we might predict that critical thresholds to organism movement will differ not only between species, but also within species occupying different environments.

Stream ecosystems provide a strong testing ground for exploring issues of environmental heterogeneity and the influence of habitat patchiness on animal movement behaviour (Downes *et al.*, 1993; Palmer *et al.*, 2000; Malmqvist, 2002). Streams contain predominantly mobile rather than sessile invertebrates, and their dispersal is determined by many factors, including intrinsic characteristics of the species (e.g. mobility, resource specificity, perceptual range) and/or features of the environment (Mackay, 1992). In streambed landscapes, or benthiscapes, the pervasive action of flowing water is a dominant physical force that generates habitat patches located in highly variable, near-bed current dynamics (Pringle *et al.*, 1988; Statzner *et al.*, 1988; Hart *et al.*, 1996; Hart & Finelli, 1999). As a result, benthic insects experience considerable physical heterogeneity during their normal activities (Palmer *et al.*, 1996). From an insect's perspective, this heterogeneous benthic landscape represents a 3D mosaic of both habitat patchiness and flow conditions, which can influence the connectivity or permeability of the landscape (*sensu* Taylor *et al.*, 1993) to movement, foraging activities and likelihood of colonisation (e.g. Hart & Resh, 1980; Poff & Ward, 1992; Lancaster, 1999; Palmer *et al.*, 2000; Wellnitz *et al.*, 2001). Variation in dispersal ability means that patchiness of resources and habitat availability is likely perceived differently by stream insects (Olden *et al.*, 2004b), resulting in differential patterns of movement and habitat occupancy (Hoffman *et al.*, 2006). Consequently, benthiscapes are an ideal setting for addressing the influence of landscape-processes on species' distributions, and thus searching for critical thresholds to animal movement in response to landscape connectivity.

The physical complexity of stream benthiscapes necessitates equally complex models that aim to unravel the threshold relationships between landscape structure and animal movement. These statistical approaches need to be able to model nonlinear relationships among variables, account for variable interactions and provide high predictive power without sacrificing explanatory power. One such technique that has gained greater attention in recent years is the application of artificial neural networks; a machine learning approach increasingly used by ecologists (Olden *et al.*, 2008). An artificial neural network is an information-processing paradigm inspired by the way biological nervous systems, such as the mammalian brain, process complex information (McCulloch & Pitts,

1943). The key element of this paradigm is the novel structure of the information-processing system, which is composed of a large number of highly interconnected elements called neurons, working in unity to solve specific problems. Neural networks are being used in greater frequency by ecologists because of their perceived predictive ability, although their explanatory utility is equally powerful but continues to be under-appreciated (Olden & Jackson, 2002; Olden *et al.*, 2004c). Here, I used neural networks primarily as an explanatory tool to understand the direct and interactive roles of habitat abundance and current velocity for shaping threshold-responses of animal movement in complex benthiscapes.

In this study I report on the results from a series of stream-side experiments investigating the influence of habitat abundance and current velocity on the movement dynamics of two ubiquitous herbivores. The study organisms are a caddisfly larva (*Agapetus boulderensis* Milne, 1936) and a freshwater snail (*Physa* sp.); benthic grazers that differ substantially in their body morphology and mobility and are therefore likely to perceive and respond differently to landscape structure (Figure 18.1). Given these intrinsic differences in mobility and the manner in which they perceive and interact with extrinsic elements on the landscape (Kawata & Agawa, 1999; Olden *et al.*, 2004b), this study addresses the following questions: How do habitat abundance and current velocity interactively shape animal movement, including the rate of movement, pathway sinuosity and directionality in relation to the direction of flow? Do critical thresholds to movement exist? If so, does the location and slope of this threshold differ between study organisms and/or differ within species for individuals exposed to different current velocities? Furthermore, does the critical threshold vary consistently across current velocities or vary similarly for the study organisms? This study uses an artificial neural network approach with recent methodological advances for exploring their explanatory capabilities to unravel the direct and interactive roles of habitat and current velocity for shaping herbivore movement, and to facilitate the detection of critical thresholds in complex benthiscapes.

18.2 Methods

18.2.1 Study organisms

Agapetus boulderensis (Trichoptera: Glossosomatidae) is a slow-moving herbivorous caddisfly that inhabits streams of western North America. *Agapetus* larvae (hereafter called *Agapetus*), or more generally members of the family Glossosomatidae, hatch in late spring, grow through five instars over the next three months, enter the pupal stage and emerge relatively synchronously over a period of approximately one month. Larvae construct and occupy hemispherical cases composed of sand grains cemented together with silk. A case contains two ventral openings located along the major axis, through which the larvae extend their thoracic legs and anal claws to grasp the substrate while moving and grazing algae (Wiggins, 1996). Although the mineral case provides effective protection against many predators, its cost is the substantial energetic expense of carrying

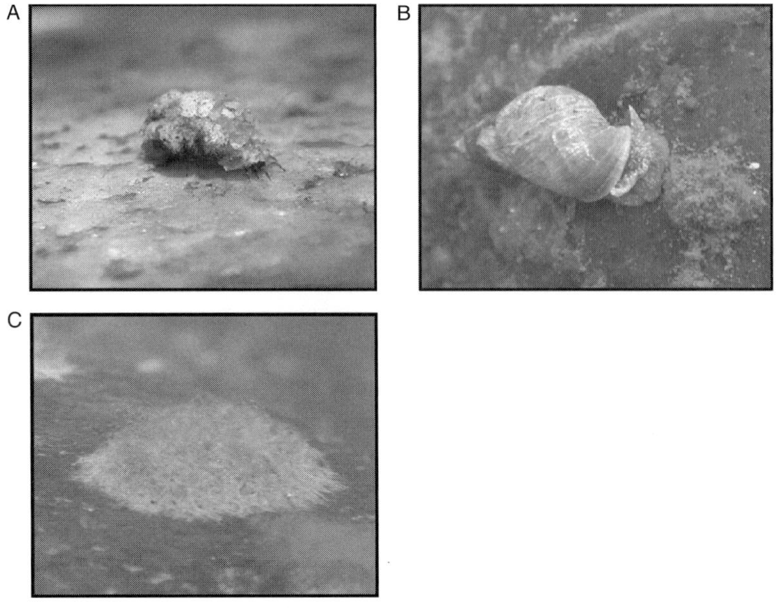

Figure 18.1. (A) The cased-caddisfly *Agapetus boulderensis* and (B) freshwater snail *Physa* sp. in their natural stream benthiscape containing retreats of the chironomid larva, *Pagastia partica* (C) that represent high-profile structure elements that are separated by smooth areas containing low-profile algae and diatom mats. Photographs courtesy of Jeremy B. Monroe, *Freshwaters Illustrated*.

it around while being opposed by the drag force exerted by the flowing water, as well as the frictional force of the case against the substrate (Waringer, 1993; Otto & Johansson, 1995). Consequently, *Agapetus* mobility is greatly limited and is restricted to smooth surfaces of substrates (Poff & Ward, 1992). The freshwater snail *Physa* sp. (Pulmonata: Physidae) (hereafter called *Physa*), in contrast, is a highly mobile herbivore that is most often found on stones along stream margins in slow current velocity habitats. Individuals use a large muscular foot to slide over the rock surface and secrete mucus from specialised glands to lubricate their path (Dillon, 2000). Like *Agapetus*, hydrodynamic drag on the shells of freshwater snails influence their mobility and directionality in running waters (Huryn & Denny, 1997).

High-profile structural elements on the streambed that influence *Agapetus* and *Physa* movement include the silken retreats of the midge larva, *Pagastia partica* (Roback, 1957) (Diptera: Chironomidae) that are colonised by filamentous algae (Monroe & Poff, 2005). The retreats of *Pagastia* are effective barriers to *Agapetus* movement, whereas the spaces between the retreats are typically dominated by a low-profile matrix of algae and diatoms upon which *Agapetus* larvae can readily move and forage. In contrast, *Physa* individuals are better able to negotiate stands of thick *Pagastia* retreats and filamentous algae, albeit with more difficulty compared with the smooth, diatom matrix (J. D. Olden, personal observation).

18.2.2 Experimental design

Movement experiments were conducted during the summer of 2001 on artificial arenas placed in a streamside channel (185 cm × 60 cm × 10 cm) adjacent to the Upper Colorado River, Colorado, USA (40°11′N, 105°52′W) (Figure 18.2). The arena consisted of a 40.6 cm × 40.6 cm frame upon which 196 unglazed porcelain tiles (each 2.54 cm × 2.54 cm) were arranged into a square matrix consisting of 14 rows and 14 columns. The streamside channel received water directly from the Upper Colorado River, and current velocity was controlled using six evenly spaced hoses that discharged water at equal rates to ensure consistent velocities. Water temperature was measured every 15 min in the experimental channel over the entire study period.

To mimic the mix of diatom patches and *Pagastia* retreats found in natural benthiscapes, porcelain tiles were used to create two patch types that imitated the form, size and complexity of these habitats. Some porcelain tiles were cultured in flow-through channels in the stream at a current velocity of 70 cm/s to produce uniform, low-profile algal mats. These tiles were considered to represent 'habitat patches' because they provided a good medium for both movement and potential foraging (Poff & Ward, 1992; Kawata & Agawa, 1999). The remaining tiles were placed in the experimental streamside channel under slow velocities, and were subjected to the addition of hundreds of retreat-building *Pagastia* larvae that were collected from the adjacent stream. Silken retreats woven by the larvae were subsequently colonised by thick filamentous algae, and were similar in both appearance and size to those in the stream (Olden *et al.*, 2004b). Tiles containing *Pagastia* retreats were considered 'nonhabitat' patches because they impeded movement of both *Agapetus* and *Physa*, although to a varying degree.

18.2.3 Movement experiments

A 2 × 2 factorial design experiment was conducted to examine the effects of habitat abundance and current velocity on the movement dynamics of *Agapetus* and *Physa*. Two random arrangements of the experimental arenas were examined for each of five treatment levels of habitat: $p_{\text{habitat}} = 0.2, 0.4, 0.6, 0.8, 1.0$, where p_{habitat} is the proportion of smooth habitat patches or tiles, and two treatment levels of current velocity: low velocity = 5–15 cm/s and high velocity = 20–30 cm/s. These treatments were based on streambed surveys of *Pagastia* retreat densities (Monroe & Poff, 2005) and the range of velocities commonly experienced by insects on the streambed (Wellnitz *et al.*, 2001). Current velocity was measured across the entire experimental arena using a Schiltknecht current probe (Schiltknecht Messtechnik AG, Zurich, Switzerland) to provide an integrated measure of velocity between 0 and 10 mm from the channel bottom.

Experimental animals were collected from the streambed minutes prior to performing a movement trial and placed at the centre of the experimental arena using a soft-bristled paintbrush. The movement pathway of 10 individuals of each species for each treatment combination of five p_{habitat} levels, two current velocities, and two replicates of the

Figure 18.2. Experimental channels (top panel) and arenas (bottom panel) used to examine the movement behaviour of *Agapetus boulderensis* and *Physa* sp. Smooth tiles (i.e. habitat patches: white) and *Pagastia* tiles (i.e. nonhabitat patches: black) were used to create arenas with varying proportions of smooth patches: p = 0.2, 0.4, 0.6, 0.8, 1.0, (where p represents the proportion of smooth tiles).

experimental arena (i.e. 2 different random tile arrangements for a given p_{habitat}) was recorded every 3 min for 1 h for *Agapetus* and every 1 min for 20 min for *Physa* on a recording map. Spatial coordinates were taken for the entire observation period (20 time steps) or until the individual left the arena via crawling or accidental dislodgement. The temporal scale of measurement intervals was based on a series of preliminary experiments and was chosen to reflect the time scale of *Agapetus* and *Physa* movement.

The influence of food resources on *Agapetus* and *Physa* movement was examined by comparing algal ash-free dry mass and algal species composition from scraped habitat (diatom) and nonhabitat (*Pagastia*) patches prior to the commencement of the experiments. Ash-free dry mass of nonhabitat patches was an order of magnitude greater ($\overline{X} = 2.68$ mg/cm^2, n = 4) than habitat patches ($\overline{X} = 0.22$ mg/cm^2, n = 4), whereas both patch types contained high concentrations of diatoms (see Olden *et al.*, 2004b for more details). These results support the structural differences and food similarities of the two patch types, and emphasise that the patch types differed only in that the *Pagastia* or nonhabitat patches contained extensive filamentous algae that impede insect movement.

18.2.4 Movement metrics

The x–y coordinates for each movement pathway were spatially referenced by digitising the recording map in ArcView GIS (ESRI Institute 2000). I calculated second-order statistics (i.e. individuals as replicates and not time steps within an individual pathway) of four movement metrics using the appropriate mathematical protocols for angular data following Batschelet (1981): (1) net displacement; (2) movement rate – calculated as the sum of the distances travelled for each time step divided by the total time of the pathway; (3) mean vector length (r) as a measure of sinuosity – calculated as $\sqrt{(\overline{\sin\theta})^2 + (\overline{\cos\theta})^2}$, where θ is the turning angle between successive time intervals (r = 0.0 represents random dispersion of turning angles between successive steps and r = 1.0 represents a perfectly straight line); and (4) homeward component as a measure of upstream orientation – calculated as $r\cos(\overline{\varphi} - \theta_0)$, where $\overline{\varphi}$ is the mean angle and θ_0 is the direction of flowing water (0 degrees). The homeward component measures how close the mean direction is to the 'homeward' (upstream) direction, and it ranges from 1.0 (precisely upstream) to –1.0 (precisely downstream). Together, this suite of movement metrics was chosen to represent the rate, direction and tortuosity of an individual's movement pathway in response to landscape structure (Turchin, 1991).

18.2.5 Multi-response artificial neural network

A multi-response artificial neural network (MANN) was used to test for the effects for habitat abundance (five p_{habitat} levels) and current velocity (two levels) on *Agagetus* and *Physa* movement behaviour as depicted by the four movement metrics. The architecture of these networks (known as a multi-layer perception) consisted of a single input, hidden and output layer. The input layer contained two neurons representing p_{habitat} and current velocity. The number of hidden neurons in the neural network was chosen to minimise the trade-off between network bias and variance by comparing the performances of different networks (see below). The output layer contained four neurons representing each of the movement metrics.

Each neuron in a multi-layer perception is connected to all neurons from adjacent layers by axons that are assigned a connection weight that dictates the intensity of the signal they transmit. The 'activity level' of each input neuron is defined by the incoming signal (i.e. values) of the independent variables, whereas the state of the other neurons is evaluated locally by calculating the weighted sum of the incoming signals from the neurons of the previous layer. The mathematics of the neural network can be expressed as:

$$y_k = \varphi_o\left\{\beta_k + \sum_j w_{jk}\varphi_h\left(\beta_j + \sum_i w_{ij}x_i\right)\right\} \quad (1)$$

where x_i are the input signals, y_k are the output signals, w_{ij} are the weights between input neuron i to hidden neuron j, w_{jk} are the weights between hidden neuron j and output neuron k, β_j and β_k are the biases associated with the hidden and output layers, and φ_h and φ_o are activation functions (in this case logistic functions) for the hidden and output layers. I refer the reader to Olden *et al.* (2008) for a more detailed description of the methodology.

I used a two-phase training approach to account for the fact that different parameter optimisation algorithms will perform best on different problems. In the first phase I used the back-propagation algorithm (Rumelhart *et al.*, 1986) and in the second phase I used the conjugate gradient descent algorithm (Hestenes & Stiefel, 1952); both of which train the neural network by iteratively adjusting the connection weights with the goal of finding a set of weights that minimises the error of the network (in this case the sums-of-squared error function). During network training using the back-propagation algorithm, observations are sequentially presented to the network and weights are adjusted in a backwards fashion, layer by layer, in the direction of steepest descent in minimising the error function. Learning rate (which controls the step size when weights are iteratively adjusted) and momentum parameters (which adds inertia to the learning motion through weight space) were included during network training to ensure a high probability of global network convergence.

In the second phase, I used conjugate gradient descent, which is a batch update algorithm that calculates the average gradient of the error surface across all cases before updating the weights once at the end of each epoch (Bishop, 1995). This algorithm can be regarded as a form of gradient descent with momentum that constructs a series of line searches across the error surface. However, instead of taking a step proportional to the learning rate as performed by the back-propagation algorithm, this algorithm projects a straight line in that direction and then locates a minimum along this line in error space (Bishop, 1995). For both algorithms I used a maximum of 500 epochs to determine the optimal axon weights. Prior to training the network, the independent variables were converted to z-scores to standardise the measurement scales of the inputs into the network.

I conducted an iterative search for the optimal neural network. One hundred networks based on different random initial connection weights were conducted for each network configuration with 1 to 10 hidden neurons (increasing by increments of 1) and random shuffling of presentation order was evoked during the back-propagation training. From this list of candidates, the top 20 networks that produced the greatest predictive performance were retained and compared with the ensemble network that combined the predictions from all the models. A single network that was deemed representative of the ensemble network based on model predictions and variable contributions was then selected as the final network. In a previous study, Olden *et al.* (2004b) showed no effect of replicate arena on movement rates, and therefore the two landscape replicates were pooled for all subsequent analyses. All neural network analyses were conducted using the Data Miner toolbox of Statistica (v. 7.1, StatSoft Inc., Tulsa, Oklahoma, USA) and computer macros written in the Matlab programming language (v. 7.0, The MathWorks, Natick, Massachusetts, USA).

Given the importance of connection weights in assessing the relative contributions of the independent variables, during the optimisation process it is necessary that the network converges to the global minimum of the fitting criterion (e.g. prediction error) rather than one of the many local minima. Connection weights in networks that have converged to a local minimum will differ from networks that have globally converged, thus resulting in drastically different variable contributions. The iterative search for the optimal and most representative neural network employed in this study (see above) ensured the greatest probability of network convergence to the global minimum. Here, I used three complementary approaches to explore variable contributions: neural interpretation diagrams to visualise direct and interactive variable effects, a connection weight method to quantify the relative contributions of the variables to network predictions, and a sensitivity analysis to explore the influence of the predictor variables across their entire range and test for the existence of critical thresholds.

18.2.6 Quantifying variable importance in neural networks

Given the repeatedly demonstrated predictive power of artificial neural networks, recent efforts have focused on the development of methods for understanding the explanatory contributions of the predictor variables in the network (Olden & Jackson, 2002). This was, in part, prompted by the fact that neural networks were coined a 'black box' approach to modelling ecological data. Recent studies in the biological sciences have provided a variety of methods for quantifying and interpreting the contributions of the independent variables in neural networks (see reviews by Gervey *et al.*, 2003; Olden *et al.*, 2004c). All of these approaches rely on the fact that the connection weights between neurons are the linkages between the inputs and the output of the network, and therefore are the link between the problem and the solution. Consequently, the relative contribution of each independent variable to the predictive output of the neural network depends primarily on the magnitude and direction of these connection weights. Input variables with larger connection weights represent greater intensities of signal transfer, and therefore are more important in predicting the output compared to variables with smaller weights. Negative connection weights represent inhibitory effects on neurons (reducing the intensity or contribution of the incoming signal and negatively affecting the output), whereas positive connection weights represent excitatory effects on neurons (increasing the intensity of the incoming signal and positively affecting the output).

18.2.6.1 Neural Interpretation Diagram

The Neural Interpretation Diagram provides a visual interpretation of the connection weights among neurons and was first presented by Özesmi & Özesmi (1999). Tracking the magnitude and direction of weights between neurons enables researchers to identify individual and interacting effects of the input variables on the output. Variable relationships are determined in a two-step manner. Positive effects of input variables are depicted by positive input-hidden and positive hidden-output connection weights, or negative input-hidden and negative hidden-output connection weights. Negative effects of input variables are depicted by positive input-hidden and negative hidden-output connection weights, or by negative input-hidden and positive hidden-output connection weights. Therefore, the multiplication of the two connection weight directions (positive or negative) indicates how each input variable influences each output variable. Interactions among predictor variables can be identified as input variables with opposing (antagonist) or similar (synergist) connection weights entering the same hidden neuron.

With even small networks, however, the number of connections to examine and interpret in a network can be extremely large. For example, a network containing 10 input neurons, 7 hidden neurons and 3 output neurons would have a total of 91 connection weights to examine. Bishop (1995) suggested removing small weights from the network to ease interpretation; however, deciding the threshold value below which weights should be eliminated from the network is unclear. Recently, Olden & Jackson (2002) developed a randomisation test to address this question. This approach randomises the response variable, then constructs a neural network using the randomised data and records all input-hidden-output connection weights (product of the input-hidden and hidden-output weights). This process is repeated 1999 times to generate a null distribution for each input-hidden-output connection weight, which is then compared with the observed values to calculate the significance level. The randomisation test provides an objective pruning technique for eliminating connection weights that have minimal influence on the network output.

18.2.6.2 Connection weight approach

A formal quantification of variable importance (as visually depicted in the Neural Interpretation Diagram) was presented by Olden & Jackson (2002). In this approach, variable contributions are quantified by calculating the product of the input-hidden and hidden-output connection weights between each input neuron and output neuron, which are summed across all hidden neurons. Positive values represent positive associations between input and output neurons, whereas negative values represent negative associations. The relative contributions of the variables are calculated by dividing the absolute value of each variable contribution by the grand sum of all absolute contributions. The connection weight approach is deemed to be one of the most appropriate methods as it has been shown to exhibit substantially higher sensitivity for identifying variable importance compared with other approaches (Olden *et al.*, 2004c).

18.2.6.3 Sensitivity analysis

Sensitivity analysis is a commonly used approach for exploring variable contributions in statistical models. Quite simply, it involves varying each input variable across its entire range (from its minimum to maximum value) while holding all other input variables constant, allowing the individual contributions of each variable to be assessed. This approach has been used for neural networks in the past (e.g. Lek *et al.*, 1996) and here I constructed response curves for habitat abundance by varying p_{habitat} across its range while holding flow velocity at each of its two treatment levels.

18.3 Results

The multi-response artificial neural network exhibited high accuracy (R^2) and precision (RMSE) for predicting *Agapetus* and *Physa* movement as a function of p_{habitat} and current velocity (Table 18.1). Net displacement and movement rate were predicted with the greatest success for *Agapetus*, whereas movement rate and upstream homing were the most predictable for *Physa*. Next, network connection weights were examined to quantify the direct and interactive importance of p_{habitat} and current velocity for predicting each of the movement metrics. In the Neural Interpretation Diagram, the set of input-hidden connection weights reflect the universal importance of p_{habitat} and current velocity for influencing overall movement dynamics, whereas the hidden-output connection weights modify the first set of weights by reducing, enhancing or even reversing the signals to maximise the model fit for each movement metric. Prior to interpretation, non-influential connection weights that were non-significant based on the randomisation test ($p > 0.05$) were pruned from the diagram.

The Neural Interpretation Diagram shows the positive influence of p_{habitat} on the net displacement, movement rate, mean vector length (i.e. increasingly straight pathways) and upstream homing of *Agapetus* through neurons D and I (Figure 18.3A). These connections exceed the relatively small negative influence of p_{habitat} on net displacement through neurons A and B. In contrast, high current velocity was associated with decreased net displacement and slower movement rates via neuron I and less convoluted pathways via neuron J. Synergistic and antagonistic interactions between p_{habitat} and current velocity were evident through a number of hidden neurons. Most notably, increasing p_{habitat} and current velocity acted together to cause greater declines in net displacement (neuron E) and greater upstream homing (neuron J); but they acted in opposition for movement rates where increasing current velocity suppressed the positive influence of habitat transmitted through neuron I.

Table 18.1. *Predictive performance of the multi-response artificial neural network for predicting the movement metrics as a function of habitat abundance and current velocity. Reported values represent model accuracy (Pearson correlation coefficient, R) and precision (root-mean-square-error, RMSE).*

Organism	*Accuracy*		*Precision*
Movement metric	R	p	RMSE
Agapetus			
Net displacement	0.67	<0.001	4.07
Movement rate	0.60	<0.001	0.17
Pathway sinuosity	0.50	<0.001	0.22
Upstream homing	0.41	<0.001	0.35
Physa			
Net displacement	0.46	<0.001	4.96
Movement rate	0.77	<0.001	2.25
Pathway sinuosity	0.50	<0.001	0.20
Upstream homing	0.53	<0.001	0.48

Movement dynamics of *Physa* were influenced predominantly by the direct effects of p_{habitat} and current velocity. Movement rates were positively associated with p_{habitat} and current velocity through neurons B-D and A, respectively (Figure 18.3B), although these factors interacted negatively through hidden neuron B where high current velocities decreased the positive influence of habitat on movement rate. Similarly, increasing p_{habitat} was related to greater upstream homing (neuron E) and current velocity was positively correlated with mean vector length (i.e. lower pathway sinuosity) via neuron A. Synergistic interactions between habitat and current velocity were not evident.

According to all axon connections in the network the total contribution of p_{habitat} and current velocity for predicting the movement metrics of *Agapetus* and *Physa* is presented in Figure 18.4. For both species, p_{habitat} was the most important predictor of net displacement, sinuosity and upstream homing, whereas current velocity was the top predictor for movement rate. Although the best predictors were the same for both species, the contributions of p_{habitat} and current velocity were much more balanced for *Physa*, particularly for movement rate and pathway sinuosity. For *Agapetus*, p_{habitat} was positively related and current velocity negatively related to all movement metrics, with the exception of upstream homing where current velocity exhibited a weak positive influence. In contrast, both p_{habitat} and current velocity were positively associated with all movement metrics for *Physa*.

Given the direct and interactive effects of p_{habitat} and current velocity for movement behaviour, the results from the sensitivity analysis provide valuable insight into the shape and possible threshold-response of the study organisms. Critical thresholds are represented in the response curves as sharp changes in a movement metric in relation to a small

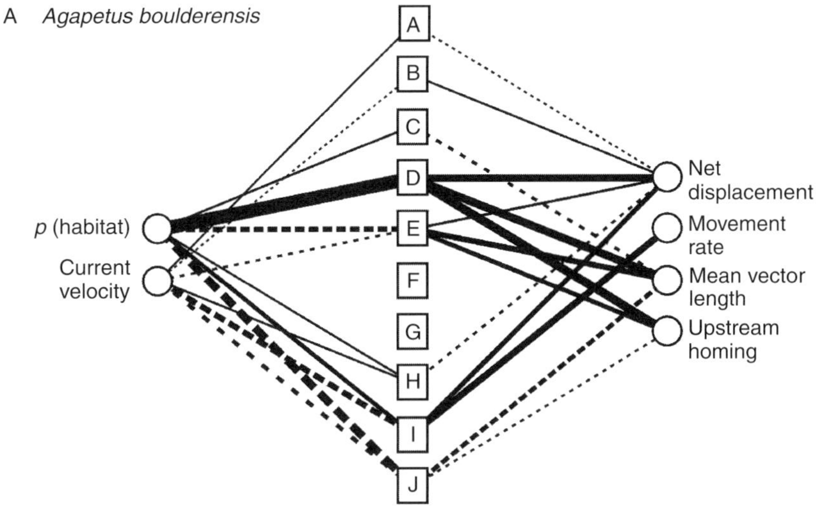

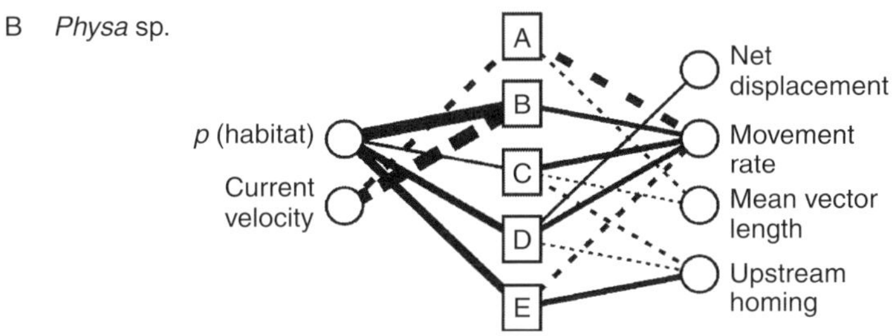

Figure 18.3. Neural Interpretation Diagram illustrating the contributions of habitat abundance (p_{habitat}) and current velocity for predicting the movement metrics of (A) *Agapetus* and (B) *Physa*. In this diagram, the relative magnitude of each connection weight is represented by line thickness (i.e. thicker lines representing greater weights) and line pattern represents the direction of the weight (i.e. solid lines represent positive, excitatory signals and dashed lines represent negative, inhibitor signals).

change in p_{habitat}. In low current velocities the net displacement of *Agapetus* responded to increasing p_{habitat} in a two-stage manner; a punctuated increase in displacement at intermediate levels followed by a gradual increase (Figure 18.5A). Movement rate similarly increased with p_{habitat} under low current velocity, but the response was not initiated until intermediate p_{habitat}; after which movement rates increased exponentially (Figure 18.5B). In contrast, net displacement and movement rate showed a minimal response to p_{habitat} in the high current velocity treatment. Mean vector length increased immediately in response to increasing p_{habitat} under low current velocity (i.e. straighter movement pathways), showing a critical threshold at $p_{\text{habitat}} = 0.2$; whereas the threshold response was delayed until $p_{\text{habitat}} = 0.6$ in high current velocity (Figure 18.5C). Upstream

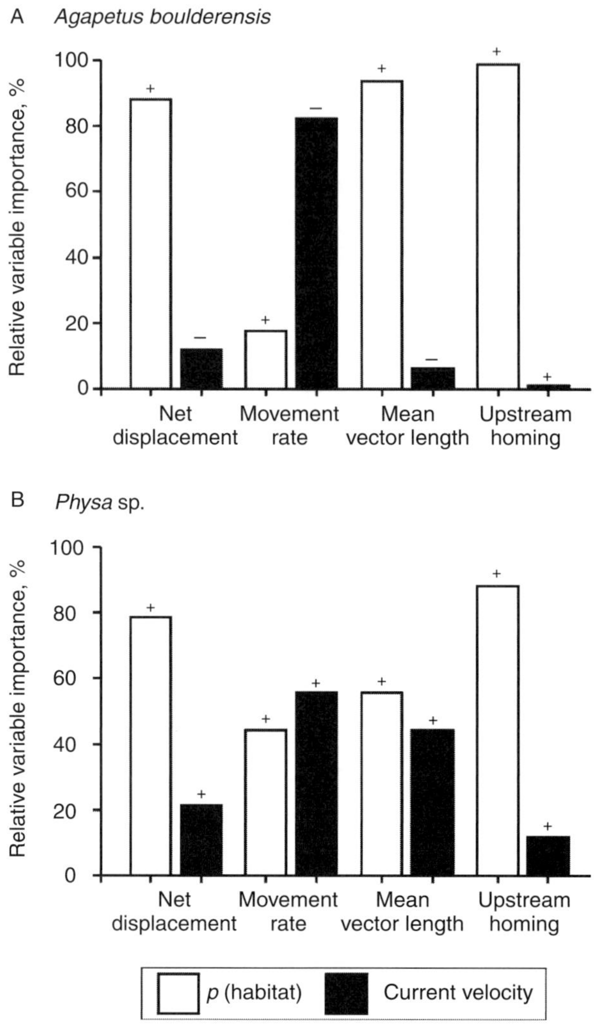

Figure 18.4. Relative variable importance of habitat abundance (p_{habitat}) and current velocity for predicting the movement metrics of *Agapetus* and *Physa*. '+' and '–' indicates an overall positive or negative influence of the variable, respectively.

homing exhibited a threshold response to p_{habitat} in both current velocity treatments, where peak homing was observed at slightly lower p_{habitat} in high current velocities (Figure 18.5D). In both cases, upstream homing then gradually decreased at intermediate p_{habitat} and showed a similar levelling-off response.

Net displacement of *Physa* showed a logarithmic response to increasing p_{habitat} in both low and high current velocities; gradually increasing at low to immediate levels of p_{habitat} and then stabilising at higher levels (Figure 18.6A). Movement rates exhibited

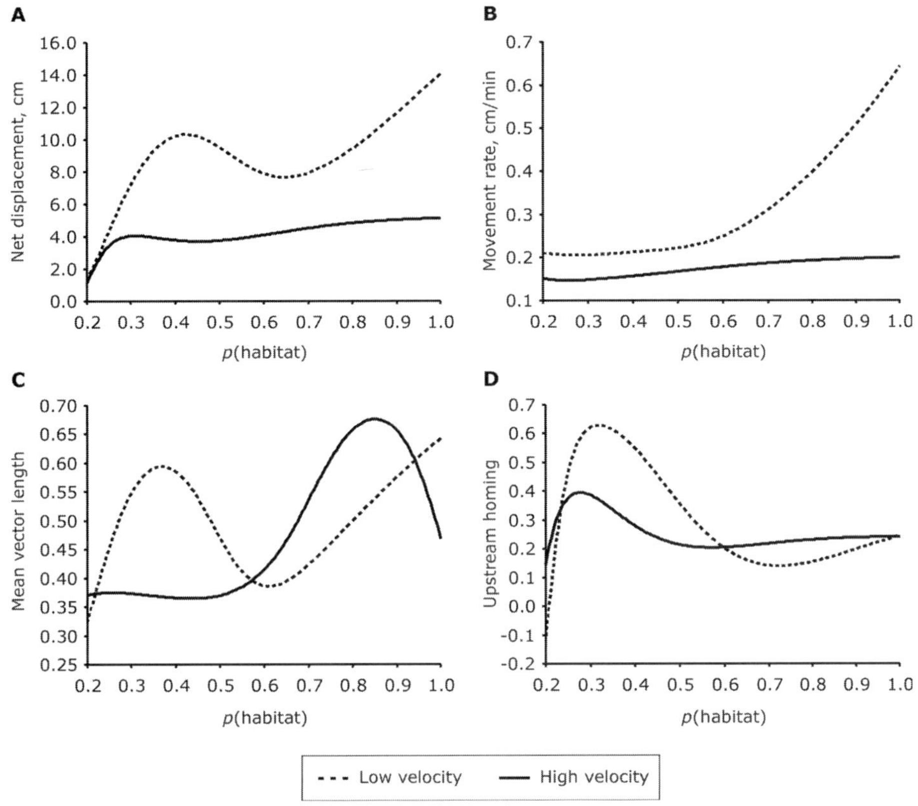

Figure 18.5. *Agapetus* response curves according to the sensitivity analysis for (A) net displacement, (B) movement rate, (C) mean vector length (0.0 represents random dispersion of turning angles between successive steps and 1.0 represents a perfectly straight line); and (D) upstream homing (1.0 represents precisely upstream direction of movement and –1.0 represents precisely downstream direction of movement).

comparable response curves in both velocity treatments at $p_{\text{habitat}} < 0.8$, after which the curves diverged sharply with elevated movement rates in high current velocities (Figure 18.6B). Mean vector length showed a positive linear relationship with p_{habitat} in low current velocities (i.e. decreased sinuosity), whereas in high current velocities a gradual increase and levelling-out of sinuosity at intermediate p_{habitat} was followed by a sharp threshold response at $p_{\text{habitat}} > 0.8$ (Figure 18.6C). Mean vector length converged to similar values once $p_{\text{habitat}} = 1.0$. Upstream homing exhibited a strong logarithm response to p_{habitat} in low current velocities and a bimodal response in high current velocities (Figure 18.6D).

Comparison of movement behaviours shows that net displacement and movement rate of *Agapetus* were distinctly higher in low current velocity; the magnitude of this difference generally increased with p_{habitat}. This differed from the movement behaviour of

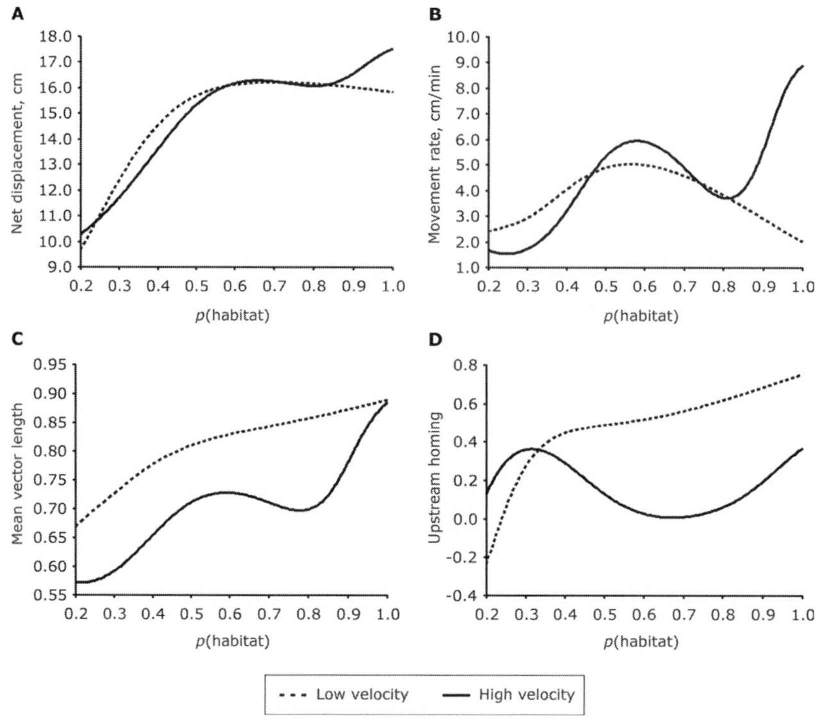

Figure 18.6. *Physa* response curves according to the sensitivity analysis for (A) net displacement, (B) movement rate, (C) mean vector length (0.0 represents random dispersion of turning angles between successive steps and 1.0 represents a perfectly straight line); and (D) upstream homing (1.0 represents precisely upstream direction of movement and –1.0 represents precisely downstream direction of movement).

Physa where net displacement was very similar between the velocity treatments, and movement rates in high current velocity were greater in high p_{habitat} landscapes. Response curves for *Agapetus* showed lower pathway sinuosity (i.e. greater mean vector length) and upstream homing in low current velocity until $p_{\text{habitat}} > 0.6$, at which the curves intersected and after which values were greater in high current velocity. For *Physa*, however, mean vector length and homing were almost always greater in low current velocity. Interestingly, the shape of the *Agapetus* response curves for net displacement, sinuosity and upstream homing in low current velocity corresponded closely to the same response curves for *Physa* but in high current velocity.

18.4 Discussion

Of the many challenges that face landscape ecologists in the coming decades, understanding critical thresholds and how they influence animal movement is an important topic of research (Turner, 2005). However, because habitat fragmentation can have nonlinear effects, it may be difficult to predict the consequences for animal movement until a threshold is exceeded. For this reason, critical threshold phenomena have been identified as a 'major unsolved problem facing conservationists' (Pulliam & Dunning, 1997). We might expect that threshold behaviour will be affected by several interacting factors rather than one particular causal agent, and that these interactions might produce complex patterns of species' responses that are seemingly unpredictable. In an effort to make such investigations somewhat more tractable, at least in a statistical sense, this study couples the strong pattern recognition ability of artificial neural networks with recent methodological advances for exploring their explanatory capabilities.

The present work illustrated that the movement behaviour of two benthic herbivores, *Agapetus boulderensis* and *Physa* sp., varied nonlinearly according to the abundance of habitat patches, current velocity and the interaction of the two. *Agapetus boulerensis*, an insect with limited mobility and flow-constrained perceptual range (Olden *et al.*, 2004b), showed increasing net displacement and movement rates with habitat abundance in low current velocity. In high current velocity, however, *Agapetus* movement was unaffected by changes in habitat abundance. These results are corroborated by instream observational studies showing that *Agapetus* larvae exhibit lower rates of movement on smooth substrate surfaces under high current velocities (Poff & Ward, 1992; Becker, 2001). This observation supports the hypothesis that the magnitude of current velocity and associated shearing force that a larva experiences as it moves across the streambed (see Waringer, 1993) is a critical factor shaping movement behaviour (Olden *et al.*, 2004b). Patterns of near-bed shear stress on natural streambeds can be highly variable, in response to an irregular bed topography that interrupts flow moving through the stream channel and creates areas of decelerating and accelerating local velocity (see Davis & Barmuta, 1989). Fine-scale structural elements, such as *Pagastia* retreats or physical irregularities in stone surfaces can interrupt the local flow field and create patches of slow near-bed current velocity, thereby shielding organisms from the highly erosive force of high current velocities (Lancaster, 1999). As a result, while increasing proportions of diatom habitat may promote greater movement in low current velocities, it is likely associated with reduced flow refugia in high current velocities, effectively decreasing the amount of habitat that is considered 'suitable' (i.e. lower risk to erosion and downstream drift) for *Agapetus* movement. Functional connectivity, therefore, is observed to decrease with increasing habitat; a result that is seemingly counterintuitive. The question of 'how much habitat is enough?' (Fahrig, 2001), in the case of *Agapetus* (and likely other benthic insects; see Lancaster, 1999) in high velocities, is perhaps better posed as 'how much habitat is too much?'

The movement rate of the freshwater snail *Physa* sp., a more mobile herbivore with a broader perceptual range (Kawata & Agawa, 1999), also responded positively to habitat abundance, but unlike *Agapetus* was generally insensitive to changes in current velocity. This finding is supported by the observation of Poff & Nelson-Baker (1997) that snail movement is relatively independent of current velocity (but see Hutchinson, 1947). Interestingly, *Physa* exhibited greater movement rates in high velocity when the landscape was completely composed of diatom habitat, suggesting that high current velocities may elicit an escape response.

Animal orientation is constrained by strong external physical gradients (Fraenkel & Gunn, 1940). Directional stimuli associated with particular physical gradients, such as flow, wind or sunlight, play an important, but commonly overlooked, role in shaping an animal's perceptual range and influencing their movement behaviour (Olden *et al.*, 2004a). Schooley & Wiens (2003) found that cactus bugs (*Chelinidea vittiger*) were more likely to orientate and move toward cactus patches located upwind compared with those located crosswind or downwind. This study showed that physical habitat and current velocity interact to influence the movement directionality of *Agapetus* and *Physa* with respect to the direction of flow; a rheotaxis phenomenon supported by Poff & Ward (1992) and Poff & Nelson-Baker (1997). Increasing current velocity and abundance of low-profile diatom habitat lead to a punctuated and nonlinear increase in upstream homing; a finding supported by a biomechanical response to the fluid dynamics of water. For example, Waringer (1993) examined the resistance of six cylindrical-case caddis larvae to different current velocities and found that for dead larvae in their cases the critical entrainment velocity (i.e. flow to be dislodged) ranged from 3.0 to 70.5 cm/s in the frontal position and only 2.2 to 20.8 cm/s in the lateral position. This suggests that *Agapetus* can reduce their erosion probability by orientating themselves with the direction of flow. Similarly, Huryn & Denny (1997) showed that upstream movements by snails are a function of torque on the snail's foot generated by hydrodynamic drag on the shell. By positioning their shell in the upstream direction, snails were able to reduce torque and stabilise their orientation to the force of water.

The manner in which high-profile structural elements of the benthiscape (in this case, *Pagastia* retreats) interacted with current velocity to influence the 'suitability' of habitat patches to animal movement support the general notion that the interpatch matrix (i.e. the landscape located between patches) should not be viewed as 'ecologically neutral' (Wiens *et al.*, 1993). That is, the characteristics of the landscape matrix do appear to influence grazer movement by changing the suitability of habitat patches by interacting with larger, overarching stimuli such as the physical current of water. The strong interplay between the highly patchy nature of benthiscapes (Downes *et al.*, 1993) and irregular near-bed current velocities (Hart & Finelli, 1999; Lancaster, 1999) illustrate the complex nature of both structural connectivity (i.e. the actual connection of habitat patches by corridors) and functional connectivity (i.e. the connection of habitat patches by dispersal) of stream benthiscapes. More generally, given that dispersing individuals may be polarised in certain directions within a landscape (see Olden *et al.*, 2004a), functional connectivity will not be equal in all directions of movement. Such landscape anisotropy with respect to movement is an exciting research front in landscape ecology (Bélisle, 2005), and requires explicit modelling (Pe'er & Kramer-Schadt, 2008). I argue that stream benthiscapes may represent an ideal model system for exploring this question and others regarding the temporal, spatial and biological sources of variability in the structural and functional connectivity of natural landscapes.

Do *Agapetus* and *Physa* show critical threshold responses to habitat abundance? This study suggests that critical thresholds do occur with respect to the rate, sinuosity and directionality of movement, and that these thresholds differ between species. Furthermore, a striking pattern was that the location and slope of these thresholds varied in response to current velocity. Critical thresholds to the net displacement and movement rate of *Agagetus* were observed in low current velocities, but were non-existent in high current velocity. In light of the above discussion, this result highlights the role that near-bed current velocity plays in shaping functional connectivity to *Agapetus* movement. Punctuated inclines in movement rates occurred at a lower habitat abundance for *Physa* compared with *Agapetus*; a result explained by the greater mobility of *Physa* and that the dichotomy of suitable habitat versus the inhospitable matrix is likely better applied to *Agapetus* than to *Physa*. In contrast, *Agapetus* and *Physa* were observed to exhibit similar threshold responses in their degree of upstream homing in response to habitat abundance. For both species, sudden increases in upstream homing in high current velocity occurred at lower levels of habitat abundance compared with slow current velocity. Taken together, these results illustrate that critical thresholds to animal movement are likely to depend on both intrinsic attributes of the species and extrinsic characteristics of the environment.

Results from this study support the general notion that species that vary in mobility and dispersal ability will vary in their response to fragmentation and will have different perceptions as to whether the landscape is fragmented (Doak *et al.*, 1992; With & Crist, 1995; Pearson *et al.*, 1996; Pe'er & Kramer-Schadt, 2008). Therefore, a critical threshold of functional connectivity is not an inherent property of the landscape, but in fact emerges from the interplay of species' interactions with landscape structure (With & Crist, 1995). As a result, a single critical threshold is not sufficient to describe the response of all species in a community to changes in landscape structure. This has important implications for the usefulness of the critical threshold concept in biological conservation (Huggett, 2005) and more generally, resource management (Groffman *et al.*, 2006). While this seems to necessitate a species-specific definition of critical thresholds to habitat fragmentation (With, 1994), future studies should consider the possibility of grouping species with similar perceptual ranges and vagilities of movement to facilitate a community-level quantification of landscape connectivity.

18.5 Conclusion

Detecting the occurrence of critical threshold effects in ecological phenomena is a challenge because ecological dynamics commonly occur over time- and spatial-scales that well exceed the extent of observation (Fukami & Wardle, 2005). The study of critical thresholds to animal movement necessitates the application of quantitative approaches that can model the nonlinear complexities inherent to ecological series, while at the same time shedding insight into the interacting factors responsible for such thresholds. Statistical tools for identifying thresholds, however, are still generally lacking in this regard (e.g. Friedel, 1991;

Toms & Lesperance, 2003), although artificial neural networks may provide a powerful approach. Moreover, with the growing need to model animal movements in three dimensions (Alderman & Hinsley, 2007), machine learning approaches such as neural networks may play an increasing role.

Acknowledgements

This research would not have been possible without the field help and intellectual contributions of Jeremy Monroe and Aaron Hoffman. I thank Pieter Johnson for providing valuable comments on the manuscript, and Tiffany for her invaluable assistance in the field and for always being willing to provide a helping hand.

References

Alderman, J. & Hinsley, S. A. 2007. Modelling the third dimension: incorporating topography into the movement rules of an individual-based spatially explicit population model. *Ecol Complex* **4**, 169–181.

Andrén, H. 1994. Effects of habitat fragmentation on birds and mammals in landscapes with different proportions of suitable habitat: a review. *Oikos* **71**, 355–366.

Batschelet, E. 1981. *Circular Statistics in Ecology*. Academic Press.

Becker, G. 2001. Larval size, case construction and crawling velocity at different substratum roughness in three scraping caddis larvae. *Arch Hydrobiol* **151**, 317–334.

Bélisle, M. 2005. Measuring landscape connectivity: the challenge of behavioral landscape ecology. *Ecology* **86**, 1988–1995.

Bishop, C. M. 1995. *Neural Networks for Pattern Recognition*. Clarendon Press.

Carpenter, S. R. 2002. Ecological futures: building an ecology for the long now. *Ecology* **83**, 2069–2083.

Davis, J. A. & Barmuta, L. A. 1989. An ecologically useful classification of mean and near-bed flows in streams and rivers. *Freshwater Biol* **21**, 271–282.

Dillon, R. T., Jr. 2000. *The Ecology of Freshwater Molluscs*. Cambridge University Press.

Doak, D. F., Marino, P. C. & Kareiva, P. M. 1992. Spatial scale mediates the influence of habitat fragmentation on dispersal success: implications for conservation. *Theor Popul Biol* **41**, 315–336.

Downes, B. J., Lake, P. S. & Schreiber, E. S. G. 1993. Spatial variation in the distribution of stream invertebrates: implications of patchiness for models of community organization. *Freshwater Biol* **30**, 119–132.

Fahrig, L. 2001. How much habitat is enough? *Biol Conserv* **100**, 65–74.

Folke, C., Carpenter, S., Walker, B. *et al.* 2005. Regime shifts, resilience and biodiversity in ecosystem management. *Annu Rev Ecol Syst* **35**, 557–581.

Fraenkel, G. S. & Gunn, D. L. 1940. *The Orientation of Animals: Kineses, Taxes and Compass Reactions*. Oxford University Press.

Friedel, M. H. 1991. Range condition assessment and the concept of thresholds: a viewpoint. *J Range Manage* **44**, 422–426.

Fukami, T. & Wardle, D. A. 2005. Long-term ecological dynamics: reciprocal insights from natural and anthropogenic gradients. *Phil Trans R Soc B* **272**, 2105–2115.

Gardner, R. H., Milne, B. T., Turner, M. G. & O'Neill, R. V. 1987. Neutral models for the analysis of broad-scale landscape pattern. *Landscape Ecol* **10**, 19–28.

Gevrey, M., Dimopoulos, I. & Lek, S. 2003. Review and comparison of methods to study the contribution of variables in artificial neural network model. *Ecol Model* **160**, 249–264.

Gompertz, B. 1825. On the nature of the function expressive of the law of human mortality and on a new mode of determining life contingencies. *Phil Trans R Soc* **1825**, 513–585.

Gordon, L. J., Peterson, G. D. & Bennett, E. M. 2008. Agricultural modifications of hydrological flows create ecological surprises. *Trends Ecol Evol* **23**, 211–219.

Groffman, P. M. and 15 other authors. 2006. Ecological thresholds: The key to successful environmental management or an important concept with no practical application? *Ecosystems* **9**, 1–13.

Hart, D. D. & Finelli, C. M. 1999. Physical–biological coupling in streams: the pervasive effect of flow on benthic organisms. *Annu Rev Ecol Syst* **30**, 363–395.

Hart, D. D. & Resh, V. H. 1980. Movement patterns and foraging ecology of a stream caddisfly larva. *Can J Zool* **58**, 1174–1185.

Hart, D. D., Clark, B. D. & Jasentuliyana, A. 1996. Fine-scale field measurement of benthic flow environments inhabited by stream invertebrates. *Limnol Oceanogr* **41**, 297–308.

Hestenes, M. R. & Stiefel, E. 1952. Methods of conjugate gradients for solving linear systems. *J Res Nat Bur Stand* **48**, 409–436.

Hoffman, A. L., Olden, J. D., Monroe, J. B. *et al.* 2006. Current velocity and habitat patchiness shape stream herbivore movement. *Oikos* **115**, 358–368.

Huggett A. J. 2005. The concept and utility of ‘ecological thresholds’ in biodiversity conservation. *Biol Conserv* **124**, 301–310.

Huryn, A. D. & Denny, M. W. 1997. A biomechanical hypothesis explaining upstream movements in the freshwater snail *Elimia*. *Funct Ecol* **11**, 472–483.

Hutchinson, L. 1947. Analysis of the activity of the freshwater snail *Viviparus malleatus* (Reeve). *Ecology* **28**, 335–345.

Kareiva, P. & Wennergren, U. 1995. Connecting landscape patterns to ecosystems and population processes. *Nature* **373**, 299–302.

Kawata, M. & Agawa, H. 1999. Perceptual scales of spatial heterogeneity of periphyton for freshwater snails. *Ecol Lett* **2**, 210–214.

Keitt, T. H., Urban, D. L. & Milne, B. T. 1997. Detecting critical scales in fragmented landscapes. *Conserv Ecol* **1:4**, Online at http://www.consecol.org/vol1/iss1/art4.

Lancaster, J. 1999. Small-scale movements of lotic macroinvertebrates with variations in flow. *Freshwater Biol* **41**, 605–619.

Lande, R. 1987. Extinction thresholds in demographic models of territorial populations. *Am Nat* **130**, 624–635.

Lek, S., Delacoste, M., Baran, P., Dimopoulos, I., Lauga, J. & Aulagnier, S. 1996. Application of neural networks to modelling nonlinear relationships in ecology. *Ecol Model* **90**, 39–52.

Loxdale, H. D. & Lushai, G. 1999. Slaves of the environment: the movement of herbivorous insects in relation to their ecology and genotype. *Phil Trans R Soc B* **354**, 1479–1495.

Mackay, R. J. 1992. Colonization by lotic macroinvertebrates: a review of processes and patterns. *Can J Fish Aquat Sci* **49**, 617–628.

Malmqvist, B. 2002. Aquatic invertebrates in riverine landscapes. *Freshwater Biol* **47**, 679–694.

McCulloch, W. S. & Pitts, W. 1943. A logical calculus of the ideas immanent in nervous activity. *B Math Biophys* **5**, 15–133.

McIntyre, N. E. & Wiens, J. A. 1999. Interactions between habitat abundance and configuration: experimental validation of some predictions from percolation theory. *Oikos* **86**, 129–137.

Monroe, J. B. & Poff, N. L. 2005. Natural history of a retreat-building midge, *Pagastia partica*, in a regulated reach of the upper Colorado River. *West N Am Naturalist* **65**, 451–461.

Muradian, R. 2001. Ecological thresholds: a survey. *Ecol Econ* **38**, 7–24.

Olden, J. D. & Jackson, D. A. 2002. Illuminating the "black box": a randomization approach for understanding variable contributions in artificial neural networks. *Ecol Model* **154**, 135–150.

Olden, J. D., Schooley, R. L., Monroe, J. B. & Poff, N. L. 2004a. Context-dependent perceptual ranges and their relevance to animal movements in landscapes. *J Anim Ecol* **73**, 1190–1194.

Olden, J. D., Hoffman, A. L., Monroe, J. B. & Poff, N. L. 2004b. Movement behaviour and dynamics of an aquatic insect in a stream benthic landscape. *Can J Zool* **82**, 1135–1146.

Olden, J. D., Joy, M. K. & Death, R. G. 2004c. An accurate comparison of methods for quantifying variable importance in artificial neural networks using simulated data. *Ecol Model* **78**, 389–397.

Olden, J. D., Lawler, J. J., & Poff, N. L. 2008. Machine learning methods without tears: a primer for ecologists. *Q Rev Biol* **83**, 171–193.

Otto, C. & Johansson, A. 1995. Why do some caddis larvae in running waters construct heavy, bulky cases? *Anim Behav* **49**, 473–478.

Özesmi, S. L. & Özesmi, U. 1999. An artificial neural network approach to spatial habitat modelling with interspecific interaction. *Ecol Model* **166**, 15–31.

Palmer, M. A., Allan, J. D. & Butman, C. A. 1996. Dispersal as a regional process affecting the local dynamics of marine and stream benthic invertebrates. *Trends Ecol Evol* **11**, 322–326.

Palmer, M. A., Swan, C. M., Nelson, K., Silver, P. & Alvestad, R. 2000. Streambed landscapes: evidence that stream invertebrates respond to the type and spatial arrangement of patches. *Landscape Ecol* **15**, 563–576.

Pe'er, G. & Kramer-Schadt, S. 2008. Incorporating the perceptual range of animals into connectivity models. *Ecol Model* **213**, 73–85.

Pearson, S. M., Turner, M. G., Gardner, R. H. & O'Neill, R. V. 1996. An organism-based perspective of habitat fragmentation. In *Biodiversity in Managed Landscapes: Theory and Practice* (ed. R. C. Szaro), pp. 77–95. Oxford University Press.

Poff, N. L. & Ward, J. V. 1992. Heterogeneous currents and algal resources mediate in situ foraging activity of a mobile stream grazer. *Oikos* **65**, 465–478.

Poff, N. L. & Nelson-Baker, K. 1997. Habitat heterogeneity and algal-grazer interactions in streams: explorations with a spatially explicit model. *J N Am Benthol Soc* **16**, 263–276.

Pringle, C. M., Naiman, R. J., Bretschko, G. *et al.* 1988. Patch dynamics in lotic systems: the stream as a mosaic. *J N Am Benthol Soc* **7**, 503–524.

Pulliam, H. R. & Dunning, J. B. 1997. Demographic processes: population dynamics on heterogeneous landscapes. In *Principles of Conservation Biology* (ed. G. K. Meffe & C. R. Carroll), pp. 203–232. Sinauer Associates.

Rial, J. A., Pielke Sr., R. A., Beniston, M. *et al.* 2004. Nonlinearities, feedbacks and critical thresholds within the Earth's climate system. *Clim Change*, **65**, 11–38.

Roe, E. & van Eeten, M. 2001. Threshold-based resource management: a framework for comprehensive ecosystem management. *Environ Manage* **27**, 195–214.

Rumelhart, D. E., Hinton, G. E. & Williams, R. J. 1986. Learning representations by back-propagation errors. *Nature* **323**, 533–536.

Schooley, R. L. & Wiens, J. A. 2003. Finding habitat patches and directional connectivity. *Oikos* **102**, 559–570.

Schooley, R. L. & Wiens, J. A. 2005. Spatial ecology of cactus bugs: area constraints and patch connectivity. *Ecology* **86**, 1627–1639.

Statzner, B., Gore, J. A. & Resh, V. H. 1988. Hydraulic stream ecology: observed patterns and potential applications. *J N Am Benthol Soc* **7**, 307–360.

Taylor, P. D., Fahrig, L. & Merriam, G. 1993. Connectivity is a vital element of landscape structure. *Oikos* **68**, 571–573.

Toms, J. D. & Lesperance, M. L. 2003. Piecewise regression: a tool for identifying ecological thresholds. *Ecology* **84**, 2034–2041.

Turchin, P., Odendaal, F. J. & Rausher, M. D. 1991. Quantifying insect movement in the field. *Environ Entomol* **20**, 955–963.

Turner, M. G. 2005. Landscape ecology in North America: past, present and future. *Ecology* **86**, 1967–1974.

Walker, B. & Meyers, J. A. 2004. Thresholds in ecological and social ecological systems: a developing database. *Ecol Soc* **9**, 3.

Waringer, J. A. 1993. The drag coefficient of cased caddis larvae from running waters – experimental-determination and ecological applications. *Freshwater Biol* **29**, 419–427.

Wellnitz, T. A., Poff, N. L., Cosyleon, G. & Steury, B. 2001. Current velocity and spatial scale as determinants of the distribution and abundance of two rheophilic herbivorous insects. *Landscape Ecol* **16**, 111–120.

Wiens, J. A., Schooley, R. L. & Weeks, R. D., Jr. 1997. Patchy landscapes and animal movements: do beetles percolate? *Oikos* **78**, 257–264.

Wiens, J. A., Stenseth, N. C., Van Horne, B. & Ims, R. A. 1993. Ecological mechanisms and landscape ecology. *Oikos* **66**, 369–380.

Wiggins, G. B. 1996. *Larvae of the North American Caddisfly Genera (Trichoptera).* University of Toronto Press Inc.

With, K. A. 1994. Using fractal analysis to identify how species perceive landscape structure. *Landscape Ecol* **9**, 25–36.

With, K. A. & Crist, T. O. 1995. Critical thresholds in species' responses to landscape structure. *Ecology* **76**, 2446–2459.

With, K. A., Gardner, R. H. & Turner, M. G. 1997. Landscape connectivity and population distributions in heterogeneous environments. *Oikos* **78**, 151–169.

With, K. A. & King, A. W. 1997. The use and misuse of neutral landscape models in ecology. *Oikos* **79**, 219–239.

19

A model biological neural network: the cephalopod vestibular system

Roddy Williamson and Abdul Chrachri

19.1 Introduction

Artificial neural networks (ANNs), inspired by the processing systems present in simple nervous systems, are now widely used for the extraction of patterns or meaning from complicated or imprecise data sets (Arbib, 2003; Enquist & Ghirlanda, 2005). Although modern ANNs have progressed considerably from the early, basic feedforward models to systems of significant sophistication, some with varying levels of feedback, modulation, adaptation, learning, etc. (Minsky & Papert, 1969; Gurney, 1997; Vogels *et al.*, 2005), they rarely contain the full processing capabilities or adaptive power of real assemblies of nerve cells. Part of the problem in modelling such capabilities is that the detailed mechanisms underlying the operation of biological neural networks are not themselves fully identified or well understood, for there is a dearth of good biological model systems that possess a wide range of processing mechanisms but whose physiological processes and cellular interconnections can be fully investigated and characterised. One of the best biological model systems available is the vertebrate visual system, but even here the full range of cellular connections and interactions have not yet been characterised and hence cannot be developed into equivalent models (van Hemmen *et al.*, 2001; Wassle, 2004).

In this report we describe a real, biological model system that comprises only three cell types but nevertheless demonstrates a wide range of complex and sophisticated cellular interactions, processing mechanisms and adaptive responses. This is the cephalopod vestibular system, a peripheral sense organ whose input can be specified and controlled, and which has a large feedback control system that can be monitored or mimicked.

Cephalopods have already supplied two model systems of fundamental importance to neuroscience and the development of neural networks. The first was the squid giant axon preparation which, through the large size and accessibility of the axon to physiological and biochemical investigation, enabled the subcellular mechanisms underlying the nerve action potential to be elucidated and then described in precise mathematical formulation (Hodgkin & Huxley, 1952). Second, the squid giant synapse, again through its large size

Modelling Perception with Artificial Neural Networks, ed. C. R. Tosh and G. D. Ruxton. Published by Cambridge University Press.

and accessibility to physiological recording methods, has enabled the mechanism of neurotransmitter release at the synapse to be precisely described and modelled (reviewed by Llinas, 1999).

Cephalopods, such as squid, cuttlefish and octopuses, are fast-moving predators that compete with fish and other marine vertebrates and hence have developed motor and sensory systems of comparable performance (Packard, 1972; Hanlon & Messenger, 1996). Like all fast-moving animals, cephalopods can sense the direction of gravity (linear accelerations) as well as the speed and direction of their turning movements (angular accelerations). The vestibular, or more correctly for a mollusc, the statocyst system used to detect these accelerations shows many parallels to vertebrate semicircular canal systems in both gross morphology and function (cephalopods: Budelmann, 1977; vertebrates: Highstein *et al.*, 2004). However, as a model for the investigation of the network properties underlying the operation of a complex, but manageable neural system, the statocyst has clear advantages over analogous vertebrate systems. First, it is embedded in cartilage not bone, and hence is easily accessible; second, precise physiological recordings can be obtained from all of the cellular elements comprising the peripheral network; and third, there is a very large and varied efferent feed-back/forward system whose influence on the operation of the system can be closely observed (Williamson & Chrachri, 1994, 2004). These features are described below, as well as the likely operation of the neural network.

19.2 Statocyst and network structure

19.2.1 Gross morphology

The gross morphologies of cephalopod statocysts have been described previously (Young, 1960; Barber, 1966; Budelmann *et al.*, 1987a, 1987b) but, in brief, consist of two endolymph-filled cavities lying within the cranial cartilage, just ventral and lateral to the brain. The left and right statocysts are mirror reversed with the precise shape of the cavity being species specific, presumably influencing the hydrodynamics of the system and hence the overall response characteristics (Young, 1989; Maddock & Young, 1984). Each statocyst contains two main areas of sensory epithelium: a macula/statolith area and a crista/cupula area. The macula system (Figure 19.1A) consists of a plate of mechanosensory hair cells with an overlying statolith; the force exerted by the statolith mass upon the mechanosensory hair cells is dependent on the magnitude and direction of any applied linear accelerations, e.g. gravity. Octopuses have a single macula with a compact statolith while squid and cuttlefish have three maculae carrying numerous small statoconia. Where three maculae are present they are set in different planes thus enabling the acceleration(s) to be resolved into its directional components.

The crista/cupula system, the subject of this report, consists of a narrow strip of sensory epithelium that winds around the inside wall of the statocyst, such that it covers the three orthogonal planes (Figure 19.1A). The strip comprises mechanosensory hair cells and

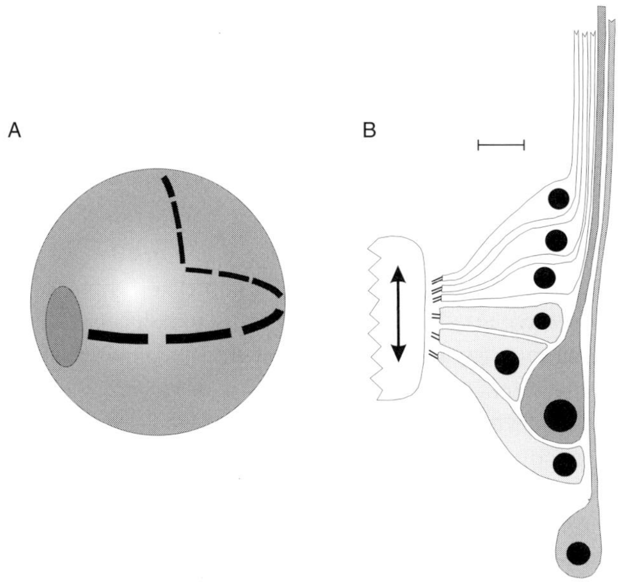

Figure 19.1. (A) Diagram of the octopus statocyst showing the ovoid plate of macula cells and the crista strip which runs around the inside of the statocyst sphere and is divided into nine segments. (B) Diagram of a greatly expanded transverse section through one of the crista segments showing the rows of primary sensory hair cells (white), the secondary sensory hair cells (light grey) and the afferent neurons (dark grey). The direction of travel of the overlying cupula is shown by the arrow. The efferent innervation is not shown. Scale bar in B = 15 μm.

afferent neurons, plus supporting cells, and is divided into segments: nine in octopuses and four in squids and cuttlefish. Each crista segment carries an overlying, sail-like cupula that is deflected during rotational movements of the animal by the flow of endolymph relative to the statocyst wall. Since the base of each cupula is in contact with the underlying mechanosensory hair cells, a cupula deflection may stimulate or inhibit these cells, depending on the direction of the cupula movement and the polarisation of the hair cells.

19.2.2 Crista morphology

A histological, transverse section through the crista epithelium (Figure 19.1B) shows that it is made up of three main cell types; these are primary sensory hair cells, secondary sensory hair cells and afferent neurons. The sensory hair cells are arranged in up to eight rows along the crista segment with each row containing only primary hair cells or only secondary sensory hair cells. Primary sensory hair cells, which have a centripetally passing axon of their own, are common in invertebrates while secondary hair cells, which have no axon but make synaptic contact with an afferent neuron, are usually only found in vertebrates (Budelmann *et al.*, 1987a) and so this mixture of the two types in a single epithelium is unique to cephalopods. The introduction of secondary sensory hair cells here is most likely an evolutionary sophistication that permits greater flexibility in signal processing, through modulation and integration of the input, in the peripheral nervous system. All of these mechanosensory hair cells are morphologically and physiologically polarised such that they are excited by a cupula deflection in one specific direction and inhibited by a deflection in the opposite direction. For the cells so far examined, the primary and secondary sensory hair cells in a single segment have the opposite polarity (Budelmann, 1977); thus a cupula movement which excites the secondary sensory hair cells will also inhibit the primary sensory hair cells in the same segment and vice versa.

The afferent neurons within a segment fall into two populations: large afferent neurons, which lie mainly beneath the secondary hair cells within the crista strip, and small afferent neurons which lie more ventrally (for the horizontally running segments), mainly at the edge, or just ventral to, the crista strip. There is morphological evidence from octopus indicating that the large secondary hair cells make synaptic contact with the large afferent neurons in a convergent manner with an estimated ratio of 4:1, whereas the smaller secondary hair cells make synaptic contact with the smaller afferent neurons in a divergent manner with an estimated ratio of 1:2 (Budelmann *et al.*, 1987a). The axons from the afferent neurons and those from the primary hair cells project via the statocyst nerves to the central nervous system within the cranium.

19.3 Physiological connections of the crista/cupula network

19.3.1 Peripheral connections and responses

Unlike the analogous vertebrate semicircular canal system (Highstein *et al.*, 2004; Eatock *et al.*, 2005), all of the afferent cells in the cephalopod crista/cupula system have their somata within the peripheral sensory epithelium and hence it is possible to make sharp electrode, intracellular recordings or whole-cell, patch clamp recordings from all of the cellular components, including the afferent neurons. Such recordings from the mechanosensory hair cells (Williamson, 1991; Chrachri & Williamson, 1998) have shown that cupula deflections perpendicular to the segment direction (e.g. in an upward direction for horizontally oriented segments) result in a depolarisation of the primary sensory hair cells and a train of action potentials in the cell axon (Figure 19.2A). Similar recordings from secondary sensory cells show that they are polarised in the opposite direction and, as seen from the imposed sinusoidal displacement of the cupula (Figure 19.2B), the response is asymmetric with much larger changes in membrane potential for displacements in the ventral direction than in the dorsal direction. This type of displacement /voltage response curve is also observed in recordings from vertebrate mechanosensory hair cells (Eatock *et al.*, 2005). Depolarising the secondary sensory hair cells in both vertebrate and cephalopod systems leads to an increased release of excitatory neurotransmitter, believed to be glutamate in both cases (Tu & Budelmann, 1994; Highstein *et al.*, 2004), and a subsequent excitation of the associated afferent neuron(s). For cephalopods, a range of

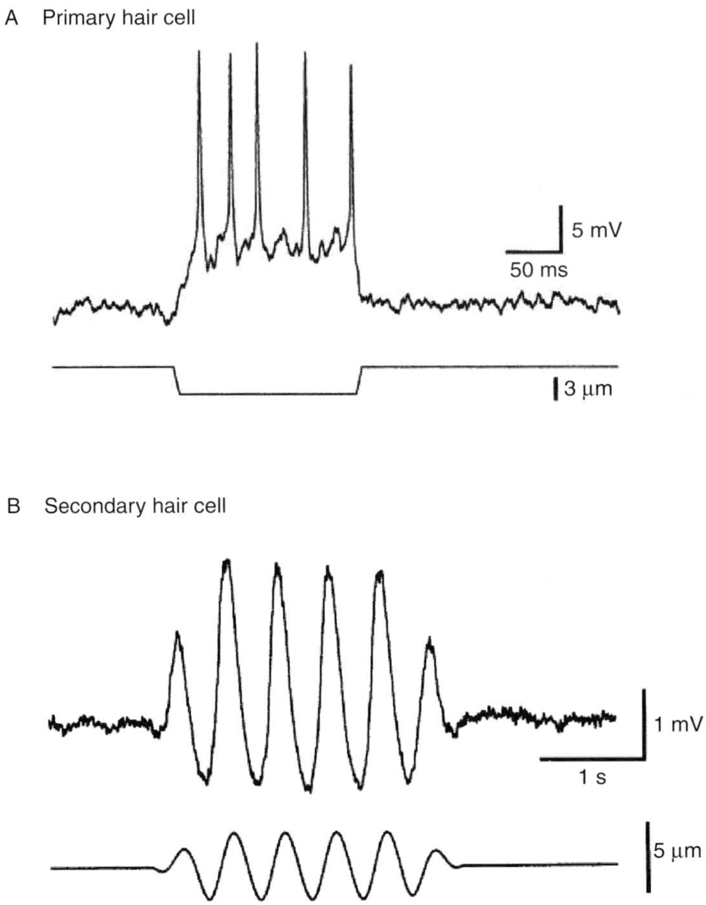

Figure 19.2. Responses of hair cells to cupula displacements.
A. Intracellular recording from a primary hair cell showing the depolarisation and action potentials resulting from a downwards displacement of the cupula. The lower trace shows indicates the time course and amplitude of the cupula displacement.
B. Intracellular recording from a secondary hair cell showing the asymmetric voltage response to an imposed sinusoidal displacement of the cupula. Note the cell is depolarised by an upward displacement of the cupula. Lower trace indicates the time course and amplitude of the cupula displacement.

neuromodulators have also been shown to influence the afferent cell activity (Tu & Budelmann, 2000a, 2000b). Note that in the vertebrate semicircular canal system all of the sensory hair cells are polarised in one direction and hence stimuli applied in the opposite direction can only be registered by a decrease in any ongoing activity. Thus, in vertebrates a steady, resting discharge in the afferents is necessary, whereas this is not the case in cephalopods and this may have significant energy-saving advantages.

Unlike the primary sensory hair cells in the crista, the secondary sensory hair cells do not have afferent axons passing to the central nervous system; however, some secondary hair cells have long, fine cellular processes extending from the cell's base along the direction of the crista segment (Williamson, 1995; Chrachri & Williamson, 1998). These processes are likely to be responsible for the spread of electrical coupling along the cells in a segment row beyond nearest neighbours, as described below, but some appear much longer than their estimated space constant and hence are likely candidates for carrying regenerative action potentials. Small action potentials have been observed in intracellular recordings from the soma of some cephalopod secondary hair cells and whole-cell patch clamp recordings from secondary hair cells have also detected an inward sodium current similar to that necessary for action potentials (Williamson, 1995). It is therefore possible that these fine basal processes carry action potentials similar to the dendritic action potentials found in some vertebrate neurons (reviewed in Häusser *et al.*, 2000), including retinal ganglion cells (Velte & Masland, 1999).

Direct recordings of the afferent nerve activity from the crista during controlled, imposed movements of the statocysts have shown that, like the vertebrate semicircular canals, the crista systems acts mainly as a detector of angular velocity (Williamson & Budelmann, 1985).

19.3.2 The peripheral efferent system

The statocyst crista/cupula system receives a very large efferent innervation from the brain, with up to 75% of the axons in the statocyst nerve being efferent axons (Budelmann *et al.*, 1987a, 1987b) and the rest statocyst afferent fibres; this proportion of efferent fibres is much larger than, for example, the 8% found in some vertebrate vestibular nerves (Goldberg & Fernandez, 1980). This statocyst efferent system forms a fine neural plexus beneath the crista ridge and innervates both the primary and secondary mechanosensory hair cells as well as the afferent neurons (Budelmann *et al.*, 1987b; Williamson & Chrachri, 1994); a single hair cell can have up to 40 separate synaptic contacts from the efferent fibres (Budelmann *et al.*, 1987b).

Activation of the efferent system has been found to enhance or depress the afferent input from the statocyst (Williamson, 1985; Chrachri & Williamson, 1998) and this is due to direct depolarising and/or hyperpolarising effects on the primary and secondary hair cells, and the afferent neurons (Williamson, 1989b; Chrachri & Williamson, 1998). Thus, as shown in the intracellular recordings from secondary hair cells (Figure 19.3), the efferent input to a cell can produce a depolarisation, increasing its activity, a hyperpolarisation decreasing its activity, or even a mixture of the two. Here the efferent fibres have been activated by direct electrical stimulation of the small crista nerve which contains some of the efferent fibres travelling from the brain to the crista. This is clear evidence for the existence of at least two populations of efferent fibres innervating the cephalopod statocyst, some excitatory and others inhibitory, and it is also apparent that individual cells receive multiple efferent contacts. An analogous excitatory and/or

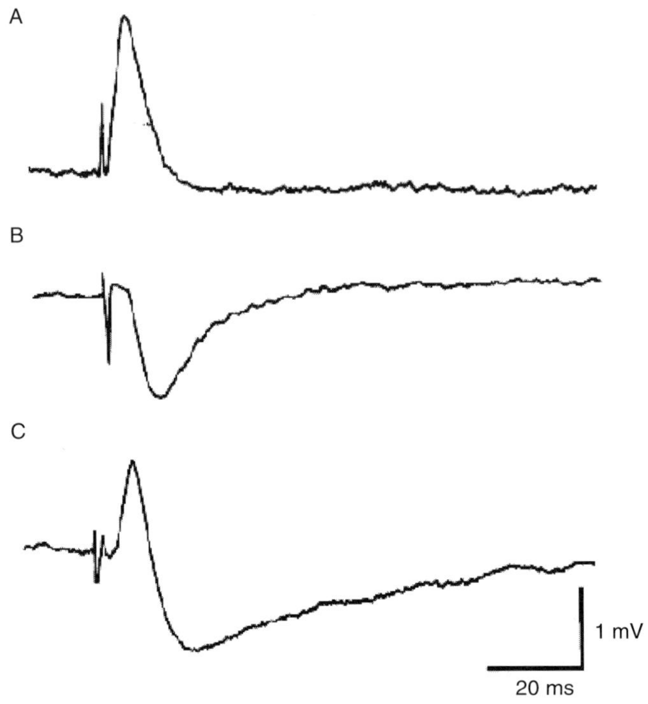

Figure 19.3. Intracellular recordings from three different secondary sensory hair cells showing the different efferent responses evoked by electrical stimulation of the crista nerve. (A) Depolarisation of the cell, i.e. excitation. (B) Hyperpolarisation of the cell, i.e. inhibition. (C) Mixed depolarisation followed by a hyperpolarisation.

inhibitory efferent innervation is also present in the vertebrate semicircular canal system (Brichta & Goldberg, 2000) but here it is much less extensive or influential. Note that because the cephalopod crista has primary and secondary sensory hair cells polarised in opposite directions, then complex permutations of inhibiting or exciting the cells polarised in opposing directions are possible.

The efferent inhibition is most likely achieved through activation of a cholinergic system (Auerbach & Budelmann, 1986) whereas the excitation is through a catecholaminergic system (Budelmann & Bonn, 1982; Williamson, 1989c), however, a variety of other neurotransmitter and neuromodulator substances, including GABA and various peptides, have also been found to influence the statocysts' activity (Tu & Budelmann, 1999, 2000a, 2000b; Chrachri & Williamson, 2004) and it may be that some or all of these are released from the efferent system, possibly as co-transmitters.

Recordings of the efferent fibre activity in semi-intact preparations of the statocysts and lower brain centres (Williamson, 1986) have shown that, with the statocysts at rest, most efferent fibres cells display only low levels of spontaneous firing activity; however, during imposed sinusoidal movements of the statocysts the efferent fibres fire bursts of activity and these are either synchronised in phase with the movement or in anti-phase with the movement. This evoked activity is presumably driven by the input from the statocysts themselves, through feedback and/or feedforward pathways, and/or from other mechanoreceptors.

19.3.3 Extensive electrical coupling within the crista network

Simultaneous intracellular recordings from pairs of cells within the crista sensory epithelium (Figure 19.4) have shown that groups of the cells are physiologically coupled through electrical synapses (Williamson, 1989a; Chrachri & Williamson, 1993, 1998). Thus, the secondary sensory hair cells along a row, within a single crista segment, are electrically coupled to their neighbours, with coupling ratios of up to 0.6. The electrical coupling ratio between cells is the ratio of the voltage change seen in Cell 2, divided by the voltage change imposed or observed in Cell 1. This observed electrical coupling in the crista extends not only to an individual secondary hair cell's immediate neighbours, within a segment row, but also some distance along the row as has been confirmed by the persistence of the coupling even after the immediate neighbouring cells have been ablated (Williamson, 1989a); this extended coupling probably arises because some of the secondary hair cells have fine processes extending along the crista row and these appear to make contact with multiple hair cells along the segment row (Chrachri & Williamson, 1998).

Electrical coupling is also present between neighbouring primary sensory hair cells along a segment row, with coupling ratios here of up to 0.4 (Chrachri & Williamson, 1993, 1998). These primary sensory cells produce action potentials when mechanically stimulated and such potentials are seen as subthreshold depolarisations in neighbouring primary hair cells (Chrachri & Williamson, 1998).

Although chemical synapses form the principal connections between the secondary sensory hair cells and the afferent neurons, low levels of electrical coupling have also

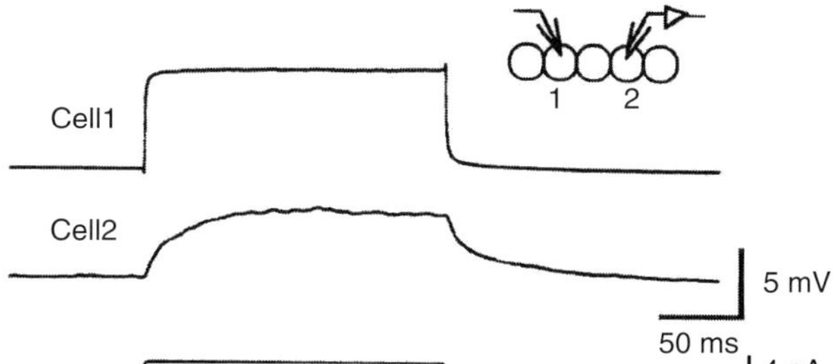

Figure 19.4. Electrical coupling between cells. Intracellular recordings from two secondary hair cells within a crista segment row showing the responses of both cells to the injection of a small, depolarising current into Cell 1. The synchronous depolarisation in Cell 2 indicates that these cells are electrically coupled.

been detected between these groups of cells (Chrachri & Williamson, 1998). Coupling ratios of up to 0.18 have been detected between these cell types and although this alone is unlikely to have a major influence on the activity of the afferent neuron, it has been found that the coupling effect can bring the afferent to fire action potentials if the membrane potential of the afferent neuron is just below threshold (Chrachri & Williamson, 1998). Unlike the electrical coupling found between the other crista cell groups, the secondary hair cell to afferent neurons coupling is rectifying in that no electrical coupling was detected in the reverse direction between the afferent neurons and the secondary hair cells. A low level of electrical coupling was also found between neighbouring, small afferent neurons, with coupling ratios of up to 0.3 (Chrachri & Williamson, 1998), but as these cells do not lie in regular rows, as is the case with the hair cells, it is not clear how far along the segment this coupling extends.

Finally, no electrical coupling was detected between primary and secondary sensory hair cells across a crista segment but such a finding would be surprising as these two types of hair cells are polarised in opposite directions and hence coupling here would act to diminish or cancel out any evoked responses.

19.3.4 Modulation of electrical coupling

The strength of the electrical coupling between sensory hair cells has been found to vary, with coupling ratios of between 0.01 to 0.6 (Chrachri & Williamson, 1998), but in addition, the coupling strength between individual pairs of cells can be modified by the application of a number of pharmacological agents, such as cAMP, forskolin and cGMP (Chrachri & Williamson, 2001). Recent experiments have shown that bath application of the cholinergic agonist carbachol can reduce the electrical coupling between crista primary sensory hair cells (Chrachri & Williamson, unpublished) and a possible mechanism for this has been demonstrated in that acetylcholine has also been shown to modulate calcium entry into the hair cells and hence probably the intracellular calcium concentration (Chrachri & Williamson, 2004); changes in intracellular calcium concentration have been shown to change cell coupling ratios in other systems (Chanson *et al.*, 1999) and such a modulation of electrical coupling is also present in the vertebrate retina network (McMahon & Mattson, 1996). Similarly, some of the pharmacologically active agents already reported to influence statocyst crista nerve activity, e.g. nitric oxide and the catecholamines, are also known to influence cell coupling in other systems (e.g. Rorig & Suttor, 1996). These findings introduce the likelihood that the statocyst efferent innervation may act to modulate dynamically the strength of electrical coupling between groups of cells, as well as having a direct effect on cell membrane potentials.

19.3.5 The central afferent and efferent projections

Nerve tracing studies have shown that the axons from the afferent neurons and from the primary sensory cells project through the statocyst crista nerves to a number of centres within the brain with the most prominent projections being to the anterior and posterior pedal lobes, as well as the ventral areas of the brachial and magnocellular lobes (Colmers, 1982; Plaen, 1987). Similar tracer studies on the statocyst efferent system have shown that the efferent axons originate from anterior pallioviscerai lobes and the pedal lobes of the central brain complex. With the main afferent projections and efferent somata being co-localised, in the main, within the suboesophageal lobes of the brain (see figure 4 of Williamson & Chrachri, 2004) it is very likely that there are significant interconnections between these two statocyst systems, and a number of appropriate inter-lobe connecting tracts have already been identified. Although the efferent fibres to the statocyst arise from at least two separate areas within the brain, there is no evidence, as yet, for functional divisions, e.g. the excitatory efferent fibres arising from one area and the inhibitory efferent fibres from another.

19.4 Network structure and operation

We have now identified the operation and many of the interconnections between the various cell types within the statocyst crista, e.g. the connection strengths and types, and can hence draw an outline network diagram (Figure 19.5) and begin to speculate on how the network properties may influence the operation of the system.

19.4.1 The primary sensory hair cell network system

The primary sensory hair cell network system appears the simplest in the crista epithelium, with the sensory hair cells' input being driven by the mechanical movements of the

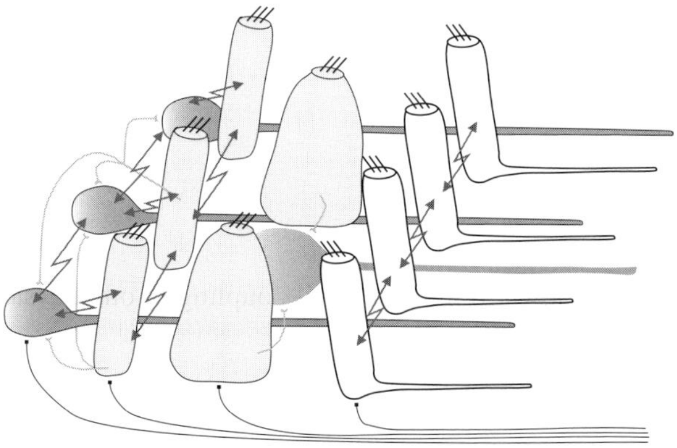

Figure 19.5. Outline network diagram of the crista epithelium showing the primary sensory hair cells (white), the secondary sensory hair cells (light grey), the small and large afferent neurons (dark grey) and the two types of synaptic connections: the electrical synapses (arrows) and the chemical synapses. Note the efferent nerve fibres, which innervate all of the cell components, coming from the crista nerve.

overlying cupula to produce either a hyperpolarisation or depolarisation of the cells and hence a reduction or increase in any ongoing afferent activity; where spike firing is evoked then, depending on the strength of coupling between neighbouring hair cells, the spiking activity in groups of neighbouring cells is likely to be in synchrony. Synchronous firing of receptor cells is also found in the vertebrate retina (Hu & Bloomfield, 2003) and there it is said to preserve high-resolution spatial signals and compress information for efficient transmission across the limited capacity of the optic nerve. However, the primary hair cell axons in cephalopods are small and slow conducting and hence the likely improvement in signal-to-noise ratio produced by the ensemble firing may be more important than the temporal signal. Synchronous firing of groups of cells from different areas may also be involved in a central 'binding' of related signals (O'Reilly & Busby, 2001).

The direct effect of the efferent input on the hair cell membrane potentials could clearly act as a gain control, improving sensitivity during slow turning movements and reducing sensitivity during rapid movements, such as during jet-propelled escape response – where the entire sensory system is in danger of saturation. This type of usage could be incorporated into either a feedback or feedforward system and the bursting activity recorded from the efferent nerves suggests a dynamic action of the efferent system.

A more specialised use of efferent fibres is found in the vertebrate auditory system where activation can change the physical responses of the outer hair cells and hence the mechanical properties of the basilar membrane and its filter characteristics (Fettiplace, 1999). Although there is no evidence that cephalopod sensory hair cells can actively change in length when stimulated by the efferent system, the possible change in coupling between the hair cells may result in significant changes in the apparent membrane capacitance and resistance of the cells, and hence the time constant of the responses produced by cupula stimulation. Thus, the frequency response or tuning characteristics of the receptor system could be increased or decreased by the impedance changes resulting from the changes in cell coupling.

Finally, the direct electrical coupling between the hair cells may also help remove uncorrelated noise from the cells' responses, as has been proposed for the vertebrate retinal cone cell network system (Smith, 2002), and hence modulation of the cell coupling coefficients may further change response sensitivity.

19.4.2 The secondary sensory hair cell network system

The membrane potentials of the secondary sensory hair cells are also modulated by the movements of the overlying cupula, responding to a cupula displacement with either a depolarisation or a hyperpolarisation depending on the direction of cupula movement and the direction of polarisation of the hair cells. These hair cells do not have axons passing to the brain and hence the effect of the membrane potential change will be an increase or decrease in the rate of transmitter release from the synapses onto the afferent neurons and hence a change in the rate of spiking discharge of these cells. However, some secondary hair cells have fine processes extending along the crista segment and it seems likely that these, as well as allowing the electrotonic spread of current, can also carry dendritic action potentials and these may act to synchronise the hair cells' excitability along lengths of a segment by, for example, providing near-synchronous depolarising inputs into groups of hair cells. Propagation of dendritic action potentials, particularly when involved in coincidence detection, has also been implicated in the promotion of synaptic plasticity (reviewed by Häusser *et al.*, 2000) and it may be that these dendritic action potentials have a similar modulatory function here. Of course such Hebbian modulation of hair cell and afferent neuron synapses is likely to occur anyway during development, and after hatching during growth, for the statocysts continue to grow in both size and in cell numbers as the animal ages.

As with the primary hair cells, the electrical coupling between the hair cells could effect an averaging of the input signal, with a resulting improvement in signal-to-noise ratio and decrease in frequency sensitivity. Here again, the large efferent input onto the hair cells can act to increase or decrease the sensitivity or possibly modulate the strength of coupling between neighbouring hair cells.

The convergence of the larger hair cell outputs onto the large afferent neurons may also improve detection sensitivity here by reducing the effect of synaptic noise on the spike generation, as has been postulated to occur in the vertebrate retinal network (Demb *et al.*, 2004). The divergence of the signal from the smaller secondary hair cells onto the small afferent neurons may act to produce correlated, or even synchronous, firing in these afferents; this phenomenon has been observed in a number of sensory systems (Usrey & Reid, 1999) and although its functional significance is not fully understood it may also lead to ensemble averaging in the higher processing centres. This correlated or synchronous firing of the afferent neurons may also be strengthened or weakened by modulation of the electrical coupling between cells via the efferent system.

19.4.3 System complexity and flexibility

We have shown that there are only three neuronal cell types present in the statocyst crista epithelium (primary and secondary sensory hair cells and the afferent neurons) and that the system receives a very large efferent innervation from the brain which can excite and/or depress the activity in the individual cells, as well as possibly modulate the strength of electrical coupling between the cells. Despite the limited classes of neurons involved here, the resulting network complexity and plasticity appears to rival that seen in more extensive neural processing networks, such as the vertebrate retina, but it is not immediately clear why such a level of sophistication is required here. A possible reason for this may be that the mechanical input signal to the neural network, i.e. a sail-like cupula driven backwards and forwards by fluid movement, is not as simple as first appears. The statocyst cupulae are known to be rather soft and gelatinous and therefore may flex or twist during turning movements that are not precisely perpendicular to the direction of the crista segment. Thus, nearly all movements are likely to cause a twisting or flexing of each of the four or nine cupulae within a statocyst and hence transmit a differential or uneven mechanical force to the sensory hair cells within the underlying crista segments; a specific head turn may then induce a pattern of differential excitation and inhibition across a single crista segment as well as across the multiple segments within both statocysts.

19.4.4 Network considerations

We have argued here that the cephalopod statocyst system presents a model biological network of relatively simple architecture, fully accessible to physiological recording techniques but which shows a range of dynamically modulated interconnections that transform a complex pattern of fluid flow, within the statocyst cavity, into a neural input signalling direction and magnitude of body movement. Clearly, this processing system cannot be modelled by a simple feedforward network for we know that it receives a large, complex and dynamically changing efferent input. At its simplest, the efferent activity could be interpreted as a recurrent signal, driven directly by the afferent input. However, this seems at odds with the magnitude and all-encompassing nature of the efferent innervation and also, because the efferent axons are mainly small, the system may be too slow for such reactivity. A more likely scenario is that the multi-layered efferent signal also contains an efference copy signal, based on the motor output, and representing the predicted, or re-afference, signal due to the animal's own intended movements. The effect of this could range from a simple reduction in input signal gain when a jet-propelled escape movement was imminent, or an increase in signal gain during fine hovering manoeuvres, to a finely tuned cancellation of the entire 'expected' input. Such efference copy systems have been already described in the mammalian vestibular system (Roy & Cullen, 2004) and the fish electrosensory system (Bell, 2001). The ability of the efferent system to modulate the level of electrical coupling between sets of cells would also enable the frequency response of the system to be adjusted and hence tuned, or matched, to the input.

A more speculative use of the efferent system would be to provide targeted co-activity along specific input pathways, thereby using Hebbian learning to increase or decrease the efficiency of specific neural pathways in an adaptive manner. Such a capability would enable a dynamic re-wiring of the network circuitry.

Finally, most neural network models consider that information processing results primarily from the properties of synapses and connectivity of individual neurons. However, we have shown here that single neurons can be non-spiking or carry dendritic spikes, both of which increase the likelihood of local signalling within a neuron, i.e. subsections of a neuron may be involved in separate, independent signalling pathways. In addition, we have also shown that groups of neurons may be dynamically connected through electrical coupling such that they act synchronously. Thus, the view of the single neuron as the basic element of the network must be revised to include the possibility of both sub-neuronal and supra-neuronal elements and these being dynamically interchangeable.

Acknowledgements

The experimental work incorporated above was supported by the Wellcome Trust and BBSRC. We would also like to thank Dr Sue Denham and Dr Guido Bugman for helpful discussions on the operation of this network system.

References

Arbib, M. A. 2003. *Handbook of Brain Theory and Neural Networks*. Bradford Book, 2nd Edn. MIT Press.

Auerbach, B. & Budelmann, B. U. 1986. Evidence for acetylcholine as a neurotransmitter in the statocyst of *Octopus vulgaris*. *Cell Tissue Res* **243**, 429–436.

Barber, V. C. 1966. The fine structure of the statocysts of *Octopus vulgaris*. *Z Zellforsch Mikroskop Anat* **70**, 91–107.

Bell, C. C. 2001. Memory-based expectations in electrosensory systems. *Curr Opin Neurobiol* **11**, 481–487.

Brichta, A. M. & Goldberg, J. M. 2000. Responses to efferent activation and excitatory response-intensity relations of turtle posterior-crista afferents. *J Neurophysiol* **83**, 1224–1242.

Budelmann B. U. 1977. Structure and function of the angular acceleration receptor systems in the statocysts of cephalopods. *Symp Zoo Soc Lond* **38**, 309–324.

Budelmann, B. U. & Bonn, U, 1982. Histochemical evidence for catecholamines as neurotransmitters in the statocyst of *Octopus vulgaris*. *Cell Tissue Res* **227**, 475–483.

Budelmann, B. U., Sache, M. & Staudigl, M. 1987a. The angular acceleration receptor system of the statocyst of *Octopus vulgaris*: morphometry, ultrastructure, and neuronal and synaptic organization. *Phil Trans R Soc B* **315**, 305–343.

Budelmann, B. U., Williamson, R. & Auerbach, B. 1987b. Structure and function of the angular acceleration receptor system of the statocyst of octopus with special reference to its efferent innervation. In *The Vestibular System: Neurophysiologic and Clinical Research* (ed. M. D. Graham & J. L. Kemmink), pp. 165–168. Raven Press.

Chanson, M., Mollard, P., Meda, P., Suter, S. & Jongsma, H. J. 1999. Modulation of pancreatic acinar cell to cell coupling during ACh-evoked changes in cytosolic Ca2+. *J Biol Chem* **274**, 282–287.

Chrachri, A. & Williamson, R. 1993. Electrical coupling between primary hair cells in the statocyst of the squid, *Alloteuthis subulata*. *Neurosci Lett* **161**, 227–231.

Chrachri, A. & Williamson, R. 1998. Synaptic interactions between crista hair cells in the statocyst of the squid *Alloteuthis subulata*. *J Neurophysiol* **80**, 656–666.

Chrachri, A. & Williamson, R. 2001. cAMP modulates electrical coupling between sensory hair cells in the squid statocyst. *J Physiol Lond* **536S**, 133–134.

Chrachri, A. & Williamson, R. 2004. Cholinergic modulation of L-type calcium current in isolated sensory hair cells of the statocyst of octopus, *Eledone cirrhosa*. *Neurosci Lett* **360**, 90–94.

Colmers, W. F. 1982. The central afferent and efferent organization of the gravity receptor system of the statocyst of *Octopus vulgaris*. *Neuroscience* **7**, 461–476.

Demb, J. B., Sterling, P. & Freed, M. A. 2004. How retinal ganglion cells prevent synaptic noise from reaching the spike output. *J Neurophysiol* **92**, 2510–2519.

Eatock, R. A., Fay, R. & Popper, A. (Ed.) 2005. *Vertebrate Hair Cells. Springer Handbook of Auditory Research, Vol. 27.* Springer.

Enquist, M. & Ghirlanda, S. 2005. *Neural Networks and Animal Behavior.* Princeton University Press.

Fettiplace, R. 1999. Mechanisms of hair cell tuning. *Ann Rev Physiol* **61**, 809–834.

Goldberg, J. M. & Fernandez, C. 1980. Efferent vestibular system in the squirrel monkey: anatomical location and influence on afferent activity. *J Neurophysiol* **43**, 986–1025.

Gurney, K. 1997. *An Introduction to Neural Networks.* UCL Press.

Hanlon, R. T. & Messenger, J. B. (Ed.) 1996. *Cephalopod Behaviour.* Cambridge University Press.

Häusser, M., Spruston, N. & Stuart, G. J. 2000. Diversity and dynamics of dendritic signaling. *Science* **290**, 739–744.

Highstein, S. M., Fay, R. & Popper, A. N. (Ed.) 2004. *The Vestibular System. Springer Handbook of Auditory Research, Vol. 19.* Springer.

Hodgkin, A. L. & Huxley, A. F. 1952. A quantitative description of membrane current and its application to conduction and excitation in nerve. *J Physiol Lond* **117**, 500–544.

Hu, E. H. & Bloomfield, S. A. 2003. Gap junctional coupling underlies the short-latency spike synchrony of retinal ganglion cells. *J Neurosci* **23**, 6768–6777.

Llinas, R. R. *The Squid Giant Synapse: A Model for Chemical Transmission.* Oxford University Press.

McMahon, D. G. & Mattson, M. P. 1996. Horizontal cell electrical coupling in the giant danio: synaptic modulation by dopamine and synaptic maintenance by calcium. *Brain Res* **718**, 89–96.

Maddock, L & Young, J. Z. 1984. Some dimensions of the angular-acceleration receptor systems of cephalopods. *J Mar Biol Assoc UK* **64**, 55–79.

Minsky, M. & Papert, S. 1969. *Perceptrons: An Introduction to Computational Geometry.* MIT Press.

O'Reilly, R. C. & Busby, R. S. 2001. Generalizable relational binding from coarse-coded distributed representations. In *Advances in Neural Information Processing Systems.* (ed. S. Becker), pp. 75–82 MIT Press.

Packard, A. 1972. Cephalopods and fish: the limits of convergence. *Biol Rev* **46**, 241–307.

Plän, T. 1987. *Functional Neuroanatomy of Sensory-motor Lobes of the Brain of* Octopus Vulgaris. PhD Thesis, University of Regensburg.

Rorig, B. & Sutor, B. 1996. Regulation of gap junction coupling in the developing neocortex. *Mol Neurobiol* **12**, 225–249.

Roy, J. E. & Cullen, K. E. 2004. Dissociating self-generated from passively applied head motion: neural mechanisms in the vestibular nuclei. *J Neurosci* **24**, 2102–2111.

Smith, R. G. 2002. Retina. In *The Handbook of Brain Theory and Neural Networks* (ed. M. A. Arbib). MIT Press.

Tu, Y. J. & Budelmann, B. U. 1994. The effect of L-glutamate on the afferent resting activity in the cephalopod statocyst. *Brain Res* **642**, 47–58.

Tu, Y. J. & Budelmann, B. U. 1999. Effects of L-arginine on the afferent resting activity in the cephalopod statocyst. *Brain Res* **845**, 35–49.

Tu, Y. J. & Budelmann, B. U. 2000a. Inhibitory effect of cyclic guanosine 3′, 5′-monophosphate (cGMP) on the afferent resting activity in the cephalopod statocyst. *Brain Res* **880**, 65–69.

Tu, Y. J. & Budelmann, B. U. 2000b. Effects of nitric oxide donors on the afferent resting activity in the cephalopod statocyst. *Brain Res* **865**, 211–220.

Usrey, W. M. & Reid, R. C. 1999. Synchronous activity in the visual system. *Ann Rev Physiol* **61**, 435–456.

Van Hemmen, L. J., Cowan, J. D. & Domany, E. (ed.) 2001. *Models of Neural Networks, Vol. IV. Early Vision and Attention*. Springer Verlag.

Velte, T. J. & Masland, R. H. 1999. Action potentials in the dendrites of retinal ganglion cells. *J Neurophysiol* **81**, 1412–1417.

Vogels, T. P, Rajan, K. & Abbott, L. F. 2005. Neural network dynamics. *Ann Rev Neurosci* **28**, 357–376.

Wassle, H. 2004. Parallel processing in the mammalian retina. *Nature Rev Neurosci* **5**, 747–757.

Williamson, R. 1985. Efferent influences on the afferent activity from the octopus angular acceleration receptor system. *J Exp Biol* **119**, 251–264.

Williamson, R. 1986. Efferent activity in the octopus statocyst nerves. *J Comp Physiol A* **158**, 125–132.

Williamson, R. 1989a. Electrical coupling between secondary hair cells in the statocyst of the squid *Alloteuthis subulata*. *Brain Res* **486**, 67–72.

Williamson, R. 1989b. Secondary hair cells and afferent neurones in the squid statocyst receive both inhibitory and excitatory efferent inputs. *J Comp Physiol A* **165**, 847–860.

Williamson, R. 1989c. Electrophysiological evidence for cholinergic and catecholaminergic efferent transmitters in the statocyst of octopus. *Comp Biochem Physiol* **93**C, 23–27.

Williamson, R. 1991. The responses of primary and secondary sensory hair cells in the squid statocyst to mechanical stimulation. *J Comp Physiol A* **167**, 655–664.

Williamson, R. 1995. Ionic currents in secondary sensory hair cells isolated from the statocysts of squid and cuttlefish. *J Comp Physiol A* **177**, 261–271.

Williamson, R. & Budelmann, B. U. 1985. The responses of the octopus angular acceleration receptor system to sinusoidal stimulation. *J Comp Physiol A* **156**, 403–412.

Williamson, R, & Chrachri, A. 1994. The efferent system in cephalopod statocysts. *Biomed Res* **15**, 51–56.

Williamson, R. & Chrachri, A. 2004. Cephalopod neural networks. *Neurosignals* **13**, 87–98.

Young, J. Z. 1960. The statocyst of *Octopus vulgaris*. *Proc R Soc B* **152**, 3–29.

Young, J. Z. 1989. The angular-acceleration receptor system of diverse cephalopods. *Phil Trans R Soc Lond* **325B**, 189–238.

Index